Harriers of the World

Oxford Ornithology Series
Edited by C. M. Perrins

1. *Bird Population Studies: Relevance to Conservation and Management* (1991)
   Edited by C.M. Perrins, J.-D. Lebreton, and G.J.M. Hirons
2. *Bird-Parasite Interactions: Ecology, Evolution, and Behaviour* (1991)
   Edited by J.E. Loye and M. Zuk
3. *Bird Migration: A General Survey* (1993)
   Peter Berthod
4. *The Snow Geese of La Pérose Bay: Natural Selection in the Wild* (1995)
   Fred Cooke, Robert F. Rockwell, and David B. Lank
5. *The Zebra Finch: A Synthesis of Field and Laboratory Studies* (1996)
   Richard A. Zann
6. *Partnerships in Birds: The Study of Monogamy* (1966)
   Edited by Jeffrey M. Black
7. *The Oystercatcher: From Individuals to Populations* (1996)
   Edited by John D. Goss-Custard
8. *Avian Growth and Development: Evolution within the Altricial-Precocial Spectrum* (1997)
   Edited by J.M. Starck and R.E. Ricklefs
9. *Parasitic Birds and Their Hosts* (1998)
   Edited by Stephen I. Rothstein and Scott K. Robinson
10. *The Evolution of Avian Breeding Systems* (1999)
    J. David Ligon
11. *Harriers of the World: Their Behaviour and Ecology* (2000)
    Robert E. Simmons

Harriers of the World

THEIR BEHAVIOUR AND ECOLOGY

ROBERT E. SIMMONS
Percy FitzPatrick Institute
of African Ornithology,
University of Cape Town,
Rondebosch, South Africa

Illustrated by John R. Simmons

OXFORD
UNIVERSITY PRESS

OXFORD
UNIVERSITY PRESS

Great Clarendon Street, Oxford OX2 6DP

Oxford University Press is a department of the University of Oxford.
It furthers the University's objective of excellence in research, scholarship,
and education by publishing worldwide in

Oxford New York

Athens Auckland Bangkok Bogotá Buenos Aires Calcutta
Cape Town Chennai Dar es Salaam Delhi Florence Hong Kong Istanbul
Karachi Kuala Lumpur Madrid Melbourne Mexico City Mumbai
Nairobi Paris São Paulo Singapore Taipei Tokyo Toronto Warsaw

with associated companies in Berlin Ibadan

Oxford is a registered trade mark of Oxford University Press
in the UK and in certain other countries

Published in the United States
by Oxford University Press Inc., New York

First published 2000

A catalogue record for this book is available from the British Library

Library of Congress Cataloging in Publication Data

(Data applied for)

ISBN 0 19 854964 4 (Pbk)

Typeset by EXPO Holdings, Malaysia
Printed in Great Britain
on acid-free paper by
Biddles Ltd,
Guildford & King's Lynn

To our parents
for 160 years of love and wisdom

Preface

This book is the fruit of a ten-year love affair with harriers spanning two continents: North America and Africa. My introduction to these birds, however, started in eastern England when, as a volunteer RSPB warden in the late 1970s, I was tasked with observing some of the first European Marsh Harriers to breed in East Anglia. Their mystique and rarity were immediately captivating, and later led me to spend many years studying their cousins on a trail that led first west, then south of the equator. In North America I was given a chance of a lifetime—to spend three years studying Northern Harriers breeding on a windblown marshland at the top corner of the Bay of Fundy in eastern Canada. As a serious student with a masters degree to pursue I was entirely absorbed by an enigmatic mating system that appeared to be maladaptive for some individuals. A second chance to study their counterparts in the subtropics took me further into the fascinating world of behavioural ecology. In the course of my studies it dawned on me that I could answer, first-hand, questions about subtropical species and their different lifestyles that were being sought with experimental techniques in temperate zones. Not only would the answers test the generality of ecological models being devised in the minds and at the desks of European and North American ecologists, but I could compare other life history traits of subtropical species to answer some pretty fundamental questions about reproduction that had rarely been tackled from the southern end. I cannot claim to have deliberately chosen harriers to study, but once in Africa it became apparent that the large body of research now available from Europe, North America and Australia made such north–south comparisons easier than in virtually any other group. I hope that this book conveys the excitement of a behavioural ecology approach to answering such questions and bridges the gap between raptor biologists and evolutionary ecologists.

During the course of this work I have benefited from, and enjoyed the friendship, encouragement and stimulation of, many colleagues. It is easiest to thank them in chronological order, beginning in Canada. Among them, Michael Simmons and Peter Smith (Acadia University, Nova Scotia) first gave me the chance to study harriers in Canada and, the late Cyril Coldwell, and Canadian Wildlife Service (CWS) officials assisted in many ways during my days on the Tantramar Marshes of New Brunswick. Some of the unpublished data included herein were collected jointly with three other harrier enthusiasts: Bruce MacWhirter (now with a doctorate on the same population), Gay Hansen of Mount Allison University, New Brunswick and Phoebe Barnard. I am grateful to them for generously sharing data and answering my queries. I was assisted in the field by Mark Forbes, Al Smith, Peter Barkhouse, Myrtle Bateman, Peter Hicklin, and Colin McKinnon.

I was equally fortunate to meet Professors Hugh Paterson and Miles Markus on moving to Africa; they allowed me to pursue my interest on African harriers unhindered. I thank my friends and academic adversaries in Johannesburg, for stimulating and sometimes unsettling arguments on adaptive and nonadaptive explanations of the universe. Alan Kemp's quiet wisdom on all matters of African raptor ecology was of considerable value during these times. During my days in the stunning Wilderness Lakes area of South Africa I was greatly assisted by conservation officers of the southern Cape Province, particularly André Boshoff who suggested the study area and helped at all stages, and Guy Palmer who supplied me with guinea pigs. Saner interludes were provided by Keith and Jane James, Alison Evans, and Nicole Palmer. The exacting task of identifying prey from harrier pellets was expertly undertaken by Margaret and Graham Avery of the South African Museum. I am especially grateful to John Mendelsohn who introduced me to the only trap I know that regularly caught harriers. Many who have commented on the papers from which some of this work is drawn are specifically thanked in those papers. However, I would like to add special thanks to Alan Kemp, Peter Frost, John Mendelsohn, Ian Newton, Penny Olsen, Alistair Robertson, Warwick Tarboton, Chris Brown, and particularly Phoebe Barnard for keeping my academic feet on the ground when my extrapolations exceeded their elastic limit. Reviewers and editors Steve Redpath, Beatriz Arroyo, Chris Perrins, Kate Kilpatrick, and Andy Radford made numerous suggestions that clarified my presentation. Others who have helped at a distance in various ways include the late Richard Brooke, Dave Baker-Gabb, Peter Chadwick, Morné du Plessis, Igor Fefelov, Anita Gamauf, Anders Møller, Alice Jarvis, Hans Kalländer, Penn Lloyd, Jeremy Midgley, Terry Oatley, Tony Robertson, Peter Steyn, and Simon Thirgood. Simon Thomsett assisted with East African marsh harrier records, while the chapter on the evolution of the genus would not have been possible without the expert work of Michael Wink (Institut für Pharmazeutische Biologie, Universität Heidelberg) and his technical assistant Hedi Sauer-Gürth. We thank all those who contributed to the new understanding of this difficult genus and its surprises by sending photographs, unpublished data, blood or feathers from South America: Gladys Garay, Rodrigo Cáceres, and especially Maria Susana Bó; South Africa: Koos de Goede, Andrew Jenkins, and particularly Andy Schofield; Madagascar: Rick Watson, Ignace Randriamanga, and particularly Russell Thorstrom; Reunion, France and Spain: Jean-Marc Thiollay, Francois Mougeot, Vincent Bretagnolle, and especially Beatriz Arroyo; Russia: Anatoly Davygora and especially Igor Fefelov; and friends down-under: Dave Baker-Gabb, Kate Fitzherbert, Dick Schodde and in particular Nick Mooney. In addition, there are several people whose books and scientific papers have subtly but profoundly shaped my thoughts and ideas. I can clearly remember the excitement on first reading the early harrier and kestrel papers emanating from Serge Daan and Cor Dijkstra's 'raptor group' at the University of Groningen in the Netherlands, and Ian Newton's 1979 and 1986 classics on raptor population ecology and sparrowhawks. A special mention must be made to Beatriz Arroyo who should really be a co-author of this book for her dynamism and enthusiasm for harriers, and for the mountain of data emanating from her pen. She is tasked with writing the update in ten years' time. Also thanks goes to Staffan and Astrid Ulfstrand for providing a hugely stimulating environment at Uppsala University, Sweden, where much of the initial writing was done. Members of the seminar group, especially Dave Wiggins, Arne Lundberg and Jacob Hoglund, gave valuable input on science, book writing and life in a fridge. I also thank my employer, the Namibian Ministry of Environment & Tourism, for granting study leave from the real world of dust and flies. In Namibia, my current home, visitors such as Joel Berger, Carol Cunningham, Guy Cowlishaw, Ian Jamieson, Sue McRae and Dave Wiggins proved invaluable friends and sounding boards, while from afar Linda Birch (Oxford University), Bill Clark and Roger Clarke provided papers, photographs, data, and unpublished manuscripts. I owe a particular debt to John Simmons, my artist-twin, who shares my love of wild days watching harriers, and cheerfully put up with my long-distance demands and critiques of his art work. His enthusiasm and skills have greatly assisted in breathing life into the following pages. We thank Robert Prŷs-Jones of the British Natural History Museum for access to harrier specimens.

I thank my parents, for producing an artist at a critical time in my ontogeny, and above all, a climate of enquiry that ultimately laid the foundation for my studies. My greatest debt of gratitude goes to Phoebe Barnard who writes better papers (and books) than I, and constantly provides alternative evolutionary explanations for me to chew on; she has shaped virtually everything that I've ever written. For that, and much more, I am eternally grateful. To Catherine and Julia, I am now happy to return their Father.

Windhoek R. E. S.
July 1999

Contents

1 *Introduction. Changing perspectives: from natural history to behavioural ecology*

In the tangled bank, at the conclusion of his masterpiece *The origin of species*, Charles Darwin mused on the ecological forces that shape and continue to shape the species that live there:

It is interesting to contemplate an entangled bank, clothed with many plants of many kinds, with birds singing on the bushes, with various insects flitting about, and with worms crawling through the damp earth, and to reflect that these elaborately constructed forms, so different from each other, and dependent upon each other in so complex a manner, have all been produced by laws acting around us.

Clearly Darwin's perspective reflected that of a country gentleman (and genius) living in a bucolic Kentish setting. Today neither the laws nor their application have changed, although, as with the hedgerows, our perspectives may have done with our increasing realization of the complexity that Darwin hinted at. For instance now an ecologist might see in the tangled bank a pair of Hedge Sparrows, *Prunella modularis*, foraging for insects in the leaf litter, with the male strenuously protecting his paternity and mate against extra pair copulations by the bizarre behaviour of cloacal pecking. Beneath their feet, hermaphroditic worms contemplate the wisdom of exchanging gametes with a mate they don't immediately need for reproduction, while close by, dragonflies dance in tandem, males silently scooping rival sperm from their 'unfaithful' mates to avoid cuckoldry. Overhead flit the first parasite-infected swallows of the spring, checking over potential mates for asymmetrical tails and other badges of quality. And inside each, of course, swarm hordes of selfish genes and supergenes ready to parasitize and manipulate regardless of the effects on the bodies that merely house them!

This new view by the behavioural ecologist sees life as an arms race between predators and prey, conflict rather than cooperation between mates, handicaps honestly advertising quality to rivals and mates, and organisms evolving fast to keep pace with the parasites that are ready to take over (dubbed the 'Red Queen' hypothesis in honour of Alice who ran ever faster to keep abreast of the queen). Less often behavioural ecologists invoke the abiotic factors emphasised by Darwin: today the emphasis is more on conflicts with conspecifics, mates and even offspring, something Darwin was less aware of. When Herbert Spencer distilled Darwinism as nature 'red in tooth and claw' he may have been closer to the truth than the image conjured up by a quiet tangled bank in Kent. While Spencer's distillation conjures up for me predators on an African plain, it is sadly true that studies in tropical regions have added little to our understanding of ecological or evolutionary principles.

Why should this be? It certainly isn't because we are unable to answer the 'why' questions that behavioural ecologists often pose. We are living in an age where many new experimental techniques in the field and lab have solved complex problems unanswerable a decade ago. Tropical–temperate differences in breeding and survival remain unanswered mainly because the vast majority of work continues in temperate zones. Where studies are undertaken in tropical areas, they are often concerned with more applied or urgent ecological problems, such as conservation biology. Often the researchers in the tropics seem unaware or perhaps uninterested in the most interesting evolutionary or ecological problems, while those who are, either lack the funds to pursue their interest, or see it as a luxury that is unsupported by their colleagues (Barnard 1995). The arguments for this evolutionary approach are well presented by Joel Berger (1996), who shows how it can solve conservation problems among small, extinction-prone, populations. It is my hope that this book will, in part, redress this imbalance, by providing a detailed comparison of a tropical predator—the African Marsh Harrier, *Circus ranivorus*—with other well-studied harriers through the eyes of a behavioural ecologist with a Eurocentric bias.

Asking why questions

One of the aims of this book is to compare the behavioural ecology of a tropical species with its congeners in northern climes. Behavioural ecology is the science of asking why an organism behaves as it does or has evolved certain traits over others in a specific ecological setting. Questions that behavioural ecologists frequently ask are: why do female birds actively seek extra pair copulations (and how do their mates prevent it)? Why, and under what conditions, do some fish suddenly change sex? Why do some male insects evolve elaborate appendages for scooping sperm from their mates? Or more simply, why do peacocks have such elaborate tails and why should peahens prefer them? If you gain the impression from this short list that behavioural ecologists are merely interested in sex, you are right. Sexual selection—(the result of differential mating success through fertilization or mate attraction) and its many consequences (such as elaborate plumes and colours useful in attracting mates, often detrimental to survival) occupy the minds and journals of many researchers and institutions. Some will be elaborated in this book. As with any discipline it is the unusual that attracts the interest. For behavioural ecologists, individuals acting in non-selfish (altruistic) ways, act as a red flag for investigation. So, why do certain youngsters, rather than breeding themselves, stay behind in their natal territory to assist their parents in rearing additional offspring (Stacey and Koenig 1990; Jennions and Macdonald 1994; du Plessis *et al.* 1995)? Even stranger are behaviours in which offspring kill their siblings so often that second sibs (Abels) virtually always perish. This not only happens among raptors (Simmons 1988a; Gargett 1990) but a whole host of other species in other taxa, and in some cases before the sibs ever see the light of day (Mock and Parker 1997). Among the most bizarre of all is infanticide of parent Australian marsupials of the genus *Antechinus* in selectively killing their own young to manipulate the sex ratio of the brood according to its original sex bias or the age of the mother (Cockburn 1992).

Equally interesting questions with broader ecological implications also arise. For example, why are species in the tropics more brightly coloured than their counterparts in temperate zones? Why do species in the tropics consistently rear fewer young? Why are they (gram for gram) longer-lived than their temperate counterparts? Do those species in aseasonal environments exhibit similar seasonal declines in reproduction? Why do they show a greater incidence of cooperative breeding, obligate siblicide, cuckoo-like parasitic behaviour and similarly 'bizarre' trends? Most interesting and revealing of all are the studies that contrast breeding, feeding and mating strategies of tropical species with temperate species to see if the same principles and assumptions hold.

If harrier or raptor ecology is your chief motivation for reading this book, Chapters 2, 3, 4, 8, and 9 will be of greatest interest since they deal with peculiarities of harriers, ideas on reversed size dimorphism, timing of breeding, hunting behaviour and the effect of food on eggs and productivity. For behavioural ecologists, the chapters on mating systems (Chapter 5), displays and their consequences (Chapter 3), copulation and sperm depletion (Chapter 6), and the arms race between harriers and mice (Chapter 4) will be of most interest. For those interested in ecological and evolutionary differences between temperate and tropical species, Chapters 8 and 9 test the different ideas forwarded for the smaller clutches and lower breeding output of tropical and subtropical species, and assess the different selection pressures apparent between African and temperate zone rate harriers.

Why harriers?

With no less than five English-language books devoted to harriers since Henning Weis's superb photographs from Denmark in 1923, it may seem like overkill for yet another book on this group. And this among a plethora of raptor books in general! This book differs from the photographic essay of Weis (1923), the sketches and natural history observations of Watson (1977), the overview of breeding biology (Hamerstrom 1986) and behaviour and conservation of temperate harriers (Clarke 1995, 1996) in that it focuses on the behavioural ecology of the world's harriers. In particular it contrasts and compares one of the most ancient harriers (Africa's marsh harrier) with its European, American, and Australasian congeners. Interwoven throughout the text are details of each species' basic breeding, feeding and life-history details. The role of the behavioural ecologist however, is to get further under the feathers, and discover why certain species exhibit certain traits. However, harriers are at once both a difficult and a highly rewarding group of birds with which to attempt this. Their breeding habitat is also often particularly difficult to access and observe in, and always full of mosquitoes. Fortunately, many species nest in dense or loose colonies allowing relatively good sample sizes. In addition, some harriers exhibit curious traits such as a highly unusual mating system, a fine-tuned response to changes in the prey base, and a fascinating ability to alter the sex ratio of their young in response to environmental conditions. Hence once the logistical problems are solved harriers offer numerous rewards.

Before this book evolved there were 13 recognised species of harrier worldwide. Africa was blessed with three resident species: the African Marsh Harrier *Circus ranivorus*, is one of the smallest in the marsh harrier complex; the most striking, the Black Harrier *Circus maurus*, virtually endemic to South Africa inhabits the grasslands of Natal and the world's smallest and most species-rich floral kingdom, the Cape Fynbos. Observing this buoyant black raptor quartering heath-covered hillsides in the southern Cape is both breathtaking and a reminder of the extreme degree of endemism in this corner of Africa (Cowling and Hilton-Taylor 1994; Simmons *et al.* 1998). The poorest known species, the Madagascar or Reunion Harrier, *Circus macrosceles/maillardi*, is endemic to the islands of Madagascar and Reunion (Cheke 1985), and provided some genetic surprises, as we shall see. One of these species—the African Marsh Harrier—I was able to study first hand. Prior to that I began my studies of harriers with a close look at its North American congeners, the Northern Harrier *Circus cyaneus*. My studies of polygyny in Canada and the influence of food on reproduction left me with more questions than answers and it was an opportunity I couldn't forgo when the chance arose to study an African relative.

As a genus, the harriers (*Circus*) are well suited for a comparative study of ecology and behaviour. The four species in Europe (Pallid, Montagu's, Hen and European Marsh) have been the focus of four books and four doctoral theses (Weis 1923; Watson 1977; Redpath 1989; Underhill-Day 1990; Gonzalez 1991; Arroyo 1995; Clarke 1995, 1996). In addition at least one book and eight dissertations have analysed mating system, behavioural ecology, population ecology and sex ratio manipulation of North America's only harrier (Burke 1979; Linner 1980; Temeles 1980; Barnard 1983; Simmons 1983; MacWhirter 1985, 1994; Hamerstrom 1986; Bildstein 1987). Mating systems have also been the focus of research in Europe (Altenburg *et al.* 1982; Picozzi 1984a). David Baker-Gabb's (1981, 1982, 1986) studies of the Spotted Harrier *C. assimilis* and the Swamp or Australasian Harrier *C. approximans* give a southern-hemisphere perspective from Australasia, leaving only Africa's three endemic species, South America's Long-winged *C. buffoni* and Cinereous Harriers *C. cinereus*, and Asia's Pallid *C. macrourus*, Pied *C. melanoleucus* and Eastern Marsh Harrier *C. spilonotus* to be studied in detail. Such studies are just appearing in South America (Jimenez and Jaksic 1988; Bó *et al.* 1996) and Russia (Davygora and Belik 1994; Fefelov 1996). I hope this book will help stimulate other in-depth studies of the remaining as well as the newly revealed species, particularly the Longwinged, Pied, and Papuan Harriers, about which we know almost nothing.

As a tropical raptor, the African Marsh Harrier allows several avenues for assessing differences between tropical and temperate species. Raptorial birds have been extensively studied in temperate regions; they are sensitive to variations in their food resources and relatively easy to follow in the field. Fortuitously, African birds exhibit similar body weights and diet to Northern Harriers, reducing the number of variables in ecological comparisons. Outstanding observational and experimental studies of kestrels, sparrowhawks and owls are rapidly advancing our understanding of the ecological principles governing the timing of breeding, lifetime reproductive success and mate choice in northern zone species. However, the generality of the proposed explanations needs to be tested on organisms under different ecological regimes. For example, while several harriers in the northern hemisphere are polygynous, those in the southern hemisphere are apparently not. The proposed evolution of such mating patterns can therefore be investigated in light of possible ecological differences between northern and southern harriers. The aerial 'courtship' displays of these species may also be revealing since northern birds re-pair each year and southern birds tend not to: what then is the function of so-called courtship displays in birds with long term pair bonds, and where the display advertises phenotypic quality, what effect might additional food have on the rate and intensity of displays?

In all raptors that catch live prey, females are larger than males—the so-called Reversed Size Dimorphism ('Reversed' because it is the opposite of the usual dimorphism apparent in birds). Are males smaller in order to catch particularly agile and difficult prey as traditionally seen; or are males smaller to enhance the agility of their spectacular displays, as recently proposed for shorebirds with similar dimorphism? Is it the female who has increased in size instead? Data presented here for the first time assist in teasing these aspects apart. In the same fashion I examine prey provisioning by males and the influence of so-called 'courtship feeding' on reproduction; there are several expected correlations derived from temperate raptors, but do they apply to the slower breeding schedules of tropical species? Likewise, are the seasonal declines in productivity, already mentioned above, seen in the relatively aseasonal climates apparent in the tropics?

Raptor ecology offers excellent opportunities for comparing selection pressures within and between continents. Raptors hunt live prey—prey that can flee, hide in impenetrable habitats and ultimately evolve complex anti-predation strategies to avoid being dined on. The hunter must become smarter to catch its dinner, but for the mouse the dinner means a life lost. Prey should always be ahead in the arms race between predator and prey because it has more to lose. One would predict, therefore, that mouse behaviour should be strongly dictated by the hunting patterns of its main predator—this in turn should lead to behavioural adaptations by the harriers to counter their prey's changes in behaviour ... and so on. This fascinating contest is dealt with in detail for harriers which hunt mice in somewhat unexpected ways.

A more recently uncovered aspect of raptor ecology (perhaps re-discovered is a better term since it was known 50 years ago), is the very high copulation rate inherent in this group. Birds may copulate up to 400 times just to fertilize one clutch of eggs, a process that takes just one copulation in sparrows and larks. The proposed—and largely accepted—reason for this is that high copulation rates enhance the chances of paternity for a group of birds that are unable to guard their mates. A closer look at these proposals for harriers and other raptors, however, reveals a few curious exceptions suggesting other interpreta-

tions. Chapter 9 also presents some of the first field evidence for possible sperm depletion as a hidden reproductive cost for polygynous male harriers servicing several females simultaneously.

I also look at the effect that food has on reproduction. Birds of prey are prime examples of species in which sex roles during breeding are well defined: in most species females are totally reliant on males for their food from pre-egg laying through to early nestling feeding (Newton 1979). Harriers are no exception. Thus, so-called courtship feeding is likely to play an essential role in the breeding outcome. But is this feeding really courtship? Is it really necessary? If females are judging their future food supplies on this rate, how accurate is it? Can they use it to assess the quality of their mates? Does it have the same intimate relationship with clutch and egg size as found in some terns (Nisbet 1973; Wiggins and Morris 1986). Are polygynous males in those species that regularly attract more than one mate better providers than monogamous species, or are they simply better at fooling their prospective mates? Why do females join already-mated males if their reproduction is impaired by doing so? Tests with Northern Harriers provide some unexpected answers to these questions.

Finally, using simple experimental manipulations of brood size and food resources similar in design to those employed on temperate raptors, I assess how natural selection has shaped the breeding strategies of African harriers (and the constraints on those strategies), to find why this species, like so many others, lays smaller clutches and produces fewer young than its temperate counterparts. Because of the use of some simple field experiments in a subtropical setting, this work provides one of the first direct tests of the idea that birds lay smaller clutches in the tropics because of constraints imposed by limited food resources.

Adaptation, constraint and the role of field experiments

An individual faces many hazards and constraints in its life and, as social commentator Bill Bryson brilliantly relates, just being born is the end point of a miraculous journey where the odds are heavily stacked against success:

> Did you know that each time your father ejaculated (and he frankly did it quite a lot) he produced roughly twenty-five million spermatozoa—enough to repopulate Britain every two days or so? For you to have been born, not only did you have to be among the few batches of sperm that had even a theoretical chance of prospering—in itself quite a long shot—but you then had to win a race against 24,999,999 or so other wriggling contenders, all rushing to swim the English Channel of your mother's vagina in order to be the first ashore at the fertile egg of Boulogne, as it were. Being born was easily the most remarkable achievement of your life. And think: you could just as easily have been a flat-worm. (Bryson 1996)

This is the essence of what behavioural ecologists term sperm competition, although the sperm is usually from different males in promiscuous or polyandrous species. Once born, every individual faces a barrage of predators, parasites, and siblicidal and cannibalistic conspecifics, all ready to eat or use its body. Drought, severe food shortages, competition for food, mates and a space to live are all additional selection pressures honing individuals to their environment. There are two possible pathways facing an organism over evolutionary time in the face of such selection. Individuals can be pushed toward an adaptive solution (aptus = towards a fit), or they can be constrained from reaching such a solution, and adopt a 'best of a bad job' compromise—or go extinct. Nevertheless, a bird may adopt the 'strategy' of laying three rather than four eggs, say, because such individuals leave more copies of themselves than birds laying four eggs. We must clearly identify the selection pressures, and particularly the constraints, before we can claim that an organism is well adapted to its way of life. For example, tropical animals may rear fewer young per breeding attempt for many reasons (more predators, longer lifespan, greater parasite burden), foremost of which is the possibility that they are constrained by environmental factors. But how would you test between these possibilities?

Consider the case of a female kestrel laying eggs in the middle of the breeding season. Researchers studying her clutch size find, as expected, that it contains one egg less than that of females laying two weeks earlier. I say 'expected' because seasonal declines in clutch size are a common temperate zone phenomenon in birds and mammals. The adaptationist, who sees every behaviour or reproductive decision as an adaptive response to maximize lifetime reproduction, might argue that the kestrel laid one egg less because she cannot expect to rear as many young as early-laying females. Possibly her late-fledging young may be unable to compete with already independent young who have filled all the empty territories and developed their foraging skills. Perhaps the extra effort of rearing another nestling may retard the female's future abilities to rear other young. She is being prudent and adaptive. However, a student who favours ideas of constraint would argue that a bird laying mid-season is *unable* to lay a larger clutch because of age or inexperience or because she is laying eggs at a time of reduced food resources in the environment.

To distinguish between the two hypotheses is not possible from pure observation—yes, young females may always lay small clutches and yes, the food in the environment does decrease, but these do not *disprove* the adaptive notion that female kestrels are being prudent by laying fewer eggs. The essential test is experimental: give the young female extra food just before she lays and compare her clutch with same-age females laying simultaneously. One removes the suggested food constraint—thus the female should lay more eggs. If the result from randomly chosen birds is that clutch size stays the same as in unfed controls, then a second experiment is needed: supplement the diet of both young *and* old birds laying mid-season. If young females are physiologically constrained they should lay the same number of eggs, but older females provided with extra food should lay an extra egg. If the clutches stay the same for both age classes the adaptationist can rightly declare that the seasonal decline is inexplicable by a lack of food alone and the smaller clutches are best explained by an adaptive decline in clutch size by prudent kestrels.

One can invoke similar arguments to claim that breeding birds lose body weight over the breeding season as a cost imposed by their energetically demanding foraging schedules to feed hungry nestlings. The adaptive alternative is that a female who gained weight to lay eggs and incubate effectively may then reduce her mass to forage more efficiently for insects or mice. She may also lose weight to become more agile to reduce the risk of predation. Both ideas explain the data set well but only a neatly designed experiment can distinguish the adaptive response from the constraint. Many such experiments are now being performed and indeed these examples are taken from ones that have already been done (e.g. Drent and Daan 1980; Norberg 1981; Meijer *et al.* 1988; Hillström 1992).

What are these constraints? Constraint may limit an organism to its particular life history or behaviour, either through the effects of the past (phylogenetic constraints), genetic make-up, present ecological setting (environmental constraint), or its present body plan and physiology (physiological/allometric constraint) (Stearns 1992). Each of these may prevent an organism from achieving the optimal reproductive avenue, despite constant selection from the environment, conspecifics, or predators. Thus those who argue that adaptive solutions evolve to each constraint are correct only up to a point—organisms can often come close to an optimal solution but close scrutiny may reveal a poorer fit than expected. One can generally distinguish such constraints by comparing what an animal does with an ideal model of what it could do. Indeed one school of thought recommends that adaptive explanations need only be sought when the more parsimonious explanations such as allometric scaling of features, like body size or age at first breeding, deviate from the expected (Williams 1966; Gould and Lewontin 1979). For morphological traits this is a fairly easy task—one simply plots body size against brain size, for example, for as many species as data allows, and look for species that lie above or below the line (Clutton-Brock and Harvey 1980). Behavioural traits are a little more difficult but still tractable. An African antelope performing on a lek to attract mates may be more successful the longer he stays, but the longer he does the more susceptible he is to losing condition and being eaten by a predator. The perfectly adapted male would long ago have evolved sufficient reserves to stay on the lek for months at a time; a physiological constraint. So as Endler (1986) has put it, 'natural selection is sufficient for evolution, but it is neither necessary for, *nor does it guarantee evolution*' (my italics). Very often, if change does not occur in the face of selection, many species, particularly specialized ones, go extinct. We will discuss several examples of such constraint when dealing with the foraging activity of harriers and particularly the evolution of smaller clutch sizes in regions close to the equator. The important point is that many behavioural ecological and evolutionary problems have more than one solution: some may have adaptive solutions or may be the result of insurmountable constraints; the best way to tease the two apart to find the ecological or physiological constraints is through field experimentation.

Overview of study species

This book is not a monograph detailing every aspect of the lives of marsh harriers and their kin. Instead I have concentrated on aspects that have intrigued me, and added other life-history details. Because my studies have focussed on two species: the African Marsh Harrier and the Northern Harrier of North America, much of the detailed work refers to them. However, I have also scoured the literature on other harriers and where appropriate added their details as well as supporting (or contradictory) evidence from other raptors and other birds. In order to provide an introduction to both study species, as well as my study areas and methods, they are briefly outlined here.

All of the world's marsh harriers occur in wetlands characterized by reeds and open water, while the smaller dry-country harriers occur over grasslands and steppe habitat. Only the Reunion Harrier regularly occurs over forests where it hunts birds (Clouet 1978). However, North America's Northern Harrier is somewhat atypical in that it prefers wetter areas of grasslands and marshes and appears to occupy the niche that a marsh harrier might take if such occurred. Since it is the only species of harrier in North America in contrast to the two (Cinereous and Long-winged) in South America, the two in Australia (Spotted and Swamp), the four in Europe (Pallid, Montagu's, European Marsh and Hen), and the two in southern Africa (Black and African Marsh) it may well have had an unconstrained choice of habitat. This does not mean that every-

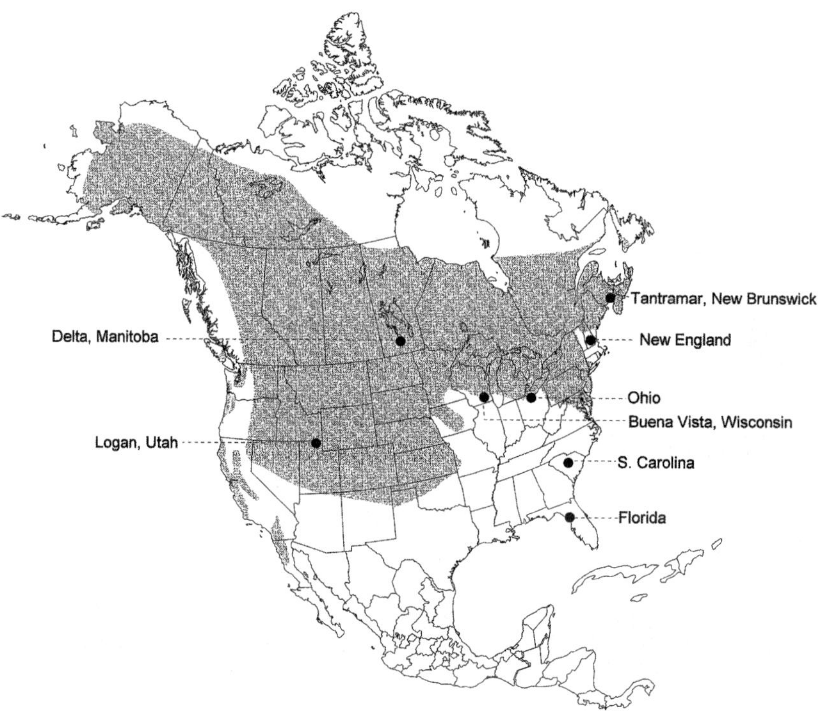

Fig. 1.1 The breeding distribution of the Northern Harrier in North America and the location of the Tantramar Marsh study site. Other study sites mentioned in text are shown, including some winter sites from the southern United States.

where in North America is it found breeding in wet locations, since some prairie habitats and upland marshes in western North America offer few such sites (Sealy 1967; Duebbert and Lokemoen 1977; MacWhirter and Bildstein 1996).

Study areas

In my study area, on the Tantramar Marsh in eastern New Brunswick, Canada (45° 53′ N, 64° 20′ W; see Figs. 1.1, 1.2), Northern Harriers occurred at relatively high densities on a reclaimed marshland, extensively dyked and protected from the extreme tides associated with the Bay of Fundy. Most nests occurred in water-logged areas, amongst sedges, *Typha* reeds and the damper areas of rough cow pastures. Some of these areas were managed as waterfowl habitat by the hunting concern Ducks Unlimited. However, particularly when densities were high, some birds nested high and dry in hayed meadows comprised of Timothy *Phleum pratense* grasses, just as some Hen, European Marsh and Montagu's Harriers in France and Spain, and the Black Harrier in South Africa have taken to nesting in cereal crops, with sometimes hazardous consequences during harvest (van der Merwe 1981; Burneleau *et al.* 1989; Arroyo 1996a). In the study area, covering about 60 km^2, no nests were destroyed by human activity during haying since birds had fledged before haying began (Fig. 1.2). Feral cats and dogs were absent and the main predators were White-necked Ravens

Fig. 1.2 The Tantramar Marsh study area in New Brunswick, eastern Canada, 60 km^2 of rough cow pastures, hayed fields, and impoundments constructed for duck management. Once flooded by the high tides of the Bay of Fundy, the whole area has been extensively dyked and drained for over 300 years.

Corvus cryptoleucus, raccoons, and occasionally foxes. One change did occur between my initial studies and those undertaken later—coyotes *Canis latrans* moved into the area, coincident with a dramatic change in prey fluctuations. These are treated briefly in Chapter 4. My studies there spanned 1980 to 1983, with simultaneous and subsequent studies by Phoebe Barnard and Bruce MacWhirter (on provisioning strategies and their consequences), and later on sex ratios and parental investment in sons and daughters by Bruce Macwhirter when he returned for his doctoral studies. With assistance provided by Gay Hansen and Matthew Saunders, the population was studied continuously from 1980 to 1985 and again from 1990 to 1993; a total of 10 years.

Precipitation is relatively even throughout the year, with snow lying on the marsh from about October to April. During this time most birds, except a few males, migrate south to the lower United States, Mexico, and Central America (MacWhirter and Bildstein 1996). Breeding is highly synchronized within a window that begins as temperatures rise in early May; egg laying is over in one month and all young are fledged by mid-August, whereupon young migrate south prior to the first snows of the year. The breeding season is thus about 120–145 days long. Some replacement nests occur if the attempt fails early on, but they form less than 10% of all nests (Simmons 1984), and may not occur at all in areas further south (Hamerstrom 1969). This relatively tightly constrained breeding schedule followed by southward migration is typical for most temperate breeding harriers (Watson 1977; Cramp and Simmons 1980; del Hoyo *et al.* 1994).

About 12 000 km to the south-east, on South Africa's Cape peninsula, the African Marsh Harrier reaches its southernmost extent (Fig. 1.3). It extends north into equatorial regions of central Africa and is largely sedentary throughout its range. I studied African Marsh Harriers from 1984 through 1986 with a short follow-up in 1987. The Wilderness Lakes system, a slightly brackish series of lakes and wetlands, lies in a beautiful valley wedged between the Outeniqua mountains and the Indian Ocean on South Africa's south coast (Fig. 1.4). The area is Mediterranean in climate, lying as it does at 34° 00′ S, 22° 40′ E, and experiences year-round ('winter') rains, unusual for areas further north in Africa. Rainfall, rather than temperature, plays a major role in the breeding schedules of these birds. Temperatures varied from below zero in the austral winter to above 40° C when hot adiabatic ('berg') winds blew from the interior. However, in general there was low variation in both monthly rainfall (annual mean was 815 mm ±260 mm) and in the maximum and minimum monthly temperatures. Although African Marsh Harriers are found breeding throughout this area, it is not necessarily their prime habitat. Recent distribution data published by the Avian Demography Unit at the University of Cape Town (Harrison *et al.* 1997) show, for example, that the reed and papyrus swamps of the Okavango delta in northern Botswana are the stronghold for the species in southern Africa. Trying to study harriers in such wild areas, however, would be difficult given the inaccessibility of nests and the risks of encounters with formidable predators. The Wilderness Lakes were rewarding because the low vegetated fossilized dunes surrounding them provided ideal vantage points from which to observe nests and

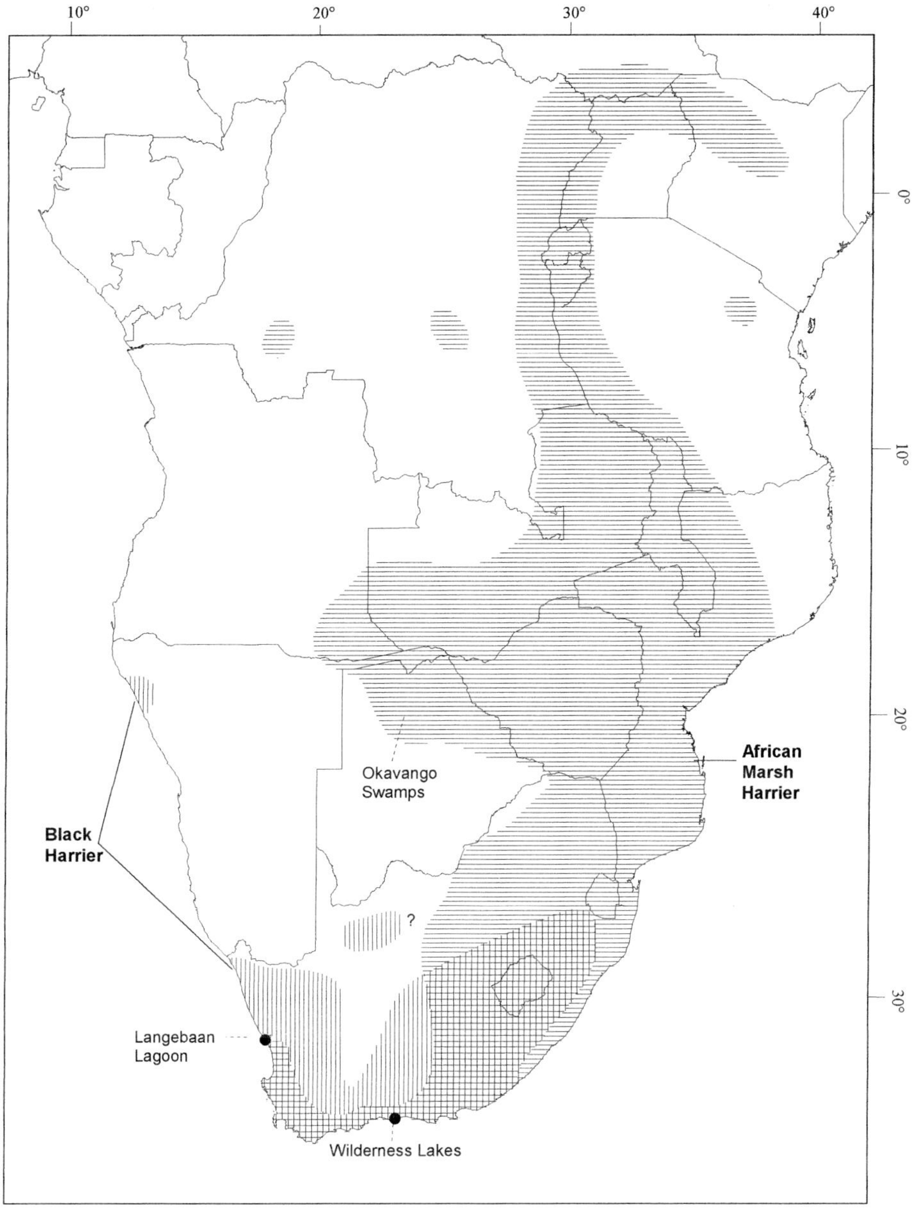

Fig. 1.3 The distribution of the African Marsh Harrier throughout south-central Africa and the Wilderness Lakes study area at the southern end of its range. The Okavango Swamps in Botswana are the centre of distribution of this species in its southern range according to recent atlas data. The world breeding distribution of the Black Harrier is also shown, as is the site of the only study of its breeding ecology (Langebaan Lagoon).

Fig. 1.4 The Wilderness Lakes study area and the lake margins favoured by African Marsh Harriers.

foraging birds. As an added bonus the whole area falls within the smallest, and most species-rich floral kingdom in the world: the Cape Floral Kingdom or Fynbos (literally fine-leaved bush). Indeed, so rich is this Mediterranean-like heathland that Carl von Linne (Linnaeus) described the area in glowing terms following the collections sent to him by his (often ill-fated) students: 'There is no place in the world with so many rare plants, animals, insects and other wonders of Nature as Africa, and it seems as if they have been concentrated to the Cape.'

My main study area, of 11 km^2, was three interconnected shallow water bodies, surrounded by reeds and drained by the Serpentine River which provided ideal floodplains for foraging harriers. Some data were also collected from two freshwater lakes 10 and 12 km east, supporting large reed beds of *Cladium mariscus* and *Typha latifolia*. Fossilized and vegetated dune ridges separate the lakes from the Indian Ocean, and support alien willows and indigenous 'Fynbos' vegetation. The immediate lake surrounds support *Scirpus littoralis* and large beds of *Phragmites australis* reeds. Bordering these normally damp areas, and extending into floodplains, are thick *Stenotaphrum* grass and *Juncus kransii* habitats, frequently foraged by the marsh harriers.

Overview of research methods

Unlike my studies in Canada, where one of the research goals was to understand population dynamics in relation to vole fluctuations, my studies of African Marsh Harriers sought to determine how *individuals* behaved. This arose partly through the constraint apparent in many tropical areas, of high predation rates in field studies, disallowing constant visits to ground nests (in some studies all nests are destroyed by predators: Skutch 1985). Instead, my objective from the outset was to gain an intimate knowledge of the behaviour of a several pairs of birds each year, and non-intrusively perturb their breeding schedules with some simple manipulations to gauge their behavioural and evolutionary responses. To minimize the impact of my research on nesting pairs, I used only one path into the nest, frequently jumping over large obstructions (e.g. banks of sedge) to make my trail discontinuous. As an added precaution I laid mongoose-sized traps on my path in wet vegetation, which were baited with either whole eggs or dried egg shells to leave no scent. In two years I caught one Egyptian Mongoose *Herpestes ichneumon*, several smaller water mongooses *Atilax paludinosus*, and one large spotted genet *Genetta tigrina*. Had these animals continued on their path they would almost certainly have taken the harrier's eggs and young. Indeed some nests were destroyed despite my precautions, so while mammalian predators are thought not follow human scent to nests (Gotmark 1992) they were probably the main predators of nests in this study. Mammalian predators of Northern Harriers were not trapped in Canada, but naphthalene 'moth balls' were laid on trails to nests in an attempt to dissuade predators.

Nest observations: assessing provisioning performance

Between 1984 and 1986 in southern Africa I observed 19 harrier nests as well several non-breeding but territorial birds. Since these harriers were monogamous, 'nests' are synonymous with 'pairs' observed. Birds were resident and often site and mate faithful (unlike Northern Harriers: Hamerstrom 1969, 1986), so some birds occurred in more than one year. In 1987, I paid a brief visit to record clutch and egg sizes early in the season. Most observations took place from simple hessian and wood hides placed 60–100 m away from nests and placed either high in pine trees or on dune ridges overlooking the marsh. These positions allowed views for over 1 km across the marsh yet were close enough for me to identify prey being brought to nests. Intensive day-long observations started up to a month before harrier breeding and continued throughout the breeding season for a total 2203 bird-hours. In Canada, observations began just prior to breeding, and continued through to fledging of young. Precise observations to record provisioning rates for a minimum of nine males per year were shorter (562 hours) than for African Marsh Harriers. In total about 900, 700, and 900 hours were spent on the marsh in 1984, 1985, and 1986, respectively. For all nest watches, I recorded copulation behaviour, provisioning and foraging performances, advertisement displays, adoption behaviour, and behaviour of young birds. To allow detailed comparisons between pairs and years, all behavioural data was recorded in a series of two-minute intervals. This method of focal animal sampling (Altmann 1974) is a simple and efficient means of analysing and comparing proportions of different behaviours, between individuals, seasons, and years. Throughout I have kept statistical analyses (and lists of references) to a minimum for the sake of readability, although interested readers are referred to the primary literature.

Analysing food provisioning was of great importance because, like other raptorial birds (Newton 1979), female harriers are completely reliant upon their mates for food from about one week before egg laying until nestlings are roughly half grown. This means that food provisioning performance was expected to play a major role in the clutch size and breeding success of the nest, as found in temperate breeding harriers (Simmons *et al.* 1986a). Its effect on copulation rates, aerial displays, and other breeding parameters was also assessed, and in some cases checked with experimental food supplements to various pairs.

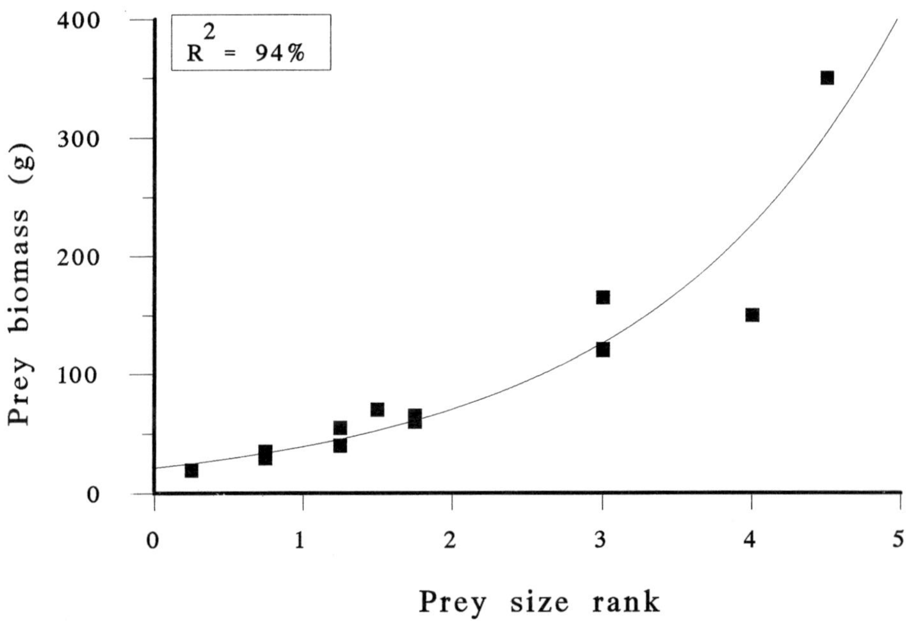

Fig. 1.5 The relationship between prey biomass and foot size rank for harriers. Foot size rank was a visual estimate of the prey size based on the harrier's foot, such that prey completely enclosed by the foot was ranked 1.0 and so on.

The rate at which food was provisioned was determined by dividing the number of deliveries by the total hours observed for each male at each stage of breeding. Recording the number of items alone, however, is inappropriate if the prey types differ substantially in size or type, so I used two measures to assess prey biomass. I happened on these methods while observing African sparrowhawks in another study (Simmons 1986a), and they were independently and simultaneously used by Dutch researchers (e.g. Masman 1986) studying kestrels. By providing birds with known weight items (or in the case of kestrels, stealing food items from the cache site) and ranking their size in the feet of the prey-carrying male, I could accurately gauge food biomass (Fig. 1.5). Several researchers have tried other methods such as timing how long an adult takes to feed on a known weight prey item. This method, at first glance the most accurate of the two, over-emphasized the mass of bird prey because raptors first pluck bird prey but rarely mammal prey, and pick over bones too large to swallow, for long periods. Thus prey ranking was better than timing the meal for harriers (Simmons 1986a). For items that could not be ranked relative to the bird's foot, consumption time was a useful guide to prey size because the two were significantly related ($r = 0.79$, 86 df, $P < 0.00001$). Thus I converted consumption time to foot size rank using the equation: Size = 0.082 + 0.121 [min]. Prey biomass and foot size rank were significantly ($r = 0.97$, $P < 0.00001$) exponentially related (Fig. 1.5) by the following equation: Biomass (g) = e(3.08 + 0.586[size]).

Experimental supplementation of food

Experimental manipulation of factors thought to influence life-history tactics and behaviours of any organism is the best way to tease apart the critical factors, and one way I tried was to supplement the diets of pre-breeding harriers. If food is critical in determining egg sizes, clutch sizes, and body condition of harriers one not only predicts that egg and clutch sizes of supplemented birds should increase, but enhanced body condition of fed birds may increase their display rates or vigilance relative to unfed controls. To supplement selected pairs of African Marsh Harriers, 1–2-metre-high posts were placed at favoured perching/eating areas in their territories. All pairs readily accepted prey provided on these posts and prey included freshly caught (usually non-reproductive) *Rhabdomys* and *Otomys* mice (28%), medium sized mammals such as Mole Rats *Bathyergus spp.* and Guinea Pigs *Procavia spp.* (22%), large mammals (dissected primates: 19%), and also fish (16%), and small birds (14%). Pairs were fed once or twice daily by placing prey on top of the posts in the early morning and afternoon. For first nests, feeding started on average 30 days before clutch initiation, and for seven of the nine nests continued until clutch completion for an average of 40 ± 13 days (range 31–62 days). For two nests I undertook additional experiments: in one case, a renesting female was fed past the date of her first nest failure, but for only 6 days. Hence feeding was stopped 27 days before she re-laid. In the second case I waited until the female was obviously pregnant, and then immediately began to feed her an average 1.6 times a day (228 g/day) until clutch completion. In this way I hoped to gauge whether supplementary feeding was important because it affected the female's condition *before* she laid (the 'capital' model *sensu* Drent and Daan 1980), or because food supply *during* egg formation (the 'income' model) was the most important factor in determining egg or clutch size. Providing prey on posts thwarted food-theft by mongooses, and territories were observed to ascertain that harriers and not resident owls were taking all the prey. Indeed, as feeding progressed I found harriers perched on their posts apparently waiting for their daily supplements; prey was then taken within 10–15 min. As far as I am aware this is the first subtropical raptor to be experimentally manipulated in this way and probably the first subtropical species of any type. Most other experiments have looked at the cost side of reproduction by increasing the brood size to measure the effects on parents.

Marking birds for individual recognition

Breeding Northern Harriers were caught using Frances Hamerstrom's (1963) Dho-gaza set-up, made from a collapsible net and stuffed Horned Owl placed close to an active nest. Captured birds were metal-ringed only, so individual recognition within the study population was possible only from atypical plumage characters of individuals (for example, some males acquired distinct bloody marks on their white breasts from clutching beheaded meadow voles as they flew; others lost single feathers in their wings, or displayed similar and distinctive behaviour patterns around the nest). Since other studies of marked Northern Harriers indicated little site or mate fidelity (31% of both males and females of 173 colour marked in Wisconsin returned the following year: Burke 1979), I assumed about two-thirds of the harriers were new birds each year. For all birds that were caught, mass, wing chord, tail, culmen, and halux claw dimensions were measured.

To facilitate individual recognition of African Marsh Harriers, all intensively observed pairs were caught and wing-tagged during 363 h of trapping (10 h per adult). While perch-hunting raptors are relatively easy to trap for marking purposes, harriers pose particular problems, as these trapping rates suggest. The dho-gaza used on Northern Harriers was also employed on African birds, but with less success. Its downfall is the length of time required for set-up, and the need for relatively still, overcast days. Since harriers won't touch, and rarely see, a noose-covered Balchatri trap baited with mice or birds, I used a method proposed by John Mendelsohn. This was a simple sprung-noose trap, activated by the bird hitting prey on a treadle in the centre; it has proven effective on many African raptors, from 60 g Pygmy Falcons *Polihieirax semitorquatus* to 6 kg Martial Eagles *Polemaetus bellicosus*. Young birds were marked with metal rings, and small wing tags. Only one tag per bird was used to minimize any impact of tagging, but this precluded any analysis of tag loss. Because I could recognize birds by their behaviour and idiosyncrasies, I determined that no tag loss occurred over four years. Some birds wore their tags for eight years and the design is described by Picozzi (1971).

The colour-marking program allowed me to investigate annual mortality rates, as well as fidelity to mates and territories, without the complexities arising from survival rates determined from ring recoveries alone. These variables are difficult to study solely with metal rings, and face added complications in an African setting where ring recoveries average less than 1% (T. Oatley pers. comm.). Putting tags on my birds minimized these problems and individual recognition was possible from considerable distances. I found no circumstantial evidence that tags affected breeding or survival, and within-population comparisons would not be compromised, even if subtle effects were present, since 90% of birds had tags (the others had colour rings). Thus tagging provided an opportunity to study harrier breeding and survival as it has been in Scotland (Picozzi 1984b) and Australia (Baker-Gabb 1981). In the following chapter we will look at some of the traits that make the harriers a unique group, and characterize a genus adapted for life in open grasslands and wetlands.

2 *Evolution and peculiarities of the harriers*

Harriers occur on every continent bar Antarctica and are relatively evenly spaced around the globe. Four occur in Europe, four in Africa, two each in Asia, Australasia, and South America and one in North America. However, recent work on the genetics of this group has revealed 2 new species and the interrelationships between all 16. In this chapter we will look at these relationships and discuss the peculiarities that distinguish harriers from other raptors.

The harriers are best known for their buoyant, low quartering flight, exceptionally light wing loading (the Montagu's Harrier at 21 g/cm^2 is among the lightest of all raptors: Brown and Amadon 1968) and their specializations on voles and birds at different times of the year. However, the harrier genus (*Circus*) is seen by some as an unspecialized group of raptors whose affinities are poorly known (e.g. Brown and Amadon 1968, p. 21). While the second statement is certainly true, I am not alone (Nieboer 1973) in being baffled by the view that harriers are unspecialized. The mating system of harriers sets them apart behaviourally from other raptors—they are the only genus in which some species are regularly polygynous; this in turn has probably led to the marked plumage differences in males and females that are also unusual in raptorial birds. They also differ from other raptors in their ground nesting habits, their spectacular ability to transfer prey in an aerial food pass, and their peculiar adaptations, shared with nocturnal owls, for foraging and high resolution acoustical abilities. These characteristics suggest that harriers have adapted superbly to a life style in open grassland and wetland that is practised by no other group of raptors. We will briefly explore the evolution of some of these traits, and some of the unexplained holes in their world distribution, described by Leslie Brown as 'one of the most extraordinary ecological lacunae in the African bird faunas'.

Harrier evolution—historical perspective

Harrier phylogeny has proven problematic and is frankly a mess. Morphological similarities between skeletons suggest, for example, that the harriers' closest relatives are the *Spilornis* snake-eagles of southeastern Asia and the *Polyboroides* harrier-hawk (Gymnogene) of subSaharan Africa (Brown and Amadon 1968; Baker-Gabb 1984). These genera occur mainly in the southern hemisphere, and the similarities are based on skeletal likeness. This superficial similarity is misleading. Circus and Accipter share a common ancestor, as we shall see (below).

Within the genus *Circus,* most authorities have generally lumped marsh harriers from all corners of the globe as subspecies of the European Marsh Harrier (Streseman 1924), giving it a definite—and misleading—eurocentric slant. Taking a more systematic approach, Nieboer (1973) undertook an extensive morphological comparison of the harriers and recognised ten species, one more than previous estimates. In doing so he agreed with Streseman (1924) who recognized five subspecies of marsh harriers, except in one important way. Nieboer differentiated the small African Marsh Harrier from the remaining four subspecies—the European, Eastern, Reunion, and Australasian harriers respectively). The African Marsh Harrier (erroneously named *ranivorus* 'the frog-eater') was seen by Nieboer (1973) and Baker-Gabb (1984) as the most primitive of the marsh harriers, because of its monogamous mating system and monochromatic plumage. Since it exhibits some traits of the dry country (steppe) harriers including a diet of small mammals (Kemp and Dean 1988; Simmons *et al.* 1991), foraging that takes it over dry grasslands, and tail/wing ratios that are intermediate between *aeruginosus* and the steppe harriers (Nieboer 1973), it was seen as the link between dry and wet country harriers (Nieboer 1973). The origin is mainly vindicated in the new genetic evidence below, but *ranivorus* is not the link between the steppe harriers and the marsh harriers as we shall see.

Fifteen years after Nieboer (1973), Amadon and Bull (1988) added three more species to this group, giving full species status to *spilonotus* in eastern Asia, *maillardi* in Reunion and Madagascar, and *approximans* in Australia. The general confusion surrounding the Australasian species *assimilis* and *approximans* is reflected in their species names which literally mean 'similar to other harriers' and 'approximately like the marsh harrier' respectively—singularly uninspiring names! Ornithologists in this region (Baker-Gabb 1979) and those outside (Amadon 1978) argued against a mere subspecies designation, pointing up marked plumage differences in both sexes. Because these differentiations are made by taxonomists chiefly on the basis of plumage and morphological differences, they are subject to vagaries introduced by convergent evolution making two unlike groups appear very similar. On the other hand, other selection pressures can produce dissimilar plumage types in species later found to be genetically close. A recent example of this is the finding that the pied African Hawk Eagle *Hieraeetus fasciatus* (and indeed all the Hieraeetus group) are close genetic relatives of the essentially monochromatic *Aquila* eagles and should be included in that genus between the Golden and Black Eagles *A. chrysaetos* and *A. verreauxii* (M. Wink pers. comm.).

Table 2.1 The harriers of the world and their presently accepted and alternative English names, after Brown and Amadon (1968), del Hoyo *et al.* (1994) and M. Wink. I use the 'common' names throughout this book. Recognized subspecies are also given. Three new species (*) are recognized: the Madagascar Marsh Harrier is no longer seen as a subspecies of the Reunion Harrier, the Hen and Northern Harrier are separated, and the Papuan Harrier from New Guinea is separated from the Eastern Marsh Harrier.

Specific name	Common name	Alternative name	Geographic name
C. aeruginosus	European Marsh Harrier	The Marsh or Western Marsh Harrier	Europe, western Russia
C. a. harterti	European Marsh Harrier	dark subspecies of above species	North Africa
C. ranivorus	African Marsh Harrier		South-central Africa
C. spilonotus	Eastern Marsh Harrier	Spotted Marsh Harrier	Asia
C. spilothorax	Papuan Harrier*	Spotted backed Harrier	New Guinea
C. approximans	Swamp Harrier	Australasian or Pacific Marsh Harrier	Australia, New Zealand
C. maillardi	Reunion Harrier	Reunion Marsh Harrier	Reunion Island
C. macrosceles	Madagascar Marsh Harrier*	Madagascar Harrier	Madagascar, Comores
C. buffoni	Longwinged Harrier		South America, Central America
C. assimilis	Spotted Harrier	Jardine's or Allied Harrier	Australia
C. maurus	Black Harrier		South Africa, Namibia
C. cyaneus	Hen Harrier		N. Europe, Russia
C. hudsonius	Northern Harrier*	Marsh Hawk	North America
C. cinereus	Cinereous Harrier		South America
C. macrourus	Pallid Harrier	Pale Harrier	Russia, eastern Europe
C. melanoleucos	Pied Harrier		Mongolia, China, Myanmar
C. pygargus	Montagu's Harrier		Europe, western Russia

The steppe harriers were always easier to group than their marshland cousins and the Hen and Northern were seen as subspecies (*C. cyaneus cyaneus* and *C. c. hudsonius)*, while the other grey harriers (Pallid, Montagu's, Cinereous) were seen as closely related. The two odd harriers falling outside these two groups were the tree-nesting Spotted Harrier of Australia and the Longwinged Harrier of South America with no obvious affinities with other harriers. Some authors (e.g., Cramp and Simmons 1980) surprisingly treated the Cinereous Harrier as subspecific to the Hen/Northern group despite their obvious plumage differences and wide separation.The Black Harrier was also sometimes lumped with the Cinereous Harrier as a small southern hemisphere steppe harrier (Nieboer 1973).

Origins

Inherent in all analyses, particularly Nieboer's (1973), is that the harriers evolved in the northern hemisphere, because of the great diversity concentrated in Europe (four species). A long history of occupation, it is argued, should lead to a greater opportunity for speciation and hence species-rich areas should give a clue to the origin of the group. David Baker-Gabb, a New Zealand ornithologist who undertook extensive studies of the two Australasian harriers, proposed instead that the harriers arose in the southern hemisphere (Baker-Gabb 1984). He noted that habitat diversity may alone explain the high diversity of harriers in Europe. He also noted that the proposed ancestral plumage type (Nieboer 1973) in which males and females are monochromatic, was commonest in southern climes (Spotted Harrier, African Marsh Harrier, Black Harrier) and absent in the northern hemisphere where 'advanced' dichromatic plumages are common (Baker-Gabb 1984). He also found that the proposed ancestral mating system of monogamy, was also more common in the southern hemisphere. He concluded that the ancestral harriers arose in the south and moved into the northern hemisphere. While his conclusions may be correct (below) the two apparently independent sources of evidence (ancestral plumage and mating system) are almost certainly related (polygynous species tend to be the most dimorphic: Chapter 5) and they cannot be used as separate evidence for ancestral origins. I also find it hard to understand why monogamy is commonly seen as the ancestral mating system. The usual argument is based on the idea that polygyny is more advanced because it is a more 'sophisticated' mating system. However, avian polygyny is related to ecological factors such as habitat type (two-dimensional marshes) and abundant food resources (Verner and Wilson 1966). Hence polygyny is expected in harriers and particularly those in northern areas for ecological reasons, and is unlikely to differentiate ancestral from evolved harriers in the two hemispheres. One might also argue that birds as evolved reptiles, which typically show little paternal care and polygynous or promiscuous mating systems (e.g., Gans and Huey 1988), would be more likely to show polygyny as the ancestral mating system. So where did the harriers evolve and how can we get closer to a solution for this problem?

Genetic similarities now tell us that the harriers shared an ancestor with the *Accipiter* hawks (Wink 1998). The two genera diverged about 5–7 My (million years) ago, differing at the genome level by about 12 per cent. The harriers closest extant relatives, therefore, are the Goshawk and Cooper's Hawk (*A. gen-*

tilis and *cooperi*). This may not directly help us establish where the harriers evolved but the *Accipiter* hawks are particularly speciose in the southern hemisphere where six species are found together in southern Africa (Brown and Amadon 1968; Simmons 1986b), and seven species co-occur on the island of New Guinea (Diamond 1985). If centres of species richness point to areas of speciation then the idea that harriers arose in southern climes is strengthened.

A new phylogenetic tree for the harriers

Using blood and feathers supplied by researchers from around the world, Wink's lab extracts DNA from the mitochondria or nuclei of cells, copying the mitochondrial cytochrome b gene using PCR (Polymerase Chain Reaction) techniques, and then assesses differences in at least one thousand base pairs. The genetic similarity and thus the evolutionary history can be reconstructed by comparing the number of differences in base pairs of one species against that of another. The less they differ the more closely they are related. DNA from the nucleus of the cell is used to determine far distant evolutionary histories while mitochondrial DNA is used to judge finer scale similarities between closely related taxa. The differences are due to point mutations (Nucleotide substitutions) which are assumed to occur at a relatively constant rate and thereby give a yardstick as to how long ago the species diverged. For birds in general a 2 per cent difference in nucleotide substitutions equates to approximately one million years (My) since divergence. It is possible, however, that divergence times are double this due to the long generation times of raptors (M. Wink pers. comm.).

Mitochondrial cytochrome b genes were available from a total of 14 of the 16 harrier species (Pied and Papuan material was not available). Several statistical analyses with slightly different assumptions were used to construct the final tree—Maximum Parsimony, Maximum Likelihood, and Neighbour Joining (Saitou and Nei 1987; Swofford 1993). If concordance is achieved between the resultant trees we can be fairly sure of the robustness of the final phylogeny. This was apparent in the phylogenetic trees drawn for this genus, but only the most parsimonious phylogenetic tree (Maximum Likelihood) is shown (Fig. 2.1).

There are many surprises, and some expected outcomes. As expected the steppe harriers and the marsh harrier form two clusters—with one exception—the Montagu's Harrier is not a steppe harrier but is ancestral to the entire marsh harrier group! This highly unusual result occurred in all three phylogenetic trees and is unlikely to be erroneous. It indicates how misleading plumage types can be in reconstructing phylogenetic trees. As field observers know, immature and adult

Fig. 2.1 A phylogenetic tree of the harriers. Similarities are based on analyses of the nucleotide sequences in the cytochrome b gene for 14 harriers using 1000 base pairs. Harriers diverged from the Accipiters about 7 million years ago. The marsh harrier group shared a common ancestor, with the Montagu's Harrier and the grey (*cyaneus)* harriers arose and separated from the Longwinged Harrier of South America before this split. Harriers appear to have arisen in South America, not Europe as previously assumed. Three new species are recognized: the Northern, the Madagascar, and the Papuan Harriers (not shown). ▶

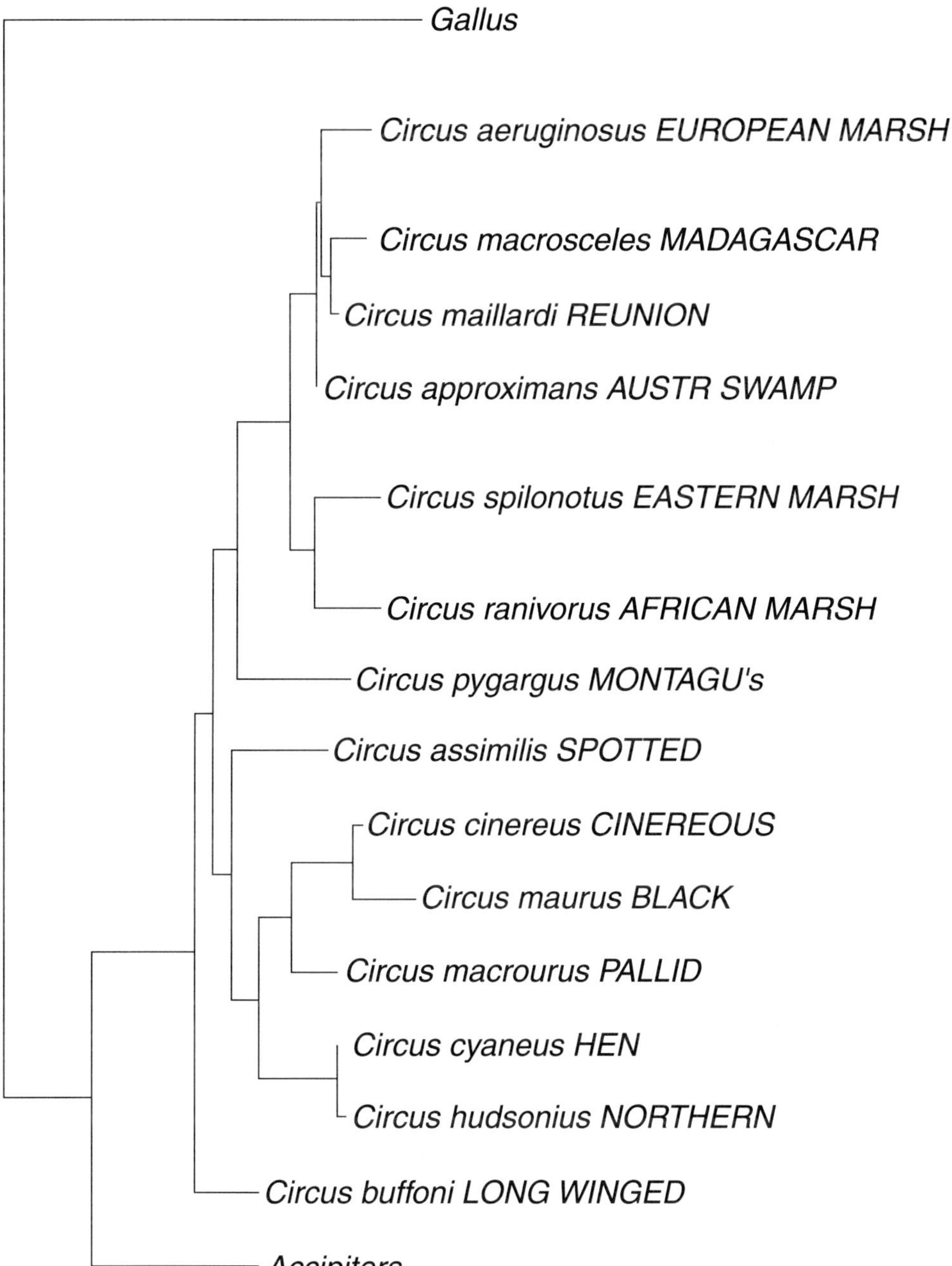

Scale: – = 1% genetic difference

Young Pallid Harrier.

Fig. 2.2 Differences in appearance of three harriers once thought to be subspecies. Males illustrated are: (a) Reunion; (b) African Marsh; (c) European Marsh.

(b)

(c)

Fig. 2.2 *Continued*

female Pallid and Montagu's Harriers are so similar they are virtually impossible to differentiate in the field. They may also hybridize in some areas of sympatry (Cramp and Simmons 1980). Yet the genetic distance between these species is calculated as 7.3 per cent, meaning they diverged more than 3.7 million years ago!

The ancestral harrier is the South American Longwinged Harrier, corroborating the idea that the genus arose in the southern hemisphere and moved north.

A further surprise is the finding that two new species have emerged. The first of these is the Reunion Harrier *Circus maillardi*, found only on the small Indian Ocean Island of Reunion, 700 km east of Madagascar. While plumage differences between this species and the Madagascar Harrier *Circus macrosceles* are not great, morphological differences between the two forms are substantial. Both were previously seen as subspecies of the European Marsh Harrier by Stresseman (1924) and Nieboer (1973). This was followed by Clouet (1978), who studied it in the field. Both harriers differ from the recently evolved *aeruginosus* by 1.8 per cent to 2.0 per cent (diverging about one million years ago). A just-completed (1998a) morphological and ecological survey of the Reunion Harrier by Bretagnolle, Thiollay, and Attie, reveals that the bird is significantly smaller in tarsus (males: 81.6 *vs.* 89.4 mm), wing (355 *vs.* 410 mm) and tail length (214 *vs.* 231 mm) than the Madagascar Harrier, for both males and females. Reunion birds (Fig. 2.2) are also darker than Madagascar birds (front cover) with males having mainly black heads, darker backs but lighter grey on upper wings. The birds on the Comores are indistinguishable from the Madagascar birds. It is a frustrating omission that body weights are not known for these harriers because they are described as highly dimorphic by Cheke (1985). This reflects the extraordinary adaptations that evolved in this harrier when it arrived on Reunion. They are the only harrier which has adapted to a forest habitat—the island was known as 'English Forest' when first discovered by British sailors in the 1620s, with forest blanketing the island down to its shores. No mammals naturally occurred there except for bats and thus the harrier probably existed entirely on birds and insects. Today the Reunion Harrier's diet comprises about 50 per cent introduced mammals, once again showing its adaptability to a changing island habitat and fauna (Clouet 1978). Unlike other harriers it regularly nests among open forest patches on the steep wooded flanks of the volcanic Reunion island. Less than 12 per cent of the 18 known nests occur in wetlands (Bretagnolle *et al.* 1998a). With a population estimated at only 72 pairs (Bretagnolle *et al.* 1998a), it is not only the world's rarest harrier, but has the smallest range of any *Circus* species. From genetic analysis of the two species we now know they differ by about 1.5 per cent and thus diverged approximately 760 000 years ago.

Island species of harrier, like other species, often exhibit very small populations and the chances of them becoming extinct would seem to be higher than for mainland populations. This may be exacerbated by a lack of genetic variation and the possibility of inbreeding. With stochastic environmental events such as fire and cyclones occasionally visiting such islands and increased human pressure it seems astonishing that small island populations persist at all. Weighted against this list of negative factors we must remember that these species have probably survived many population bottlenecks and inbreeding depression through the combination of deleterious genes. Selection may have acted constantly and eliminated the deleterious genes in such populations. Furthermore, adaptations to the island environment—or changes to it—may quickly spread through any population that is numerically small. Both the Reunion and particularly the Madagascar Harrier may be benefiting from human activity, because agriculture practices there result in clearance of forested land and in some areas rice is planted, over which the harriers hunt (R. Thorstrom pers. comm). Thus both nesting areas and foraging areas may be increasing for these harriers on Africa's islands and their populations may well be larger than in the recent past.

The second 'new' species arises from the splitting of the Northern Harrier *(hudsonius)* from the Hen Harrier of Europe. Traditionally viewed as a subspecies of the widespread Hen Harrier (AOU 1982) the Northern Harrier has been isolated from the Hen Harrier for about 410 000 years and differs by 0.8 per cent from it (Fig. 2.1). So superficially similar are these birds that detailed behavioural and ecological studies of populations on the two continents by

Table 2.2 Body dimensions and Reversed Size Dimorphism indices of the world's 16 species of harriers. Mean values after Nieboer (1973), and mass data from Nieboer, and Brown and Amadon (1968) unless stated otherwise. Measurements given in [ ] are those derived from egg volume – female mass equations found in Chapter 7. The largest values for each variable are given in **bold.**

Species	Mass		RSD	Wing chord		Tarsus		Tail		Source
	♂	♀		♂	♀	♂	♀	♂	♀	
Pallid	310	440	35	340	371	66.5	71.8	208	236	
Montagu's	265[1]	345	27 (16)	365	372	58.1	60.5	218	225	
Hen (west)[2]	346	527	41	338	376	69.3	74.5	213	243	a
Hen (east)	340	500	40	349	386	69.4	75.3	220	250	
Northern	370	546	38	346	390	73.6	81.6	213	241	b
Cinereous	338	496	38	329	358	66.1	71.2	210	233	c
Pied	290	390 [429]	29	353	369	74.2	79.1	202	217	
Black	[364][3]	[573]	—	344	371	71.0	73.0	231	254	
Spotted	465	671	36	393	440	89.7	**98.2**	**245**	**274**	d
Longwinged[4]	397	612	**43**	395	415	77.1	82.7	235	254	e
Longwinged	"	"		413	**445**	77.9	85.6	242	266	
African Marsh	430	570	28	367	378	77.6	81.6	220	235	f
European Marsh	530	720	30	393	413	84.7	88.4	224	239	
	633	836	28							g
Eastern Marsh	600	780	26	404	422	86.7	89.5	228	242	
Australian Marsh	**647**	**890**	32	406	424	**89.9**	97.0	227	240	h
Papuan	—	—	—	380	409	83.4	92.0	222	232	
Reunion	—	[573]	—	350	374	80.0	83.3	216	234	
Madagascar	600	850	34	**415**	438	89.1	93.1	235	247	i

RSD–Reversed Size Dimorphism, based on difference in mass calculated following Storer (1966): (♀–♂/[(♀ + ♂)/2].

[1] From Russia; males in France (0°) are 30 g heavier (295 g: Leroux and Bretagnolle 1996), but female are the same (345 g). RSD indices therefore decline to 16, the lowest for any harrier.

[2] (west) from 0 to 30 longitude; (east) from 110 to 160 longitude.

[3] This male mass predicted from correlations between halux claw and body mass of 11 other species (R^2 = 97%, mass = 9.36 + 0.02 [halux]). The female mass, derived from correlations with egg volume, is similar to one unpublished mass of 574 g (A. Kemp pers. comm.).

[4] Northern South America first, followed by southern South America; mass data from unknown location.

a. Cramp and Simmons 1980; b. MacWhirter and Bildstein 1996, c. Jimenez and Jaksic 1988; d. Marchant and Higgins 1993; e. Brown and Amadon 1968; f. Simmons (unpublished); g. Bavoux *et al.* 1988 – western France; h. Baker-Gabb 1982 i. R. Thorstrom (unpublished)

Simmons *et al.* (1987) concluded that the American Ornithologists' Union 1982 decision to lump them as one species was well founded. The males differ only in the lighter grey plumage of Eurasian birds (more brown feathers are retained in the Northern Harrier) and the brown speckling on the breast of North American birds. Morphologically there are significant differences between the two populations only in wing chord but not body mass (Scharf and Hamerstrom 1975) and these authors suggested they be treated as different species. Since the longer wing chord measurements (Table 2.2) may simply reflect the greater migratory move-

(a)

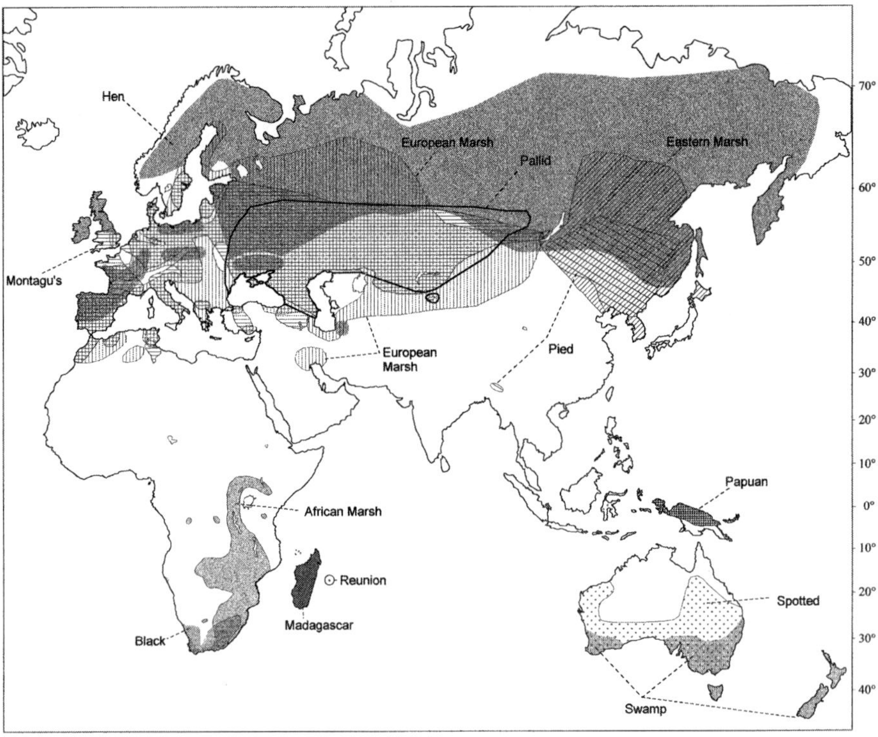

Fig. 2.3 (a) Breeding distribution of the 13 Old World harriers. The Hen Harrier has the widest distribution, while the greatest diversity occurs in western Russia where Montagu's and European Marsh (square hatching), Pallid (dark line), and Hen (grey shade) all occur. Three species overlap in western Asia: the Pied, Eastern Marsh, and Hen Harriers. The European and Eastern Marsh Harriers overlap and hybridize on the eastern shores of Lake Baikal. Four species are recognized in Africa, two of which exhibit the smallest distributions: the Black and Reunion Harriers. Two species are currently recognized in Australasia, one of which (Swamp Harrier) occurs alone on New Zealand. A third, the Papuan Harrier (*C. spilothorax*), differs from the Eastern Marsh Harrier (and resembles the Pied Harrier) in having a black not a streaked head and a spotted back. It is a resident. (b) opposite Breeding distribution of the three New World harriers. Note the general lack of breeding below 30° N to equatorial regions for species across North America, Europe, and Asia. The Longwinged Harrier has the most disjunct breeding distribution of all the harriers and breeds both north and south of the equator.

ments of Northern Harriers, this difference is adequately explained by their ecology. One can speculate that if *cyaneus* arose from *hudsonius* then the latter has changed little morphologically in the approximately half million years it has been separated because there were no pressures in the form of other harriers in North America. Despite the morphological similarities I treat them as separate species throughout this book because their genomes are more strongly divergent

(b)

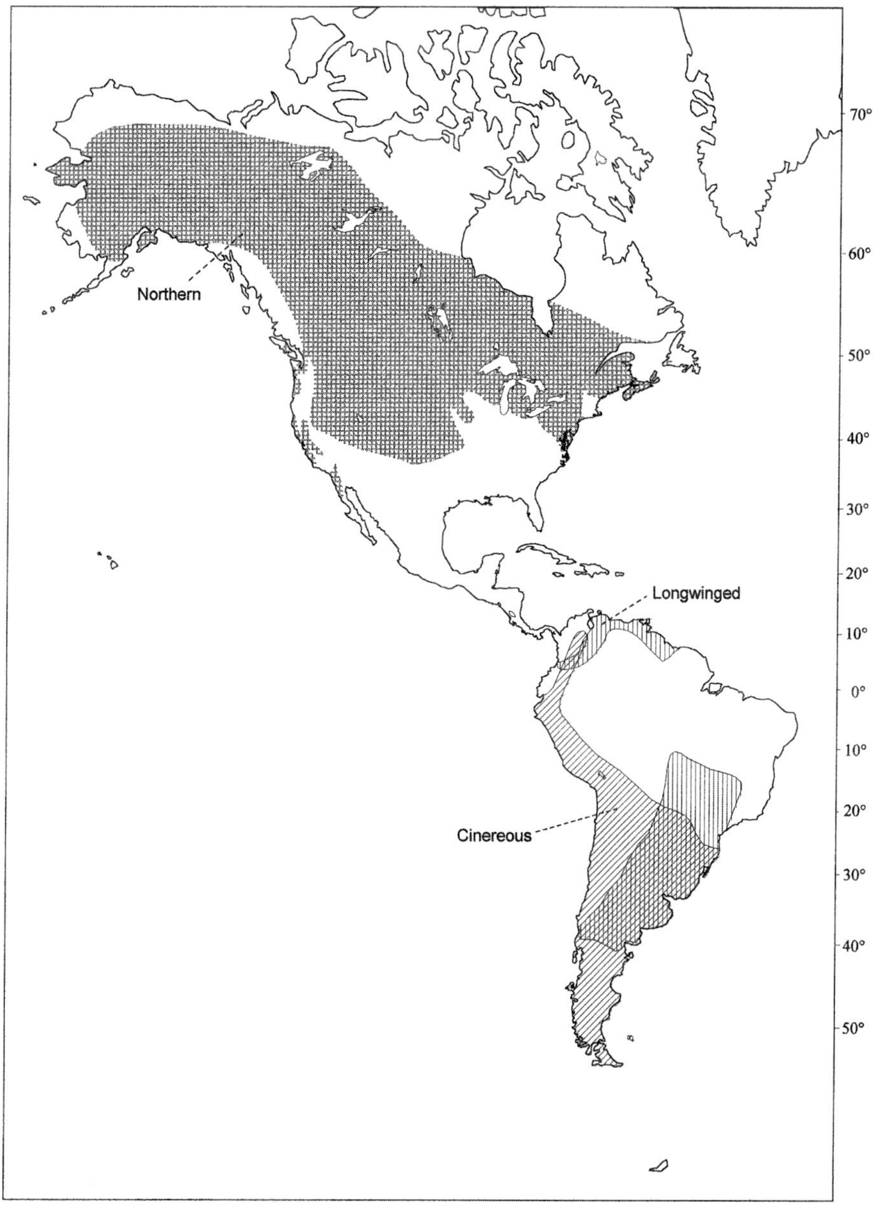

Fig. 2.3 *Continued*

than those of species already separated on other grounds. The Northern Harrier is referred to as *C. hudsonius* from hereon.

A third 'new' species is proposed here although DNA was not available for analysis. As a sedentary island species the Papuan Harrier *C. spilothorax* on Papua New Guinea has very probably diverged from its nearest geographical neighbour the Eastern Marsh Harrier *C. spilonotus*. The bird is the only truly equatorial harrier and is resident in the central highlands and eastern sections of New Guinea (Coates 1985; Beehler *et al.* 1986). This bird resembles most the Eastern Marsh Harrier (*C. spilonotus*) in plumage characters. But whereas the Eastern Marsh Harrier migrates as far south as Borneo, this species is resident on New Guinea (Coates 1985, Fig. 2.3). It is smaller and darker than the Eastern Marsh Harrier and its Latin binomial *spilothorax* suggests it is a 'spotted backed' harrier. It is heavily streaked brown on the back and neck (Brown and Amadon 1968). Two forms of males occur: the central highlands birds lack the streaked throat of lowland birds and exhibit a well defined black head and pale breast much like the Pied Harrier. Because other island harriers prove to be separable at the species level (above), its plumage differs from its nearest neighbour, it is sedentary and it is morphologically smaller and separable from the Eastern Marsh Harrier (Table 2.2), I treat it as a good species. The number of harriers worldwide therefore totals 16, three more than currently recognised. Further differentiation of the harrier group will await blood samples from other island species, such as the those on Fiji and New Zealand *(gouldi)* and the Falklands (*C. cinereus histronicus).*

The steppe harriers

The ancestor of this small group of dry-grassland harriers appears to be the Australasian Spotted Harrier. These birds diverged about 3.0 My ago. While the tree-nesting habits of this species may seem good evidence that this bird is an ancestral harrier linking it with the Accipiters, Baker-Gabb (1984) argued that this harrier was once a ground-nester and its tree nesting is a derived, not an ancestral trait. This conclusion was based on the flimsy nature of the nests and their resemblance to the ground nests of other harriers. Thus rather than the bird exhibiting ancestral nest building traits it appears to have resorted to tree nesting having once been a ground nesting bird. This conclusion is supported by the finding that its extant ancestor is the Longwinged Harrier which is also a ground nesting species (Fig. 2.1).

While the Montagu's Harrier unexpectedly falls outside the Hen Harrier group, the Pallid Harrier (*macrourus)*, is confirmed as a part of this group. It diverged from *cyaneus/hudsonius* about 2.8 My before it gave rise to the Cinereous (1.9 My) and Black Harrier (2.8 My) later. The whole group's ancestor is the Longwinged Harrier *C. buffoni* of South America. They share morphological similarities in having a light wing loading, but are as different in Reversed Size Dimorphism as any two harriers can be (Chapter 3). It is interesting to comtemplate that in both South America and Africa some of the oldest harriers, the Longwinged and African Marsh, forage and breed alongside two of the youngest, the Cinereous and Black respectively (Bo unpublished, Chadwick 1997). The affinities of the striking Pied Harrier *C. melanoluecus* are not known but as a slight and long winged steppe harrier it is probably an offshoot from the Pallid-Black-Cinereous group.

The marsh harriers

As we have seen, the marsh harrier group has swollen to seven species, and all bar the Papuan Harrier can be confirmed as full species given that their genetic distances all exceed 1.5 per cent. The African (*ranivorus*) and the Australasian (*approximans*) appear to have arisen about the same time from the Montagu's ancestor. The Australasian birds then rapidly gave rise to the African island species and also the youngest member of this clade, the European Marsh harrier (*aeruginosus*) some 815 000 years ago. One presumes that the recurrent ice ages promoted the recent speciation of these species. As expected by Nieboer (1973) the African bird is the oldest marsh harrier. One anomaly was the Eastern Marsh Harrier, which, given its geographic position between the Australasian and European birds was expected to be a link between them. However, the analysis showed it is to be more recently related to the African marsh than it is to either *approximans* or *aeruginosus*.

Morphology- and DNA-derived phylogenies compared

It is of interest to see how accurately Nieboer's purely morphological approach predicts the phylogenetic relationships of the harriers as found by genetics (Fig. 2.1). Nieboer (pp 80–1) correctly predicted the following points:

(i) the Longwinged and Spotted Harriers are 'the oldest offshoots of the harrier stem';

(ii) the African Marsh Harrier 'represents the least advanced phase [of marsh harriers]' and the European Marsh Harrier 'represents the most advanced phase in the development of the Marsh Harriers';

(iii) the Pallid Harrier 'is considered a specialised Hen Harrier'; and

(iv) the nine harriers are subdivided into three groups arranged in a progressive sequence of evolutionary development—Spotted and Longwinged Harriers followed by the small (Hen) harriers and then the marsh harriers.

Bar a few details these statements are essentially accurate. What he did not predict accurately were:

(i) the Black Harrier, because of its lack of sexual dimorphism is 'probably the least advanced of this [small harriers] group' (it is the most recently evolved);

(ii) because of its specializations Montagu's Harrier is probably the 'most advanced in the group of small harriers' (it is not now included in this group at all);

(iii) 'the small harriers have been derived from one probably Palearctic stem' (they arose from an Australian stem instead); and

(iv) 'it may be that all marsh harriers are of an African origin and have originated from the early African Marsh Harrier' (they arose from the Montagu's Harrier of central Europe).

This checklist of comparisons suggest that morphological analyses of the harriers can give a broadly accurate picture, but sexual dimorphism and morphological specializations can be misleading as guides to evolutionary affinities.

To summarise, the harriers shared ancestry with the accipiters and we know that the marsh harriers are sufficiently different to confirm species status for the African, Madagascar, Reunion, European, Eastern, Australian, and probably the Papuan harrier. The most surprising finding not evident from plumage characters is that these marsh harriers shared a common ancestor with Montagu's Harrier. This bird shared a common ancestor with the Long winged Harrier of South America—the most ancient of the harriers. This suggests that South America and not Europe was the birth place of the harriers. We can tentatively suggest, therefore, that the Longwinged Harrier simultaneously gave rise to the 'grey' steppe harriers via the southern Spotted Harrier and the marsh harriers via the Montagu's Harrier.

Nieboer (1973) noted towards the end of his analysis: '*But an answer to the question whether the present* C. ranivorus *from Africa is an African offshoot of the [European] Marsh Harrier, or whether all marsh harriers have originated in Africa, can only be found through speculation. The answer will remain hypothetical*' (p 77). Just twenty-five years on, this collaborative analysis has turned Nieboer's speculation from hypothesis into reality.

Harrier displays

Genetic evidence for differences between species might best be complemented with additional behavioural evidence. We can assume that a member of one species will not recognize another species because it does not recognize the mate-attraction displays. This arises from the view that species are members of a population which share a specific mate-recognition system. The cues used may be pheromonal, behavioural, audial, colouration, or any subtle combinations of these and other factors. This idea grew out of work by Hugh Patterson in the 1980s on Drosophila flies, subsequently applied to other sexually reproducing species. He named the system SMRS (Specific Mate Recognition System: Patterson 1985), and although not well known it has useful heuristic value because it allows us to look at subtle differences in those behaviours which are most likely to be used in mate recognition: the display.

Mate-attraction displays of the harriers are reasonably well described and observations confirm that displays in the African Marsh Harrier serve to attract mates based on displays by unmated males (but not females), and spectacular displays by paired males who have temporarily lost their mates (Chapter 3). This is seen in several other species (Australasian, Montagu's, European and Hen/Northern) and since it serves to attract females, it should qualify as a Specific Mate Recognition component. Do differences in displays support the genetic evidence for differences in the marsh harrier group? African Marsh Harrier aerial displays consist of two elements: sky-dancing, a horizontal display seen in several other species, and sky spiralling, a vertical descending display of greater vigour (Chapter 3). The latter is uncommon among European Marsh Harrier populations (Blanco *et al.* 1993) whose displays are well known (Cramp and Simmons 1980), and unknown in the Reunion Harrier (Clouet 1978; Jones 1989). The difference confirms the likelihood that the African Marsh Harrier is separable from both the Reunion or European Marsh. Further differences such as plumage and foraging behaviour mentioned previously support this conclusion. The pied Reunion species not only differs in plumage, but frequently hunts in evergreen forest, and there chases mainly small avian prey. Its display does not exhibit spiralling, which differentiates it from *ranivorus*, and differences in calls also separate it from the displays performed by *aeruginosus* (Clouet 1978). Because of the close similarity in the dark pied-like plumage of the Eastern Marsh and Reunion Harriers, it seemed plausible that they were closely related (Bretagnolle *et al.* 1998a). This we now know not to be true (Fig. 2.1). The spiralling display, common in the African but not the European birds, is similar to the display in Australasian birds described by Baker-Gabb (1981) as 'twisting like a falling leaf'. Since it also calls at the top of each U-shaped undulation like the African bird, but not like the European Marsh Harrier, their displays suggest closer affinities. Based on these characteristics, I suggested in 1991 (Simmons 1991a) that harriers evolved in the sequence African—Australasian—Eastern—Reunion—European. This is also not entirely correct as the African and Eastern birds are more closely related than either are to the Australian species. Eastern Marsh Harriers also do not exhibit spiralling but do skydance (I. Fefelov pers. comm.). Thus, it seems appropriate to suggest that sedentary populations exhibit spiralling (African, Australian, some southern European) while migratory populations do not (most European, and Eastern Marsh Harriers). Given that skydancing is absent or poorly developed in older harriers (Spotted and Montagu's Harriers) the best we can state is that skydancing and spiralling can act as a guide only to harrier relationships.

Morphological and behavioural differences between harriers

Female mass and other body dimensions for the world's harriers are given in Table 2.2. Apparent from this listing is that the marsh harriers are the heaviest of this group, the Australasian Swamp Harrier *Circus approximans* being the heaviest at almost 900g, the lightest being the dry country Montagu's Harrier at 345 g. The most dimorphic is the South American Longwinged Harrier *Circus buffoni*, a trait which we will discuss further in chapter 3 with respect to reversed size dimorphism. Surprisingly, the Madagascar Harrier (*C. macrosceles*—the appropriately named long-legged harrier) is the largest in body dimensions (Nieboer 1973) with an average wing chord (415 mm), 19% longer than its Reunion cousin (350 mm), and even 2% longer than the heaviest Australasian Swamp Harrier (408 mm). The body weight of this curious harrier has recently been recorded by Russell Thorstrom of the Peregrine Fund Programme in Madagascar at 850 g the second largest after the Swamp Harrier. This is 48% bigger than that predicted from egg volume-body mass equations for the Reunion Harrier (573 g). This is an appreciable difference for two birds that are traditionally treated as subspecies, suggesting that their treatment here as separate species is well founded. This has independently been suggested by Bretagnolle *et al.* (1998a) who studied the bird in 1997–98.

Distribution patterns and morphological variation

Harriers are distributed widely across the continents, and islands such as the West Indies, Fiji, Reunion, New Guinea, and Tasmania (Fig. 2.3). By far the widest distribution is shown by the Hen Harrier, spread across Europe and Asia, while the Northern Harrier exhibits a similar distribution across North America from east to west. The two are separated in the Alaska/eastern Siberia region, by at least 1000 km and none occur on the Aleutian Islands. The smallest continental range is that of the southern African endemic Black Harrier *C. maurus*, which breeds almost entirely within South Africa and western parts of Namibia (Simmons 1997a, Fig. 2.3). The newly recognized Reunion Harrier has the smallest island range, being restricted entirely to the 2515 km^2 island.

Clines in the body size of birds were noted by Nieboer (1973) who measured museum specimens from across the globe. Hen Harriers, and the now split European and Eastern Marsh Harriers tend to get larger as one travels from west to east (by about 3% in wing chord measurements). The largest harriers are also found in the east with the Australasian Swamp Harrier being the largest and heaviest at almost 900 g. Somewhat surprisingly, for a tropical island species, but in keeping with this trend, the Madagascar Harrier is also the largest of the African harriers (at 850 g female mass) and its wing chord is longer than any other harrier including the Longwinged Harrier of South America (Table 2.2; Nieboer 1973).

We do know, however, that Bergmann's Rule—that species become smaller as they approach the equator—is found in the harriers too. On the African continent African Marsh Harriers stretch from the southern tip of South Africa at 34° S to the Kenyan equator (Fig. 2.3). While the relative weights are unknown, their wing chord measurements decrease by between 3% (males) and 5% (females). The size difference (Fig. 2.4) has even suggested to several authors (Streseman 1924; Clancey 1964) that the two may be subspecies (*C. ranivorus ranivorus* and *C. r. equatorialis*), an idea which is not generally accepted today (Brown and Amadon 1968; del Hoyo *et al.* 1994). Another suggestion is that the larger size in the south arises from competitive interactions with the Black Harrier in southern Africa (Nieboer 1973). This seems unlikely, however, since the two rarely interact during breeding even when they do co-occur on nesting grounds (Chadwick 1997). Furthermore, the same latitude-dependent trait is shown by the Longwinged Harrier (Fig. 2.4) over the same range (5° N to 30° S) in South America. Its wing chord decreases 5% (male) to 12% (female) as it approaches equatorial regions. It is the female that changes size more rapidly than the male, a point we will return to in discussions of reversed size dimorphism (Chapter 3).

This is not so for the Montagu's Harrier, however, where the males increase in body size as one goes west and south in Europe. This geographic difference leads to changes in the degree of reversed size dimorphism further west in Europe. In Nieboer's (1973) sample which came from areas east of 22°

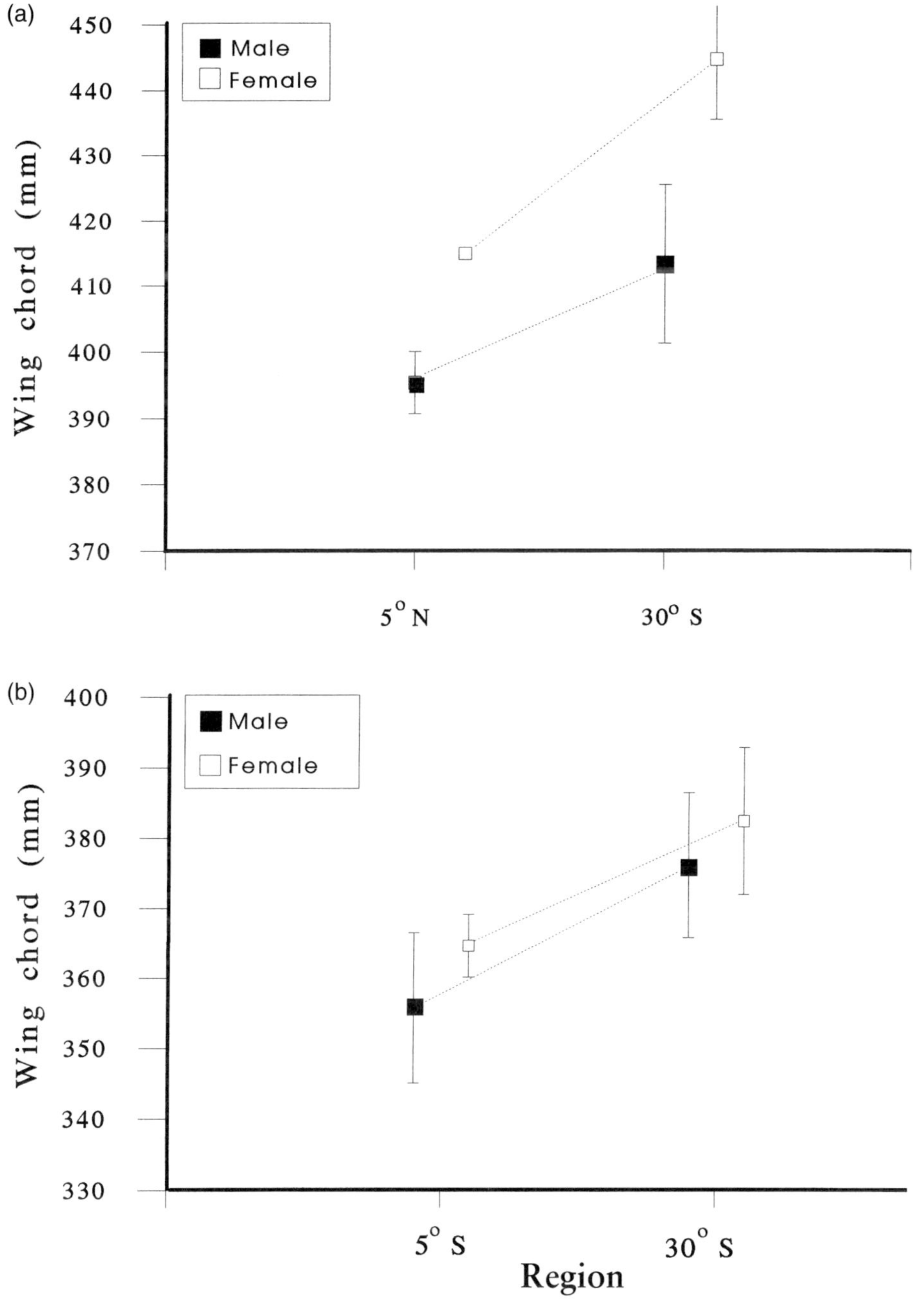

Fig. 2.4 (a) Wing chord differences for Longwinged Harriers over the same latitude range in South America (after Nieboer 1973). (b) Differences in wing chord measurements of African Marsh Harriers from equatorial regions (East Africa) and South Africa. Both species conform to Bergmann's Rule—equatorial birds are smaller than those from temperate climes.

(i.e. mainly Russia), female Montagu's were the same body mass (345 g each) as those in western France (0°), but males were 11% heavier in the west (295 g: Leroux and Bretagnolle 1996) than in the east (265 g). This makes a sizeable difference to the indices of reversed size dimorphism based on body mass which decrease from 27 in the east to 16 in France—the smallest value for any harrier (Table 2.2). We will explore the reasons for this in the next chapter.

The Hen Harrier is found farthest north in the northern hemisphere with its northern limit near the 70th parallel (Fig. 2.3). The Cinereous Harrier occurs furthest south and has been recorded on the Falkland Islands at the southern tip of South America. There, island males are similar in skeletal measures such as tarsus length (66.2 mm) to mainland birds (66.1 mm) but the flight apparatus such as wing chord (308 vs. 329 mm) has undergone a 7% reduction. The same reduction is seen in body measurements between Madagascar Harriers and their cousins 700 km east on Reunion.

The greatest diversity of breeding harriers is found in eastern Europe and, in an area near Kiev in western Russia, one theoretically could find Pallid, Montagu's, European Marsh, and Hen breeding together (Fig. 2.3). This diversity is surpassed in India where wintering harriers can include the Pied, Pallid, Montagu's, Hen, European, and the Eastern Marsh Harrier (Ali and Ripley 1968; Clarke 1995). Migratory birds, except the Hen Harrier, move into the Sahel, stretching east to west across the bulge of Africa, and there may mix with the smallest of the marsh harriers, the African *ranivorus* in East Africa.

Africa

Some curious distribution patterns occur here. There is a complete gap with no resident or breeding harriers just north of the equator in the Sahelian grasslands. The habitat cannot be inappropriate for harriers because it is the main wintering area for European, Pallid, and Montagu's for about seven months of the year (Brown *et al.* 1982; del Hoyo *et al.* 1994; Clarke 1996). Indeed it is particularly important for many Eurasian populations of harriers because rainfall in this region appears to have a profound effect on breeding populations the following year (Clarke 1996). The lower the rainfall, the less likely are locust swarms, and the less likely overwintering Montagu's Harriers are to either survive or breed the following year. So while the Sahel is vital for non-breeding harriers, it holds no breeding birds, and was described by Leslie Brown as an extraordinary lacunae in the African avifauna. How can we explain this gap?

The answer may lay in the observation that non-breeding birds are able to exist for a large portion of the year, while breeding birds do not—or cannot. If breeding is not possible there, the first clue as to why is the decreases in the clutch size found with latitude (Chapter 9). For every 10° decline in latitude, clutch size decreases by between 0.3 and 0.7 eggs for European birds. A clutch of 5 eggs in Europe (50° N) therefore will have decreased to between 3.5 and 1.5 eggs for a hypothetical harrier breeding at the equator. From the south up (for African Marsh Harriers), a similar clutch of 2.6 eggs can be computed for a bird attempting to breed at 10° N. Since brood size of successful females is smaller than clutch size by approximately 55% (for European Marsh: Underhill-Day 1984) to 50% (for Hen Harriers: Picozzi

1984a), successful pairs would theoretically be rearing only 1.7 to 0.7 young. For successful African birds whose fledged broods are about 75% of their mean clutch size, birds attempting to breed at 10° N can expect to rear less than 2 young. These small clutches and fledged broods imply, theoretically, that it may not be possible or worthwhile to attempt to breed with such low output. This approach presupposes that appropriate prey are available to such harriers. Are they?

Since several breeding harriers rely on a good supply of small mammals, early in the season, for successful mate attraction, large clutches, and future success (Butet and Leroux 1993; Hamerstrom 1979; Simmons *et al.* 1986; Chapter 4) even when their overall diet appears to be dominated by insects (Arroyo 1997), are small mammals actually available to our hypothetical harrier breeding at these latitudes? It has already been noted that the range of the African Marsh Harrier and its main prey—the Striped Mouse *Rhabdomys pumilio* coincide well (Simmons 1997a), especially in East Africa where the mouse prefers cooler highland areas (Taylor and Green 1986). Neither the bird nor the mouse are found below about 1500 m. This suggests that African birds are highly dependent on this diurnal rodent.

However, from studies of European birds wintering in this gap, we also know that their diet is different from that found in their northern quarters. For example, both Cormier and Baillon (1991) and Arroyo *et al.* (1995) determined that wintering Montagu's Harriers in Senegal, West Africa, take over 90% insects in their winter diet, much different from their breeding season diet which is dominated by small mammals early in the season (Arroyo 1995) and small passerines in general (Clarke 1996; Arroyo 1997). European Marsh Harriers which typically take larger prey such as waterfowl and rabbits in Europe (Clarke 1995), also take mainly insects in these equatorial climes (Arroyo *et al.* 1995; Clarke 1995). For one of the largest harriers to be taking mainly insects suggests that they are somewhat constrained by a lack of more appropriate prey. Where Montagu's Harriers occur further south, their diet may include more birds as suggested by Stronach (1991) in the Serengeti where harriers chase insectivorous passerines. Even further south, some small mammals, typically caught by the resident African harriers, may begin to enter the diet of European Marsh Harriers (Simmons 1988). It is pertinent to note that Lesser Kestrels, which take mainly vertebrates during breeding in temperate areas, also prey almost exclusively on insects in their wintering areas in South Africa (Anderson *et al.* in press).

It may be, therefore, that no harriers breed in this apparently vast suitable tract of Africa because appropriate mammalian prey are lacking, but this is unlikely. According to distribution maps, suitable grassland species such as the *Arvicanthis* rats and *Lemniscomys* mice (Kingdon 1978) occur throughout this region and are at least in part diurnal although mainly crepuscular. Also Blackshouldered Kites *Elanus caeruleus*, well known for their specialization on small mammals (Mendelsohn and Jaksic 1989), occur in the Sahel where they take mainly rats (J.-M. Thiollay *in litt.*). However, these species often hunt in the twilight hours, possibly to avoid excess heat, and may synchronize their hunting with activity patterns of these rats. Many other small mammals in the Sahel are nocturnal (Kingdon 1978), again possibly to escape the high daytime temperatures, and are thus not available for diurnal harriers. This suggestion is supported by Clarke's

(1996) observations from India at the same latitude, where locust prey also predominate in harrier diets. He notes that harriers finish hunting about three hours after sunrise and thereafter spend long periods perched in the shade of bushes, escaping the heat. Only during the middle to late afternoon, four hours before sunset, do harriers begin hunting again. So perhaps harriers are too heat stressed to hunt for long enough to breed. But how do Blackshouldered Kites manage to breed when harriers apparently cannot? Unlike kites, harriers only employ energetically expensive flight foraging techniques, which may preclude them hunting mid-day as found by Clarke (1996). By contrast, kites often perch hunt and expend little energy doing so (Mendelsohn and Jaksic 1989). Recent unpublished work on foraging strategies of European Kestrels *Falco tinnunculus rupicolus* resident in Kenya show similar patterns: birds subsist mainly on insects and spend much longer periods perch-hunting than their northern counterparts at equivalent times of the season (van Zyl and Amutete 1998). Thus, circumstantial evidence suggests that harriers may not be able to breed in this region because small mammal prey are not available in large enough numbers (evidenced by declining clutches), and both lack of potential prey and heat stress in foraging harriers may prevent them from hunting throughout the day. Both are easily testable in the field. Chapters 8 and 9 explore this in greater detail, showing that tropical raptors, including harriers, have a much reduced foraging yield (prey caught per unit effort) relative to their temperate congeners.

South America and Eurasia

This distribution gap in equatorial regions is not confined to African birds. Another gap at about the same latitude is apparent in the split breeding distribution of the South American Longwinged Harrier. The birds are relatively common in southern South America (i.e. Argentina), and much scarcer in northern countries such as Guyana and Trinidad where only one nest has been found (Herklotz 1961). The gap between these two breeding areas is presumed to arise from unsuitable habitat presented by the vast Amazonian rainforest. Is it possible that mice are once again mainly nocturnal or burrowing or available in low numbers and thus unavailable to the diurnal harriers in these regions? Furthermore, none of the Eurasian harriers breed below 30° N, except perhaps Pied and Montagu's Harriers, which are reported as resident and breeding at ~26° N in northern Burma (now called Myanmar) (Neufeldt 1964; Ali and Ripley 1968). Breeding by Pied Harriers has recently been confirmed in the foothills of the Bhutan Himalayas at 27° N (Narayan and Rosalind 1990), but numbers are low and a large gap exists between these populations and the main breeding stronghold in eastern Russia, north-eastern China and Mongolia at 45–50° N (Fig. 2.3). These species also winter in areas closer to the equator, and some of the world's largest roosts of harriers occur in India (15° S) where thousand of birds may gather at a single site (Clarke 1996). They, too, often eat insects there (Clarke 1995).

In support of this idea of decreasing availability of appropriate small mammals, the diet of Montagu's Harriers changes in constitution from being fairly specialized (often birds and mammals) in northern latitudes to more diverse in southern climes such as southern France and central Spain. Insect prey then becomes most important (Arroyo 1997). If higher energy prey such as mammals or birds are critical for other populations, the slow switch to more insects may also indicate a lack of more appropriate prey. But does this reflect a switch due to a lack of other prey (a constraint) or does it reflect a preference for these prey types elsewhere (adaptive response)? Both Arroyo (1997) and Bildstein and Collopy (MS) concluded from studies of availability and choice, that the diets of both Montagu's and Northern Harriers in different areas of the world are a reflection of the prey available rather than a specialization on prey type per se. (This does not preclude specialization taking place where particularly appropriate and easily caught small mammals are available.) If the lack of appropriate small mammal prey is the cause of this latitudinal shift to more insects it might also explain why even further south (below 35° N) harriers rarely breed, but can overwinter because insects are plentiful. Studies of the diet of the European Marsh Harrier subspecies *harterti* in North Africa—the southern most breeding of any Eurasian harrier—would be pertinent in this regard.

North America

A similar distributional gap is also apparent in North America south of 36° N (Fig. 2.3b); here Northern Harriers winter but rarely breed (MacWhirter and Bildstein 1996). Fortunately, diet data are available and show that for wintering birds, those farthest north feed almost exclusively on voles (*Microtus* species: 84–93%), while diets in the southern United States are variable but with median proportions of 72% mammals. Importantly, voles are replaced by bird prey, which are much more difficult to catch (Temeles 1985, 1987): frequency of bird prey increases in southern wintering areas where median values were 29% birds, relative to northern regions where the frequency is always less than 10% (MacWhirter and Bildstein 1996). How does this translate into foraging efficiency and yield? In a review of hunting harriers in three wintering areas (Ohio, 39° N; South Carolina, 33° N; and Florida, 29° N) Bildstein and Collopy (MS) found that harriers furthest south took up to 95% birds and the prey caught per hour of hunting (0.87 and 0.99) was much lower than for either wintering birds north in Ohio (1.4 prey/h) or breeding harriers even further north in Utah (2.1 prey/h) at 42° N (Linner 1980). These data also suggest that North America's only harrier may be limited in its breeding in these regions because appropriate small mammals are too scarce to allow sufficient prey capture efficiency. We will return to comparisons of prey capture yield and show in chapter 9 that subtropical harrier hunting yield lies between the limits for overwintering Northern Harriers which do not breed (1.4 prey/h) and close to that for breeding birds in Utah (2.1 prey/h) which breed at the southern extreme of their range.

Clearly studies of the diet and efficiency of harriers that breed in temperate and equatorial regions (the African Marsh Harrier in Kenya/Uganda and the Longwinged Harrier in Trinidad/Guyana) would prove particularly fruitful in answering why these distributional gaps occur.

Melanism

Before we leave geographical and species-specific variation it is useful to look at reports of plumage variation. All of the northern hemisphere species are dichromatic, with males exhibiting varying amounts of grey in their adult plumage. In the southern hemisphere, two species are entirely monomorphic in plumage—the Black Harrier in South Africa/Namibia, and the Spotted Harrier in Australia—while the Swamp Harrier and African Marsh sexes are barely distinguishable on the degree of greyness apparent in otherwise brown plumaged males. Albino birds are very rare with some reported from Scottish Hen Harriers (Watson 1977). Melanism is more often reported and has been recorded for Longwinged, European Marsh, African Marsh, Montagu's, Northern, and Hen Harriers. These are often striking looking birds with a bright yellow eye enhanced by the dark black of the head. Bill Clark (1987) was the first to draw attention to the (relatively low) incidence of melanism in European birds caught on migration in Israel and found melanism in both males and females. He noted that the likely origin of these birds was eastern Europe or western Asia—near the edge of their range. In the southernmost part of their range south of the Ural Mountains, Anatoly Davygora noted (in Clarke 1995) that perhaps 20% of the resident birds were melanistic. A dark form of the Marsh Harrier is also known in the southernmost section of its range in western Europe where it extends into North Africa. This race -*harterti*- is dark headed and smaller than the typical European Marsh Harrier (del Hoyo *et al.* 1994). A high incidence of melanism is also known in Montagu's Harriers in the westernmost portion of their range in northern Portugal and north-western Spain. Here 20–30% of pairs are melanistic, while elsewhere melanistic individuals make up less than 1% of the population (B. Arroyo pers. comm.). Interestingly these birds spend longer pairing than typical grey-coloured birds and begin their broods later than most other pairs. Since seasonal reproductive declines are apparent in Spanish populations (Arroyo 1995), this suggests one ecological reason why melanism is so rare in these populations. Research on the mate-choices and reproduction of melanistic populations would therefore be highly instructive. Finally, the only three dark individuals reported for the African Marsh Harrier were recently uncovered by Bill Clark (1997a), and true to the trend in other harriers they were found in the far northerly sections of their range in the Aberdare Mountains and other areas of Kenya where African Marsh Harriers are rare.

So why should melanistic individuals be more common towards the extremities of their ranges? Since it is not merely a north–south trend, latitudinal trends in colour morphs cannot account for it. Arroyo's observations that melanistic Montagu's Harriers may be at a reproductive disadvantage because they take longer to find mates and breed suggest that selection may decrease their incidence in areas where populations are denser and competition for mates more intense. Melanistic individuals may have a greater chance of reproducing and passing on the trait to future generations in areas of lower density at range extremes. Further research to test these ideas is needed in northern Portugal and the southern Ural Mountains.

Food passing behaviour

One of the best known traits that differentiates the harriers from all other raptors is the habitat of passing prey in an aerial transfer. Many other raptors transfer prey either on the nest or on a nearby tree or rocky ledge (Newton 1979; Cade 1982), although aerial transfers are not unknown in some falcons and hawks. The behaviour in harriers involves males passing prey to their mates and later in the breeding season, both parents passing prey to their young. In

Fig. 2.5 An aerial food transfer between male and female European Marsh Harriers. In a typical pass the male flies up at the female's closest approach, and the prey is released as he does so. The female therefore catches prey which is momentarily stationary.

many descriptions of aerial food transfers (e.g. Watson 1977), including my own interpretation (see Palmer 1988), female harriers are described as flying from their nests as the male appears (prominently trailing the prey in a lowered foot), positioning themselves slightly behind and below the male, and catching the prey in front of them as it drops the 1–2 m from their mate (Fig. 2.5). Closer observations, however, reveal that prey-carrying males fly upwards just as the female closes to within 2 m and release their prey as they do so. The effect is that the prey is momentarily moving upwards after release and is virtually stationary in the air as it is caught by the female. This appears to be a deliberate manoeuvre on the part of the male, as it makes the catching and delivery of a hard-won food item considerably easier for the female or offspring.

Although some texts imply that females routinely turn right over to collect prey (e.g. Cramp and Simmons 1980) my observation of almost one thousand food transfers suggests it is rare and occurs only when females attempt to grab prey in a foot to foot fashion. More usually, they catch prey in front of them or by tilting to one side, but some interesting and spectacular variations have been observed. Among Northern Harriers in particular, females occasionally approached their mates from in front and received the food as they swept past them, side to side. On one occasion a female stooped in from above her mate, who deftly turned on his side, legs at full stretch, and the female took the prey as she swept past. Balfour (1962) describes a similar case in Hen Harriers: the male dived in from overhead and prey was passed foot to foot to his mate. Other clumsier passes occurred when females flew head-on directly at their mates, virtually colliding with them before the female snatched the prey. The longest drops seen were of about 6 m and all (4) were caught easily by the females involved. Two northern females who fumbled their initial attempts deftly dived after the prey and caught it before it hit the ground. Hamerstrom (1986) reports that sometimes two prey are dropped by the male and the female has almost no chance of catching both. Hosking (1943) reported an unusual incident in which a *female* European Marsh Harrier twice transferred prey to her mate who delivered it to the nest. This behaviour probably arose because the female did not accept the photographic hide at the nest (like Australasian birds European Marsh Harriers are sensitive to hides near their nests), and she transferred it to the male who had accepted the hide and delivered the prey.

Some foot to foot transfers have been described by Hamerstrom (1969) in Northern Harriers in heavily DDT-affected populations in Wisconsin. In the New Brunswick cases the birds may have been food-stressed, but not apparently pesticide influenced. The habit seemed peculiar only to a few pairs; more a case of individual variation than abnormal behaviour. Similar 'talon to talon' transfers occur in *cyaneus* populations in Scotland (Balfour 1962; Watson 1977: 117) but African birds were never seen to transfer talon to talon; 1–3 m drops were more typical. This may occur because pair bonds of African Marsh Harriers are more stable than those of Northern Harriers (who pick new mates every year), and thus they have more chance to perfect their releasing and catching techniques.

Evolution of food passing—why only harriers?

Why harriers differ from other raptors in this respect has rarely been explored, but the lack of trees in their typically marshy or open grassland habitat may preclude other methods of food transfer. Most other arboreal or cliff-nesting raptors transfer prey at a convenient perch close to, or even at, the nest. We can assess the possible evolution of this trait by looking at the methods and efficiency of food transfers in two species of harrier, the African Marsh and the Northern Harrier.

African birds, with more aquatic environments than the Northern Harrier, presumably have fewer opportunities to transfer prey on the ground; one might therefore test the idea that having fewer raised transfer sites (such as trees or cliffs) has led to the evolution of aerial food passes. Do African birds transfer prey aerially more often than Northern Harriers? For each food pass I recorded whether prey was transferred aerially or on the ground (or nest), whether it was caught or dropped, retrieved or lost, the recipient (mate or offspring), and the stage of breeding.

Efficiency

Aerial food passing is highly efficient: of 305 aerial passes recorded for African Marsh Harriers, 93% were caught. Of the 21 aerial transfers that were missed, 48% were due to inexperienced fledglings failing to catch the item, 19% were dropped during the earliest transfers of the season between adult birds, and 24% were dropped by newly paired birds whose performance soon improved. Thus in 305 passes only two were dropped by experienced adults, after the first transfers of the season. In total, only about 3.7% of all prey were dropped in adult–adult passes (Table 2.3), the remainder were dropped by juveniles. Of the 21 prey dropped by African birds, 16 were retrieved from the reeds or grass below; 5 were lost, either because they lay in impenetrable reeds or below water, and once because the recipient was unmotivated to recover it. How does this efficiency compare with other harriers?

Adult Australasian Swamp Harriers never dropped prey in about 150 aerial transfers (Baker-Gabb 1982), while Jimenez and Jaksic (1988) report 2 drops in

Table 2.3 Efficiency of aerial adult-to-adult food passing among five species of harrier worldwide.

Species	Female mass	Aerial prey transfers	Number missed	Efficiency (%)	Source
Australasian	880	150	0	100	Baker-Gabb 1982
European	720	169	6	96	Fernandez and Azkona 1994
African Marsh	570	295	11	96	Simmons 1991b
Northern	546	246	10	96	Simmons 1991b
Cinereous	500	60	2	97	Jimenez and Jaksic 1988

60 transfers (3.3%) by female Cinereous Harriers *C. cinereus* in Chile. Among Northern Harriers, 246 aerial food transfers were seen, of which 96% were caught by the female (only two transfers were seen between adults and offspring). The heaviest harrier, the European Marsh Harrier, was no less efficient at catching aerial transfers and adult–adult transfer efficiency among five species of harrier never dropped below 96%—a very high rate considering the variations possible (Table 2.3).

Of the 10 prey missed by female Northern Harriers, two were caught before they reached the ground, only one was not retrieved, and four drops occurred because the recipient was already carrying prey (something never seen among African birds).

Frequency of different methods of prey transfer

While aerial transfers were common for African Marsh Harriers (49% of all transfers) and usually efficiently executed, they were by no means the only method of prey delivery. For example, on 617 occasions when male marsh harriers arrived with prey for their dependants, it was not transferred on 3% of the occasions. This was mainly because males were reticent in transferring prey at the very beginning of the season. The same has been observed in Hen Harriers early in the season (Balfour 1962–63). Females chased their prey-carrying mates, vociferously food soliciting, but were denied access to the food. There were two possible reasons for this intriguing behaviour: it was a male ploy to test the female's state of breeding readiness through her motivation, or it may have arisen from the male's conflict over giving prey up to another bird in these incipient stages of breeding. That both *cyaneus* and *ranivorus* males eventually did relinquish their prey following several minutes of chasing supports both views. Those males that continued to refuse to give up prey either stood over it for long periods, or eventually ate it themselves.

Most other prey transfers (46%) by African birds took place on the ground in the territory. The majority (66%) of the 107 ground passes that took place to females before their young were hatched occurred just before or during egg-laying (Table 2.4). In fact ground passing of prey at this stage of breeding (71

Table 2.4 Seasonal differences in various methods of prey transfer by male African Marsh Harriers to their mates and dependent young 1984–86.

Stage of breeding	Type of prey transfer					
	Aerial (succ)	Aerial (missed)	Ground pass	Cached	Not passed	Total
Pre-breed to egg-laying	50	3	71	5	9	138
Incubation	56	3	36	1	6	102
Nestling period	118	2	129	2	0	251
Flying young	60	13	49	0	4	126
Totals	284	21	285	8	19	617

transfers) outnumbered the aerial transfers ($n = 53$). Similar trends are reported for Montagu's Harriers *C. pygargus* (Pandolfi and Pino D'Astore 1990b). Either egg-laden females were more reticent to fly to accept aerial transfers at this stage, or males preferred ground passes as a prelude to copulation. Since prey delivery was not a pre-requisite for copulation in African birds (Simmons 1990), the reticence of egg-laden females to perform agile aerial manoeuvres is more plausible.

On a further 1% of all occasions, African males cached the prey, either because their mate was unwilling to rise to accept it, because their offspring could not be found (away foraging), or because they were satiated. So for African Marsh Harriers aerial food passing is about as frequent as ground passing, with a small percentage of prey not being passed at all.

Male Northern Harriers arrived with prey on 320 occasions, and on four occasions it was not passed to their mate. Unlike African birds, this was not because males refused to give it up, but because it was stolen by another female before they could reach the intended female. These 'robber' females were sometimes, but not always, within the male's polygynous harem. This was also observed in Wisconsin harriers (Hamerstrom 1986).

Aerial transfers occurred more often among Northern Harriers than among African birds, comprising 77% of all 320 transfers. Since the two species are virtually identical in body mass (546 g vs. 570 g females masses: Table 2.2), aerial agility is unlikely to explain the differences. Ground passes comprised a mere 22% (vs. 46% for African birds) of Northern Harrier passes, but changed in frequency over the years with prey fluctuations. More male visits to nestlings occurred during prey-poor years than in prey-rich years when females were on hand to collect the male's prey in an aerial transfer. Combined, the two species passed prey aerially on 58% of 937 occasions, and on the ground 38% of the time. Only 4% of all aerial prey transfers were dropped, and almost all of these were retrieved.

Evolution of aerial food transfers

We can use the foregoing results of efficiency and method to ask how the trait may have evolved. Contrary to the prediction that Northern Harriers, with more opportunity to leave their prey on the ground should ground pass more often, this species food-passed aerially more frequently (77%) than African Marsh Harriers (49%). African Marsh Harriers lost their prey more often when they dropped it (24% not retrieved) than Northern Harriers (10% not retrieved), as would be expected in reed marshes with deep water margins. Thus ground passes may be more frequent in Marsh Harriers because of the threat of losing hard-won prey during prey transfer. Nevertheless, since aerial passing is still the predominant method for both harriers why did it evolve in this genus and not others?

An interesting insight into this question is provided by the only tree-nesting harrier, the Australasian Spotted Harrier. Following the logic presented above, one would expect Spotted Harriers, with their arboreal-oriented habits, to have perhaps dispensed with aerial food passing in favour of leaving prey in trees. But not so. Baker-Gabb (1982) reports that, while early passes are more frequently passed near the nest in a tree, aerial passes predominate from incubation on. Spotted Harrier behaviour does not support the idea that a lack of raised transferral sites has led to the evolution of aerial food transfers in harriers.

There are two further possibilities: (i) aerial passing of prey may be safer than ground passing because it leaves no scent or remains for a mammalian predator to find. At the same time (ii) it may be that aerial transfers are more typical for an aerially oriented genus like harriers than for perch-and-wait predators. Two lines of evidence support the first idea. First, aerial prey transfers in Northern Harriers are three times as common as ground passes. These birds nest in 'dry' marshes and mammalian predators are significantly more likely to find (and destroy) nests in dry areas than those built in wet ones (Simmons and Smith 1985). Hence, prey or scent left in dry marshes is more likely to be discovered than those in wet marshes. Second, the Montagu's Harrier, which nests in drier (grassland) habitats than the Northern Harrier (Cramp and Simmons 1980; Arroyo 1995), passes prey aerially even more often (86%) than the Northern Harrier (77%). Only 14% of all 291 transfers recorded by Pandolfi and Pino D'Astore (1990b) were on the ground. The trend from these three harriers is for more aerial passing as the breeding

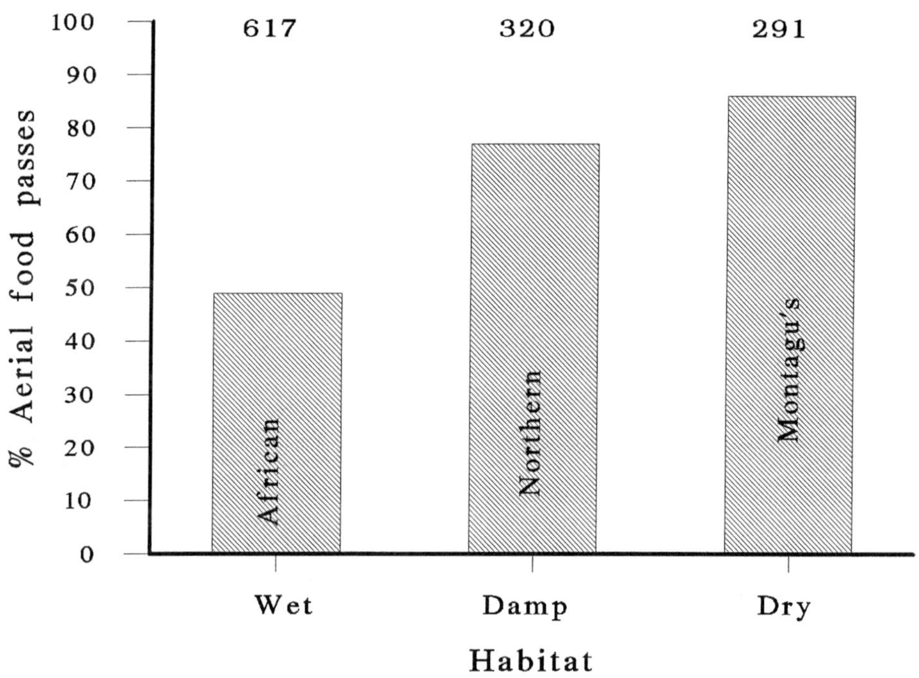

Fig. 2.6 Differences in aerial food passing frequency between harriers breeding in wet to dry habitats: African (Simmons 1991b); Northern (Simmons 1991b); Montagu's (Pandolfi and Pino D'Astore 1990b). Numbers above bars give the total number of food passes seen. Other food transfers occur on the ground.

habitat gets drier (Fig. 2.6). Hence aerial food passing seems to be associated with dry nesting habitats and the African Marsh Harrier's propensity to ground pass more frequently can be explained by a lower risk of discovery by terrestrial predators in its wetter habitat. It seems, therefore, that aerial prey transfers evolved in harriers because it was (and is) both safer and easier than ground passing.

Specializations of the harriers—ground nesting

The ground nesting of harriers also sets them apart from other raptors, except nocturnal predators such as the owls (Strigidae) with which they are often sympatric. The ground nesting of species which typically nest in trees such as Merlins *Falco columbarius* and Bald Eagles *Haliaeetus leucocephalus* (Newton 1976a; Sherrod *et al.* 1977) often arises where predation pressures are relaxed, making it a viable alternative to the normal arboreal habitat. However, clearly predators are not absent from many of the areas where harriers breed, and it is more likely that the diet of the harriers, such as wetland birds, or abundant mice in open grasslands, has drawn them to such habitats where trees were, and are, not available. It is probably this lack of availability that has driven the harriers to ground nesting.

Given that ground nesting harriers are more susceptible to predators (between 9% and 52% of all nests in seven species are destroyed by predators—Table 9.17), one might expect special adaptations for reducing predation pressures. Obligate ground nesters have evolved four methods:

(i) to nest on islands inaccessible to predators (Lack 1968);

(ii) to nest in groups to cooperatively mob and drive off predators (Kruuk 1964) or dilute the probability of predation;

(iii) to nest in close association with larger predatory species which offer protection through their presence and own nest-defence (Wiklund 1982);

(iv) to nest in particularly well concealed evenly spaced sites (Redmond *et al.* 1982).

Cooperative mobbing and well-concealed nest sites (ii and iv above) are the most likely, but at least one study noted that Northern Harriers nest on predator-free islands in Massachusetts, USA, and are not found on those with mammalian predators (Veit and Petersen 1993). Given that harriers are susceptible to predators it is surprising that only two studies have determined predation in relation to site characteristics or possible cooperative defence.

In her study of the advantages of clumped nesting in Montagu's harriers, Arroyo (1995) found birds were spaced between 30 m and 4300 m apart, with a bimodal distribution apparent either side of 600 m. Extremes in this species have been noted of 6–8 m between nests in colonies of 22 pairs in Russia (Karpov and Bekbaev 1990 in Arroyo 1995), or 20 nests in 2–3 ha in France (Cramp and Simmons 1980). The clumping in Spain was not due to differences in nest site moisture levels—all were dry sites—or vegetation types, as many sites not used one year were used the next with no difference in the habitats or vegetation density. As expected, clumped pairs attracted other birds in mutual defence of their nests, and they did detect experimentally introduced stuffed predators at nests quicker than did solitary pairs. By playing alarm calls along with the presentation of these stuffed predators, Arroyo and co-workers (1998a) were able to show that it was alarm calling that attracted other birds. In food-rich years predators were detected quicker (because parent birds were more often in the nest vicinity) and attack rates were higher, a finding common to Northern Harrier populations (Simmons *et al.* 1987). However, despite these clever experimental tests of the advantages of coloniality, birds in the clumped Spanish populations were *less successful* in the number of young fledged per year than solitary birds (Arroyo 1995)! This may have been caused by the greater aggression of males who nested together (taking time and parental investment away from defending and provisioning youngsters), or possibly because young, not old birds, tend to make up the colonies. That this species is the smallest, lightest, and only truly colonial harrier, suggests that it may be more susceptible than other, larger, harriers which have a better chance of driving off predators single-handed. Whatever the reason, clumping did not appear to be of any advantage to Montagu's Harriers nesting in central Spain. Possibly the selection pressures (such as predators) giving rise to the trait are no longer operating in modern-day Spain and the reasons will have to be sought in other populations. The predation levels of 9% of 165 nests (Arroyo 1995) are the lowest for any harrier (Table 9.17), supporting this reduced predation view.

Behavioural ecologists have considered many reasons for coloniality in various species and those apparent for Montagu's Harriers in nest defence are among the commonest cited. However, Newton (1979) suggested that raptor coloniality was associated with food resources that are variable in both space and time. Hence species such as vultures, insectivorous kestrels, and Ospreys may enhance the food finding abilities of one another, and mutually benefit from the association. Vultures are the best example of this: a foraging net may cover hundreds of square kilometres until one bird descends onto a carcass, pulling in all the others with it (Mundy *et al.* 1992). This type of information sharing has given rise to the information-centre hypothesis (Ward and Zahavi 1973). By contrast, species that hunt prey which are dispersed relatively uniformly throughout the environment are unlikely to benefit from shared information and should show territorial behaviour and defend exclusive ranges.

Few good tests of this information sharing idea have been undertaken but Ospreys, at least, are known to actively convey information to others in the colony with a special flight behaviour once they have successfully located and caught shoaling fish. The same flight is not apparent when solitary fish are located (Greene 1990). Montagu's Harriers do not appear to convey any information and are unlikely to do so because birds forage in different areas and aggressive interactions occur when they meet on foraging grounds (B. Arroyo pers. comm.). This lack of cooperation was confirmed in tests with Swamp Harriers in New Zealand which roost communally. Baker-Gabb (1982) left out food baits for birds to find and then determined on the second day whether more harriers were attracted to the baits, as would be expected if information was being exchanged. While birds foraging close by were attracted when they saw another wheel in mid-air and descend onto a bait (a form of local enhancement or direct piratical behaviour), young birds that found baits actually left the local communal roost and roosted singly within 200 m of the baits. Adult birds who discovered the baits and continued to roost communally were generally not followed by other roosters when they returned the following day. Indeed in the one case (out of 17) when a bird was followed, it flew past the baits, shed its follower, and then circled back to feed! Thus in all cases these Australian harriers actively prevented others from finding the food source by changing their behaviour—a sort of misinformation hypothesis. Thus among harriers, clumping is unlikely to arise from shared benefits in finding food and little benefit is apparent in terms of increased breeding success. In India (Clarke 1996) non-breeding birds, at least, sometimes form massive roosts on large open fields where predators would find it hard to approach undetected. Many other harriers roost in rank grassland, where the dry and crackling nature of the vegetation also makes stealthy approach difficult. The general conclusion, therefore, is that harriers probably choose their roost sites for anti-predation reasons.

However, given that success is not enhanced in colonial breeding Montagu's Harriers, why does this species nest in loose colonies? One point to note is that most advantages and disadvantages proposed for colonial living disregard sexual selection and pre-nesting behaviour. Wagner (1998) has suggested that colonies may form as a result of females clustering to seek extra-pair matings. So, rather than large colonies promoting a high incidence of extra pair copulations, it is the high probability of extra-pair copulations (EPCs) that promotes clumping of females and therefore males. This hidden lek hypothesis is a reversal of the normal ideas explaining relatively high EPCs in colonial species. It is, however, unlikely to apply to raptors for two reasons. First, as we shall see in Chapter 6, there is no correlation between EPCs and coloniality in raptors, and second, EPCs are unusually low in species such as raptors in which one would expect a high incidence because males are often absent hunting for their mates. For Montagu's Harriers, ongoing research in western France and Spain will have to consider all factors that might select for coloniality to understand its present-day maintenance. Bretagnolle and co-workers (1998b) assessed responses between three study areas in France and the Spanish site and found that the responses of birds to the decoy predators are closely related to natural predation risk in the study sites. Thus the Spanish site may have suffered relatively low levels of predation and not been the ideal area in which to test the anti-predator advantage of coloniality. An evaluation of the relative success of single and colonial pairs in all four study areas is needed.

Among the less clumped nest dispersion of Northern Harriers in eastern Canada, birds did not cooperate in defence when tested with stuffed Great-Horned Owls *Bubo virginianus*. However, significantly higher nest success was apparent for birds that nested in sites ranked as wet or very wet (29 of 33 sites successful with standing or running water), relative to 11 of 20 sites that were dry (Fig. 2.7). Neither nest concealment nor vegetation type played any role in nest success (Simmons and Smith 1985). Like differences in aerial food passing behaviour of birds in dry and wet habitats (above), the difference in success may

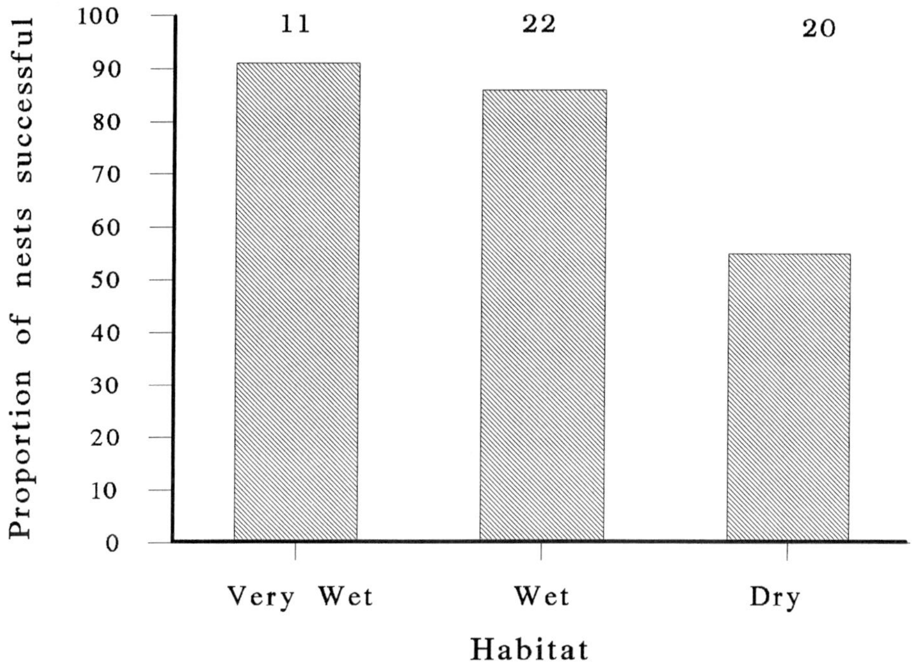

Fig. 2.7 Differences in nest success in relation to habitat moisture for Northern Harriers in New Brunswick, Canada.

also be related to terrestrial mammalian predators hunting less in wet habitat if scents are harder to follow there. By calculating the number of theoretical wet to dry sites (127:145) on the 60 km^2 marsh in New Brunswick, it was also possible to determine that Northern Harriers there were not limited in their choice of sites, but the wet sites were used more often than expected by chance alone (41:23; $X^2 = 6.6$, $P < 0.02$). Harriers in Orkney likewise preferred the wetter swales in which to nest (Balfour 1962). However, Hen Harriers in The Netherlands used 'humid or dry ground' on which to nest. This apparent switch from wet to drier habitat is associated with competition from other species in these areas. In Continental Europe Hen Harrier are sympatric with European Marsh Harriers, which prefer wetter habitats, while harriers in Orkney or North America face no apparent competition for nest sites. From several angles, therefore, wetter sites appear to be preferred because they reduce predation.

Several harriers, however, have been known to use trees and one species—the Spotted Harrier—has moved to this habitat permanently. Or did it always nest in trees? Perhaps all harriers once tree-nested if they evolved from the arboreal *Accipiters* and the present tree-nesting of the Spotted Harrier is the ancestral trait? On comparing the structure and durability of the nest with other arboreal species Baker-Gabb (1984) found that Spotted Harrier nests were unlike those of other tree-nesting species, being flat and flimsy and made of light twigs and thistle stems. Their placement was also atypical in that the birds made no attempt to weave their nests around stout upright branches but simply laid them on forking horizontal branches. Consequently, 14% of 28 nests in south-eastern Australia blew down while in use, significantly more than for other arboreal species in the same area. This suggests they once built ground nests and have simply moved to tree-nesting. This is not the only unusual feature of this species and aspects of its display and allometric relations with egg size will be treated later. Similarly flat and simple nests constructed of typical nesting material are characteristic of Hen Harriers which have taken to nesting in Sitka Spruce in Northern Ireland (Scott *et al.* 1993/4). Since these nests were up to 4.5 m high they are not birds merely nesting in bushes in marshy areas. While neither study assessed the selection pressures which may have promoted the use of tree-nesting, it is notable from Baker-Gabb's earlier work (1982) that Spotted Harriers have a lower nest-predation rate than most other harriers (Table 9.17). Studies of the persistence of tree-nesting in Northern Ireland's Hen Harriers and the levels of predation and prey relative to nearby sites, in both harriers, will be particularly interesting to follow. Whatever the outcome, it seems that harriers were probably never tree-nesters, but their main grassland/wetland prey have promoted the evolution of ground nesting and this unusual habit is accompanied by particular adaptations to avoid terrestrial predators.

Facial feathering and hearing abilities

Like most owls, every harrier has a cowl of feathers surrounding its eyes, giving it an owl-like appearance. These feathers are specialized and stiffened conch feathers in a primitive parabolic shape suggesting that hearing is enhanced to locate small mammal prey well hidden in dense grassland. A harrier quartering over favoured habitats keeps its head pointing almost vertically down (see chapter opening), rather than looking forwards as other raptors do, and often at heights no greater than 1–2 m. Foraging owls do likewise, suggesting convergence in their foraging patterns and modes of prey detection. Clark and Stanley (1976) found morphologically similar feathers in the Short-eared Owl *Asio flammeus* and Northern Harrier, suggesting harriers do possess hearing acute enough to detect prey by sound alone. Rice (1982) went further by studying Northern Harriers in the lab and field. He determined that harriers could resolve the direction of the sounds with accuracy equal to that of Barn Owls *Tyto alba* (1–2°) with their large forward-facing disc. He then tested them in the field by played synthesized vole sounds to wild birds through tiny microphones hidden under matting covered with dirt and grass. Birds successfully hit, and in one case carried off, the microphone even though the sounds were terminated when the bird was within 3 m. (Rice did this to determine how well a harrier can localize sound from this distance.) Kestrels and hawks tested at the same time yielded resolution efficiency of only 8–12°. While the owls also have asymmetrical ear-openings to assist in determining exact locations in a horizontal and vertical plane (Payne 1971; Knudsen and Konishi 1979), harriers do not. They may not need this if other (visual) cues are available to them to determine vertical components of location during their diurnal foraging. Rice concluded that the stiff facial feathers assist in the acute hearing abilities of harriers do allow them to detect and catch mice they cannot see.

While this seems fairly convincing evidence, there is an unsolved problem in Rice's conclusion. The problem was pointed out by Calford *et al.* (1985) who tested the acoustical regions within the midbrain of Australasian Swamp Harriers by anaesthetizing birds, placing electrodes in the midbrain, and then playing sounds of different frequency to them. By assessing electrical responses, they determined that optimal hearing frequency was less than 5 kHz for the harrier (and a falcon and goshawk) but up to 10 kHz for the Barn Owl. However, their most important finding was that at such frequencies the diameter of the facial ruff (which they measured at 72 mm) was too small for the optimal frequency to give any assistance with enhanced directional hearing. Only at frequencies above this would the harrier be able to accurately locate sound, yet this frequency is above that which the harriers can hear! By contrast, because the Barn Owl's facial disc is larger (~110 mm), the frequency at which the bird can accurately locate sounds is only 2.7 kHz, and the disc becomes highly directional over the frequency range of the neurons tested (~4–10 kHz). The authors concluded that the harrier's facial ruff cannot function as directional sound collector and it must serve another purpose!

How can these two surprising and opposite conclusions—harriers can localize sound much better than other hawks, yet their facial disc is not large enough to allow them to do so—be reconciled? Both were thorough and convincing studies, so were there perhaps incorrect or overlooked underlying assumptions? One possible and unexpected solution, originating far from these studies in North America and Australia, is provided by Neufeldt (1964), who worked in the southern corner of Siberia. She studied the virtually unknown Pied Harrier in Amurland in the late 1950s and early 1960s and noted an unusual aspect of the harriers which no one before, and few since, have pointed out: they are able to raise and lower their facial ruffs. She photographed an incubating female responding to noises around the nest and from her observation hide. The raised ruff (p. 301) clearly protrudes several centimetres *beyond* the normal width of the neck (see Fig. 2.8), while under normal circumstances it is typically held flat and no wider than the neck feathers. If this trait is common to other harriers, and those which are captured, restrained, or not foraging sleek their ruff feathers down, then Calford *et al.* (1985) may have under-measured the true size of the ruff in their captured birds, and wild foraging birds may capture and localize sounds in a reflector approaching the size of a Barn Owl's pronounced facial feathering. This suspicion is partly confirmed by Clarke's (1995) observation that 'feathers of the various parts of the disc are independently mobile, with their own musculature and nerves, allowing the harriers to direct sound into the ear by lifting feathers and altering the shape of the disc.' (p. 87) Measurements of the raised facial ruff have yet to be published, but they may reconcile the theoretical limits, used by Calford *et al.* (1985) to dismiss the function of the facial feathering in harriers, with the obvious abilities of harriers to localize and capture unsighted prey. While perhaps not the only explanation, it seems the most parsimonious.

Fig. 2.8 A female Pied Harrier with the facial ruff raised in response to sounds around the nest. The sketch is based on a photograph by Neufeldt (1964).

Evolution of the facial ruff

Why did harriers evolve such a specialized sound-localization characteristic in the first case? For owls it is obvious that their nocturnal foraging and the low light levels have been selective pressures allowing them to capture prey in complete darkness. Yet harriers are strictly diurnal (at best crepuscular) hunters. Rice (1982) believed the abilities evolved because they hunted prey in long grass in which visual penetration is not as great as auditory. I suggest that the detection of terrestrial prey such as mice and voles was more important than merely concealed prey because harriers specialize on these at critical times of their lives. There are two reasons for proposing this: one proximate one ultimate.

First, observing young marsh harriers on a African flood plain make one unsuccessful foray after another into thick grass in which they (and I) can hear mice scurrying and squeaking, but which they cannot possibly see, suggests it is mice not birds that have selected for their acute hearing abilities. Passerines sing from elevated perches but are generally silent on the ground unless feeding young. Bird-chasing by harriers involves a different—ambush and fast flight—strategy while flying over physical boundaries (Underhill-Day MS) with little or no use of their acute hearing. Acute hearing may be useful for locating nestling birds in concealed ground nests, but such predation typically occurs at a time of plenty and is less likely to be a major selective force. Second, while many studies report a wide variety of prey in the diet of harriers, three species that have been assessed for seasonal differences in prey choice during breeding each show a specialization on small or medium mammals during the critical courtship and egg-laying periods (*hudsonius* 92%, Barnard *et al.* 1987; *ranivorus* 82%, Simmons *et al.* 1991; ~80%, *pygargus* Arroyo 1997). Just as important, raptors in general find avian prey significantly more difficult to catch than small mammal prey (median values for birds 13%, $n = 20$ raptor species, vs. 23% for mammals, $n = 8$ species: Temeles 1985) as do harriers in particular (17% vs. 55% success, respectively, for African Marsh Harriers; Chapter 4). Such differences also suggest that harriers should find it profitable to specialize on mammals rather than birds at critical times. Thus, I would suggest that ancestral harriers specialized on mice and other small mammals in thick grass, and this became one of the chief selective pressures in the evolution of their facial ruff. That the larger marsh harriers (European, Swamp, and Madagascar) do not take many small mammals does not negate this idea, because once a ruff and directional hearing have evolved it would not disadvantage a harrier which took larger prey to retain it, albeit in reduced form. The ruff might remain as a relic of evolution past.

Support for the view that the facial ruff is chiefly an aid to detecting concealed micro-mammals is found in Clarke's (1996) observation that the facial ruff in the Hen Harrier, the best known vole specialist, encircles the ears and meets below bill. In the Montagu's Harrier, however, the ruff is confined to an arc just behind the ear openings. This is expected in a species in which birds and insects predominate in the diet in most areas across Europe (Arroyo 1997).

Behavioural peculiarities

We have thus far concentrated on morphological, distributional, and ecological aspects that help define the harriers, some of which are unique to the group. I will finish this broad sweep by focussing briefly on behavioural aspects. Harrier mating patterns are well documented and well known, and polygyny is frequently recorded in both Northern and Hen Harriers in North America and Europe, respectively. It is also known, but highly variable, in populations of European Marsh Harriers across Europe. It has also been recorded in the Australasian Swamp Harrier and may occur at low levels in Black Harrier populations. That it is not due to nest clumping alone is evident from its low incidence (< 2%) in populations of the Montagu's Harrier which breed in colonies of various density across Europe.

Why this unusual mating system occurs at all is controversial since it is well established that females within the harems reproduce poorly relative to monogamous birds. Why polygynous females continue to choose mated males has been the subject of intense speculation and modelling, and some ingenious ideas have emerged. The incidence and evolution of this unusual trait is fully explored in Chapter 5.

Polyandry is rarely reported in birds in general, and raptors in particular. It is, however, regular in the Galapagos Hawk *Buteo galapogoensis* (Faaborg *et al.*

1980), Harris' Hawks *Parabuteo unicinctus* (Bednarz 1987), Bearded Vulture *Gypaetus barbatus*, and Egyptian Vulture *Neophron percnopterus* (Heredia and Donazar 1990; Tella 1993), and has recently been uncovered in Africa's Pale Chanting Goshawk *Melierax canorus* (Malan *et al.* 1997). Three cases of apparent polyandry have been recorded in Montagu's Harriers, once each in England, France, and Spain (Khan in Cramp and Simmons 1980; Cormier 1990; Arroyo 1996). For these cases to be treated as true polyandry, both males are expected to copulate with a breeding female, rather than just feed her or her brood. Provisioning alone may rather indicate cooperative breeding (i.e. helping) rather than polyandry. In this system helpers assist with brood maintenance but are not rearing their own young, even though they may be increasing their inclusive fitness by raising related brothers and sisters. In at least two of the cases with Montagu's Harriers young males were involved and almost certainly added nothing genetically to the breeding attempt—cooperative polyandry. In the Spanish case, two adults were provisioning half-grown young but copulations had not been witnessed. Aggression was minimal between the males and both attacked a predator placed at the nest (Arroyo 1996). Since provisioning was higher at this nest as it was in Pale Chanting Goshawk polyandry (Malan *et al.* 1997), having two males certainly improves chances of success for the breeding female. Arroyo speculated that young males may be provisioning females in order to improve their chances of mating with them in future years. However, none of these studies could evaluate this.

Polyandry has also been witnessed in an interesting case involving the light and dark morphs of the Longwinged Harrier in the Buenos Aires province of Argentina. Maria Susana Bó and her associates studied sympatric breeding populations of both South American Harriers on the Laguna de los Padres Reserve in 1992–93 and 1993–94. In six Longwinged Harrier nests one female copulated with a light morph and a dark morph male and was fed by both birds. She laid a clutch of three eggs, only one of which hatched. The chick was then killed by one of the males—a possible case of infanticide driven perhaps by the uncertainty of paternity?

Three other cases of cooperation between harriers are known, one each involving European Marsh Harriers, Black Harriers, and African Marsh Harriers. The latter case allowed an assessment of the 'future benefits' hypothesis for young males, since tagged birds were followed for several years. In the former case, one female European Marsh Harrier who had lost her large nestlings to a predator switched her attention to a second pair's brood whom she fed (Buczek 1994). This case from eastern Poland suggested misdirected parental care rather than any adaptive strategy to win a future territory or mate. Short observations at a nest of Black Harriers in South Africa by Graeme Arnott revealed two females apparently brooding the chicks: a female dropped off prey at the nest and disappeared over a ridge with the male; on approaching the nest to count the chicks, Arnott disturbed a second female brooding them. In the case of the African Marsh Harrier, however, a young male (male B) took over the entire provisioning and defence of an unrelated almost grown brood following the disappearance of the brood's natural father. Not only did he supply the brood with large amounts of food for the next 36 days but the female solicited food and copulations from him as well as a third, neighbouring, male, from whom she took food to feed her young (Simmons 1992). I tagged both new males and watched their movements for a further two seasons: the newly territorial male (B) retained the territory for the next two seasons, but not the female who left to breed with another male in a different territory. This suggests that territory acquisition rather than mate acquisition was the male's primary motive, and his feeding behaviour was not sufficient to retain the experienced female he had assisted. This itself suggests deceptive behaviour—through her food solicitation—on the part of the original female and at one time she had three males feeding her young. Since male B himself bred successfully for at least the next two seasons he appeared to suffer no loss in fitness through his provisioning behaviour and he gained a productive territory. His behaviour, therefore, was neither altruistic nor maladaptive. As interesting as these cases are, they are exceptional and not the usual route to mate or territory acquisition.

We have seen that the harrier genus, comprising 16 species, has a number of specializations, such as acute hearing, aerial food passing, and polygynous mating patterns, each of which can be attributed to the open grassland or wetland habitat where they have evolved. The following chapters will explore in greater detail several of the aspects touched on here, including the mating system, aerial agility and its role in mate choice, copulation behaviour, and other ecological and evolutionary aspects.

3 *Aerial displays: mate choice and reversed dimorphism*

The father of systematic biology, Carl von Linnaeus, mistakenly lumped the harriers in with falcons when he named the first two harriers *Falco aeruginosus* and *Falco pygargus*, in 1758. It took Frenchman Lacépède to realise their differences in 1799, and he was presumably impressed enough with the display behaviour of the harriers to rename the genus *Circus*—the hawk that circles. The name captures one trait that distinguishes harriers from the falcons—their astounding acrobatic nature and uncontrolled abandon during aerial displays before and during breeding. In this chapter I will explore the different types of display, the reasons why harriers expend a great deal of energy displaying, what they communicate to other birds, and whether females can and do use the displays as a guide to male fitness.

Like other raptors, particularly the bird-eating ones, all the harriers show reversed size dimorphism (RSD) almost as extreme as that of the sparrowhawks. That is, females harriers can be half as large again as males. It is generally agreed that the degree of dimorphism is related to bird prey and the agility of that prey—bird hunters are more dimorphic than mammal hunters (Snyder and Wiley 1976; Newton 1979). For harriers, this is even more striking since plumage dimorphism is also marked between the sexes, unlike many other raptors. This has correctly been identified as linked with the polygynous nature of harriers (Newton 1979; Baker-Gabb 1982), rather than the originally proposed counter-shading needed by open country hunters such as harriers to aid prey capture (Nieboer 1973). Some authors (Jehl and Murray 1986) researching shorebird morphology have suggested that aerial agility in display is an important selective component of the dimorphism, for which there is some evidence. Associated with this, and not mutually exclusive to it, some raptor biologists have argued that females, not males, need to compete for highly competent (food providing) mates and this female–female competition has driven them, rather than males, to become ever larger (Olsen and Olsen 1987). Whether this applies to harriers is easy to check because one can measure the strength of this sexual competition through their degree of polygyny—species in which polygyny is intense means males should be more 'sought after' and hence female competition for them greater than in monogamous species.

We will explore whether bird catching ability and sexual selection interact to promote high degrees of dimorphism in the harriers towards the end of the chapter, using new data. This is probably a unique approach since no other group of raptors lends itself to an assessment of sexual selection like the mono-

gamous/polygynous harriers. This is not to say that sexual selection does not occur in monogamous species—there is still selection on males to be attractive to females, to begin breeding as early as possible or attract additional females for extra pair copulations and offspring—such selection, however, is harder to see and is of lower magnitude.

Skydancing and other displays

First, what is so special about the displays and what is their function? Eddie Balfour, long time harrier watcher on the Orkney islands, has captured some of the energy and sheer abandon of the birds, performances on spring days in northern Scotland:

> The Hen Harrier exhibits flying powers which combine great vigour with an airmastery of the first order. Generally after a bout of undulating flight he appears to abandon himself in mid-air and hurtles earthwards, seemingly out of control, wings threshing, sometimes spiralling, until he seems destined to crash to the ground. He is, however, in perfect control for when near the ground he sweeps upward and ascends with vigorous strokes to regain height. When at the top of the ascent he may execute a half-roll, or turn a somersault to fling himself into the next headlong dive. (1962–63, p. 146)

This aerial display was dubbed 'skydancing' by Frances Hamerstrom in her classic (1969) study of the Northern Harriers, *Circus hudsonius*. The undulating displays of the harriers are that much more impressive because of the counter-shaded plumage of males of most species that flash white and grey as the bird barrel-rolls at the top or bottom of each undulation. Some displays involve over 100 undulations in a single sequence, and birds end by disappearing into potential nest sites. The energetic output of the harrier display has never been measured but there are other ways of assessing the energetics involved. One that has been used on passerines such as Skylarks *Alauda arvensis* and American Bobolinks *Dolichonyx oryzivorus* is to wing clip displaying males and estimate how long they spend in aerial flight compared with unclipped controls. The results indicated reduced display flight as predicted, suggesting flights are energetically costly (Møller 1991a; Mather and Robertson 1992). Non-intrusive methods include determining display intensity in good-food and poor-food years. Better still, one can supplement feed prior to breeding and determine their display activity relative to unfed controls in the same year. The first such correlational and experimental studies for raptors are described here.

There are two distinct aerial displays performed by harriers: *skydancing* and *skyspiralling*. The first is a high (70–400 m) horizontal display in which birds circle to the desired height, often with vigorous and obvious effort, before levelling out and starting exaggerated, deep wing beats, hereafter called 'wing-rowing' flight. This serves to increase the bird's air speed before it loses height in the first tentative undulations. As the display proceeds so the dips in flight increase until the bird falls steeply through 20–50 m before swinging upwards to regain height, with flailing wings. At the peak or trough (depending on the species) of most undulations (Us), birds turn on their axis and call—a squealing rising 'oweep'—before swooping earthward again. This calling is timed differently for African and European Marsh Harriers; the African calls at the top, and the European bird at the bottom, of each undulation. Approximately 10–12 Us are performed per minute with a horizontal speed of about 1 km in 9 min. Some displays of African Marsh Harriers last up to 20 min but more usually are complete after 2–3 min. Among Montagu's Harriers the skydancing display seems relatively infrequent, and indeed birds rarely roll around their axis (Arroyo 1995). Montagu's Harriers are also unusual in that they are the only harrier in which males sometimes carry prey while displaying, but only after pairs have formed (Clarke 1996). Skyspiralling (below) is much commoner in this species. For some harriers the display calls are frequently heard, particularly it seems, among the Pied, Australasian, African, and Reunion birds, whereas at least among Northern Harriers in Canada the displaying birds are relatively quiet but not mute. Among Black Harriers, high circling seems to be commoner than skydancing per se (van der Merwe 1981) but dancing does occur and calls are heard at the top of each U. Skydancing is unknown in at least one harrier. The Australian Spotted Harrier has never been observed performing undulating flights, and high soaring and mutual following through its arboreal breeding habitat are the few displays noted (Baker-Gabb 1982). Incipient *skyspiralling* has been described once but it is obviously rare (Marchant and Higgins 1993).

The second form of aerial display was originally thought to occur only among African Marsh Harriers, and is one I dubbed *skyspiralling* because it is vertically orientated and a more vigorous display than the horizontally orientated *skydancing* (Simmons 1991). The descent may follow the horizontal display, but often starts with a soaring bird beginning wing-rowing flight, before dropping in a series of fast spiralling Us (Fig. 3.1), pulling up suddenly in the descent, and calling at each peak, before terminating in level flight close to the ground or more frequently on the nest. Indeed skyspiralling is a spiral within a spiral because as the bird falls, it also twists vertically about its own axis (Fig. 3.1) producing a wild flailing, uncontrolled display of great agility, speeding up as it nears the ground. Some birds stall completely at the end of such displays, before dropping into the nest where they call. Particularly vigorous displaying birds, mainly unmated males, immediately start soaring again to repeat the performance.

I believed that spiralling was unique to African birds because either descriptions from other species didn't match what I was observing or it wasn't described at all. Bildstein and Gollup (1988) described one Northern/Hen harrier display as 'falling like a leaf', while Baker-Gabb (1982a) described a similar display for the Australasian Swamp Harrier that ends in the territory. Robinson (1950) describes the display of the Montagu's Harrier as 'cork-screwing'. In their comprehensive reviews neither Glutz *et al.* (1971) nor Cramp and Simmons (1980) mention such a display in European birds. However, harrier researchers Blanco *et al.* (1993), working in central Spain, provide evidence that similar spiralling displays occur in European Marsh Harriers. Since then both Clarke (1995) and Arroyo (1995) have indicated that European Marsh Harriers

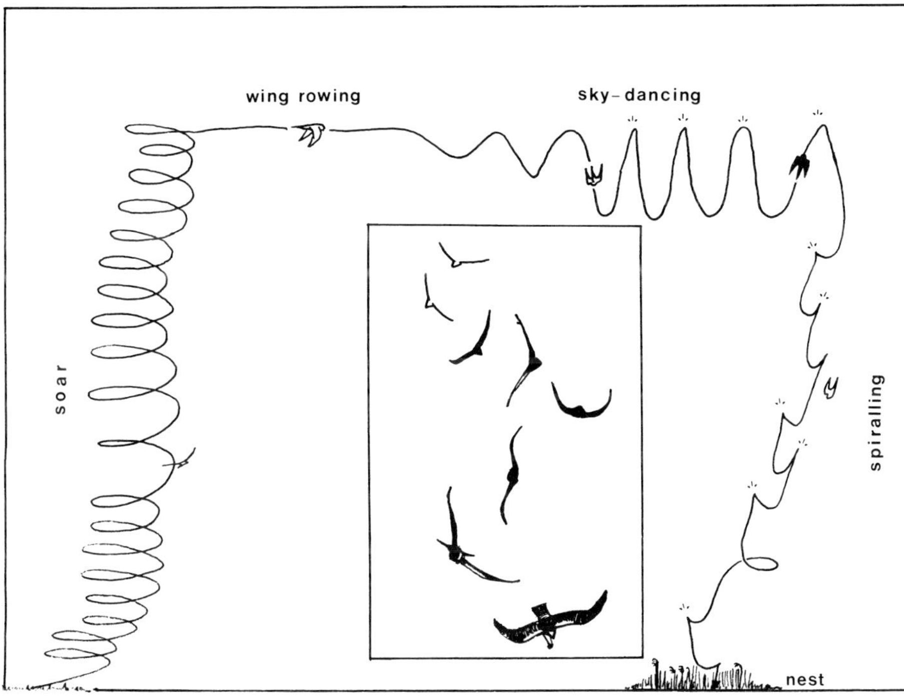

Fig. 3.1 A schematic representation of the aerial skydancing and skyspiralling displays of the African Marsh Harrier. Inset is the skyspiralling descent as seen from below.

and Montagu's Harriers perform skyspiralling displays, and for Arroyo's Spanish population, females perform spiralling displays as often as males (79:93). Skyspiralling has not been observed in the Eastern Marsh Harrier while skydancing is common (I. Fefelov unpublished) and is performed by both male and female. Females display least often but may do so with their mate or in territory defence against intruders. While displays have to be described for other harriers, it seems the two types of display occur in at least four species, and may well be found in others.

Variations seen in African Marsh Harrier skydancing displays over three years were backward loops executed by two birds, in which the bird completed a full circle (up to three times) by turning backwards at the top of each U. Frances Hamerstrom and Eddie Balfour have observed similar evolutions in Northern/Hen harriers, and recent observations of Black Harriers indicate that they too perform so-called loop-the-loop displays (Clark 1999). One wonders if they occur in birds where territories are small and males or females are forced to display in a limited aerial arena. This may explain the preponderance of vertically oriented skyspiralling displays of Montagu's Harriers (Arroyo 1995; Robinson 1950) because of their colonial nesting habits and limited territories.

Displays: courtship or aggression?

Female participation

To understand whether such displays are mate-attracting (courtship) or territorial aggression, we must determine the context of the display and the participants. Female harriers in various populations are known to perform both spiralling and skydancing displays at different frequencies. In central Spanish populations for example, Blanco *et al.* (1993) showed females performed 58% of all (65) displays recorded. Since female participation was rarely recorded in more northerly migratory populations (e.g. Witkowski 1989), they attributed differences in female displays to the sedentary nature of their population and the need to retain territories. Females also displayed in Thiollay's (1970) Camargue population in southern France, supporting their contention. The idea is that where females need to be territorial because they reside in an area year round, they perform territorial displays; migratory females do not need to since they fit into existing male territories during breeding. However, Arroyo's harrier population in central Spain are migratory yet show female participation (46% of 172 displays) so they offer no support for the sedentary female-territorial idea of Blanco *et al.* (1993).

Nevertheless, sedentary female Hen Harriers in Scotland also display, and unusually did so up to five times as frequently as males (Picozzi 1984a). Picozzi and Balfour both attributed this high display frequency to a predominance of females in the population (skewed 3:1 in some years) and the need for some females to display to attract males in the dense populations found in northern Scotland. Females land in potential sites, inducing active interest from the resident male (Balfour 1962–63; Picozzi 1984a). Since polygynous males in a dense population can afford to be choosy, males are more likely to choose a displaying female than a non-displaying one. These females are not always successful however: some late-settling females in *hudsonius* populations in Canada displayed to and attracted up to three breeding males, but the resident monogamous males aggressively drove them off. It may also be that females are exhibiting intense territorial behaviour, particularly to competing harem females in polygynous situations since, as implied by Arroyo (above) and evidence below, skydancing may be more territorial than mate-attracting.

Male participation

If males are small for reasons related to aerial display activity, as proposed by Jehl and Murray (1986), we would expect male harriers to perform more vigorous or longer displays than the females. A measure of the vigour of skydancing in harriers is the number of U-shaped undulations. Among African Marsh Harriers, this measure of vigour confirmed that male displays ($\bar{x} = 12.6$ Us) were more vigorous than those of females ($\bar{x} = 9.5$ Us). Over three years, males displayed more vigorously than females in two years, and more often in all three years (Table 3.1). However, the maximum number of continuous Us recorded was 52 by a pre-breeding but food-supplemented female in 1985. Even greater sexual differences are recorded for *hudsonius* in Canada (Table 3.1); males per-

Table 3.1 Variations in aerial skydancing/spiralling displays of Northern and African Marsh Harriers by sex and year. Prey abundance (mice caught / 100 traps) is an index of mice in the environment during breeding.

Year	Total skydancing sequences		Mean intensity (Us)		Rate (displays/100 h)	Prey abundance
	♂	♀	♂	♀		
Northern						
1980	43	5	13.7	11.0	5.3	6.6
1981	10	0	10.3	0	1.4	1.9
1982	6	0	6.2	0	0.7	0.5
Totals	59	5				
	[female participation: 8%]					
African						
1984	24	7	11.5	8.0	3.3	19.8
1985	49	25	10.6	11.5	8.1	12.3
1986	26	11	17.4	5.9	10.0	11.5
Totals	99	43	12.6	9.5		
	[female participation: 30%]					

Prey indices are not comparable between areas: Northern Harrier prey (voles) were snap-trapped, African prey (mice) were live-trapped. Correlation of prey abundance with male sky-dancing (Northern Harriers) $r = 0.99$, $n = 3$. No correlation was found for African harriers since supplementary feeding obscured any natural relationship.

formed 92% of all display sequences, and their displays were generally more intense (more Us per sequence) than females. The longest display involved 74 continuous undulations by a male (covering about 1 km), whereas the longest female display was 25 undulations. Males, especially unmated birds in highly polygynous populations in Orkney, have been recorded performing over 100 such undulations by Balfour (1962–63). The different mating patterns probably explain the greater proportion and higher intensity of displays in the Northern Harrier: 63% of the 609 Us recorded in Canada were by polygynous males, even though they comprised just 14% of the male population. The difference in female participation between migratory Northern Harriers (8% performed by females) and sedentary African Marsh Harriers (30%) supports the idea (above) that females in sedentary populations of harriers exhibit territorial behaviour, whereas migratory females do not need to. Only Montagu's Harriers in central Spain do not conform to this trend, which may therefore be a trait peculiar to this colonial nesting species.

In summary, female participation in displays seems to vary among populations depending on migratory tendencies (sedentary females may display as a means of territorial advertisement), adult sex ratio (increasing in those skewed towards females), or mating system (increased male participation because of premiums on mate attraction).

Among other species of harrier the basic display patterns are similar, and Reunion Harriers and Pied Harriers of far Eastern Asia are known to end their displays in likely nesting habitat (Neufeldt 1964; Jones 1989), indicating it may have a mate-attraction component. Once male Black Harriers have attracted a mate, their skydancing performance changes from steep dives from high altitudes to low level and widely spaced undulations at ground level (Chadwick 1997). Like African Marsh Harriers, with which they sometimes share nesting habitat in South Africa, males call at the top of their undulations and barrel-roll like other harriers. Among Australian Spotted Harriers in which skydancing is absent, a female attracted to a soaring male is led to their woodland nesting site. Once perched, males perform vocal bowing displays, tucking their head into their chest while uttering 'kittering' calls. These performances generally occur close to, or on, the nest and usually cease past early incubation (Baker-Gabb 1982). Since several other harriers begin displays by soaring, it may be that the Spotted Harriers' display is a primitive form of the skydancing now prominent in most other harriers. Unfortunately, we know nothing of the displays of South America's two species—the Longwinged and Cinereous Harriers, both of which prove to be interesting species, as we shall see below.

Conditions promoting displaying

On hot days (above 30° C), African birds soared beyond binocular vision but could be heard displaying, and were typically re-sighted as they descended in the skyspiralling display into their territories or nests. Nesting females, incubating or brooding below, often food-called in response to displaying birds. Astonishingly, they appeared to recognize their mates as suggested by an additional soft chuckling alarm if the displaying bird was a neighbour or nomad. I was able to recognize a bird as the female's mate or neighbour because each male was marked with wing tags and sometimes small windows cut into the wing feathers. While the females may also have been using the (artificial) marks I was recognizing through binoculars, the females always correctly identified their mate. Grounded males observing displaying males also called while crouching with their head lowered and bobbing, and tail fanned. This frequently precipitated a display flight of their own. This stimulatory effect could induce up to four males and their mates to skydance and call simultaneously before the display was terminated or graded into a purely territorial leg-lowering displays (Barnard and Simmons 1986). Pied, Black, Australian Swamp, European, and Eastern Marsh are all similarly induced into 'communal' displays when several males perform in an aerial amphitheatre high above the ground (Sharland 1932; Neufeldt 1964; Thiollay 1970; Baker-Gabb 1983; Clark 1999; Chadwick 1997; I. Fefelov unpublished). I suspect that rather than communal displays being peculiar to these species, it is more likely that environmental conditions were well suited for soaring and one bird stimulated others to soar and display.

Diurnal differences

Among African birds, displays occurred with increasing frequency towards midday and did not differ between the sexes in this regard. Displays per hour differed significantly from an even hourly rate ($X^2 = 39.5$, 10 df, $P < 0.005$) and the hour 11.00–12.00 saw three times the average display rate at 17.7 dis-

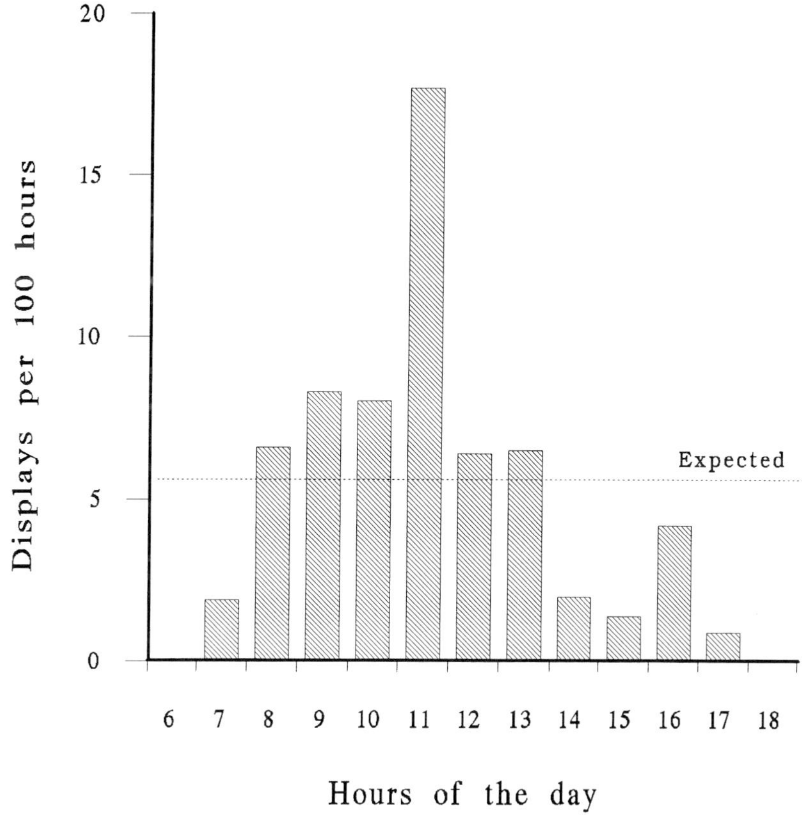

Fig. 3.2 Hourly display frequency of African Marsh Harrier adults in the southern Cape Province. Expected display frequency is based on an equal hourly rate.

plays/100 h. Another minor peak (4.2 displays/100 h) occurred between 16.00 and 17.00 (Fig. 3.2), before displays tailed off completely after 18.00 (sunset in September). Most displays took place in still conditions (58% of 146 displays) and fewest (8%) in strong wind, perhaps reflecting optimal soaring conditions before displays began. This is true also of male Pied Harriers which display most in fine windless weather (Neufeldt 1964). Displays were more frequent at temperatures between 20° and 24° C (58% of 146) but less frequent at temperatures above this. However, in August when most displays were performed, temperatures seldom rose above 24° C in the southern Cape Province of South Africa, and therefore were consistent with the idea that harriers prefer higher temperatures for the initial soaring prior to display.

Seasonal differences

Displays are performed most often just prior to breeding. Where this has been quantified most displays are performed by the male. For example in Montagu's Harriers males spent almost 8% of the time in display prior to

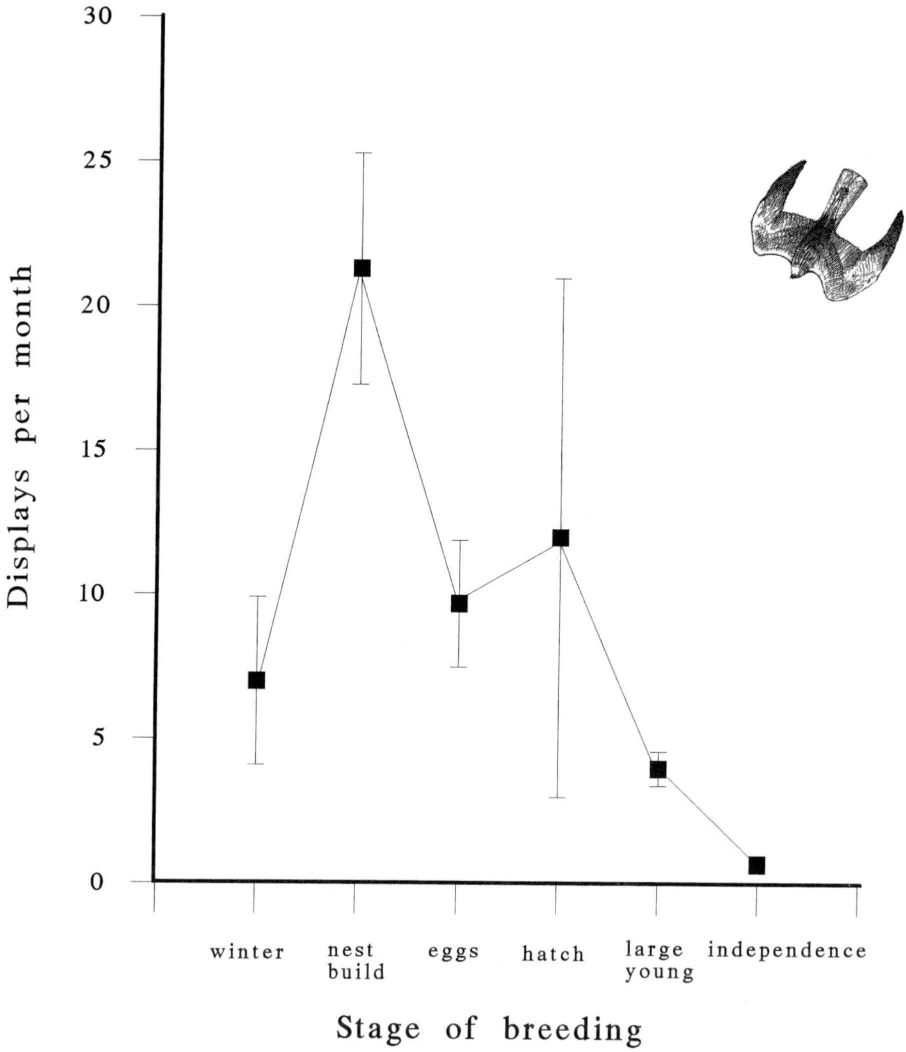

Fig. 3.3 Seasonal differences in displays performed by African Marsh Harriers. The wide variation apparent during egg laying (± 1 standard deviation) arose because of greater displaying to floating birds that came into the area.

laying while females spent half this time spiralling and dancing (Arroyo 1995). For African Marsh Harriers, most display activity was recorded prior to laying and this declined as breeding progressed (Fig. 3.3); however, a minor peak was apparent when floaters arrived as eggs were hatching causing territorial birds to display. European Marsh Harriers in Spain showed similar trends with about 15 displays/10 h prior to laying, declining to 7.5 displays/10 h between hatching and fledging and virtually no displays (0.3/10 h) thereafter (Blanco *et al.* 1993).

Display functions of spiralling and skydancing

Among the main displays, I detected no obvious differences between the contexts in which the *dancing* and *spiralling* displays were undertaken by African Marsh Harriers. However, since spiralling is a more intense display of agility, and since it terminates in a potential nest site for almost all harriers studied, it is likely that it is the main mate-attraction display, while the undulating display may be territorial in nature. There are three lines of evidence for this, mainly indirect:

1. Spiralling is the least common display of the two performed by African Marsh Harriers, as expected in a species where pair bonds are long term and attraction displays are 'not needed'.
2. Montagu's Harriers often nest communally in breeding territories so small that males would be unable to skydance over them to prove territorial ownership. This species has been described as non-territorial (Pandolfi and Pino d'Astore 1990a), although more subtle displays are known, suggesting limited territoriality (Arroyo 1995). It appears that in this species undulating (skydancing) displays are neither as common, lengthy, or aerially spectacular as in other harriers (Arroyo 1995). The spiralling display, by contrast, is much more common and clearly directed at opposite-sex birds (61% of 138 observations). This would be expected if spiralling was the mate-attraction display and the undulating skydancing display was territorial.
3. Evidence from other species of raptor in which undulating flights have been re-evaluated.

Function of undulating flights in other species

Most raptor biologists have assumed that undulating flights in general are mate-attracting displays, a view stemming, I believe, from Leslie Brown and Dean Amadon's (1968) review of displaying eagles, hawks, and falcons. In cases when this interpretation was critically re-evaluated by researchers determining the context in which they occurred, each researcher converged on an alternative conclusion. Harmata (1982), and Collopy and Edwards (1989) for example, both concluded that for Golden Eagles *Aquila chrysaetos* undulating flights were principally territorial in nature. Iribarren, in a 1975 study of the Spanish Booted Eagle *Hieraeetus pennatus*, suggested that all three major flight patterns recorded in his study were territorial and not courtship. Among African eagles Rick Watson (1989) found no evidence for courtship flights in the highly aerial Bateleur Eagle *Terathopius ecaudatus*. Like African harriers, other African species such as the Bateleur may courtship display less frequently since birds stay together for many years and there is less scope for mate attraction. Hence in all these species, re-appraisals of undulating flights concluded that they were territorial and not mate-attracting as previously thought.

This corroborates the evidence for harriers, including European Marsh Harriers in which Witkowski (1989, p. 292) states of undulating displays: '1) they form a part of territorial behaviour and 2) they are a part of mating behaviour. [but] In my opinion, their basic, if not the only function is territorial advertisement.' He suggests that undulating flights have been misinterpreted because the response of a male to an approaching female 'never included dancing'. Instead males in his Polish population (watched for seven years) pursued females in flight and attempted to make them land in the territory, sometimes pretending to attack them. While Witkowski's interpretations are in line with the evidence that skydancing is mainly territorial, the complete lack of skydancing to approaching females is unusual behaviour for European Marsh Harriers. His population was affected by pesticide DDT for the first four years of the study, and it is strikingly reminiscent of a change in behaviour found by Hamerstrom (1969) in Northern Harriers during the worst DDT years in Wisconsin. She found no skydancing (which was common before and after the DDT era) and a general breakdown in territorial behaviour during that time. In each case the context in which displays occur must be determined, preferably without pre-existing bias (and the influence of DDT).

I conclude that the undulating skydancing is indeed mainly territorial, but may attract females because it is visible for several kilometres and signals a territory-holding male. Skyspiralling by contrast, because it terminates on a nest platform, may be chiefly aimed at prospecting females (as found in Montagu's Harriers). One clear example from African Marsh Harriers came during observations of a paired African Marsh Harrier female seeking extra-pair copulations with a neighbouring male (Simmons 1990). Her old mate (of three years) then performed striking and vigorous skyspiralling displays, repeated several times, to which she was attracted and where she subsequently bred. Both displays may thus convey territorial and mate-attraction components, depending on who is viewing them. They are best seen as advertisement displays, with variations dependent on the species, nesting density, and mating patterns.

Other aerial displays

Skydancing and spiralling are not the only aerial displays seen in the harriers. In many other species, birds fly with their legs down and talons prominently displayed. Raptors with unfeathered tarsi often exhibit bright yellow, orange, or even red legs which may enhance the effect. Leg-lowering has been observed in other species and commented on several times and three basic hypotheses forwarded: raptor biologists tend to favour heat dissipation on hot days (e.g. Gargett 1990); for larger species particularly those landing (e.g. vultures) or those in turbulent winds, lowered legs may assist flight stabilization. Our own interpretation (Barnard and Simmons 1986) favoured threat displays. This interpretation was the most parsimonious of the three because most interactions between raptors in southern Africa occurred when conspecifics were close by and were apparently intruders. These birds subsequently left the area following the display. The other ideas were unsupported by our data; since leg-lowering occurred on cool and hot days with equal frequency, the feathered tarsi of most raptors are unlikely to be good heat dissipators, and both large (vulturine) and

small (*Elanus* kite) raptors displayed leg-lowering in stable and windy conditions. The common denominator was the presence of another bird.

Leg-lowering and physical contact

Leg-lowering among harriers is frequently observed and involves low-level parallel flying between neighbours with wings held at an exaggerated angle above the horizontal and the orange-yellow tarsi fully lowered with talons often lightly clasped. These displays were mainly intra-sexual (89% of 83 displays among African Marsh Harriers), and among other harriers are interpreted as aggressive interactions serving to delineate common nest territory boundaries (Baker-Gabb 1981 for Australian birds; Bildstein and Collopy 1985 for American birds; I. Fefelov for Eastern Marsh Harriers). As expected most interactions (67/83 for

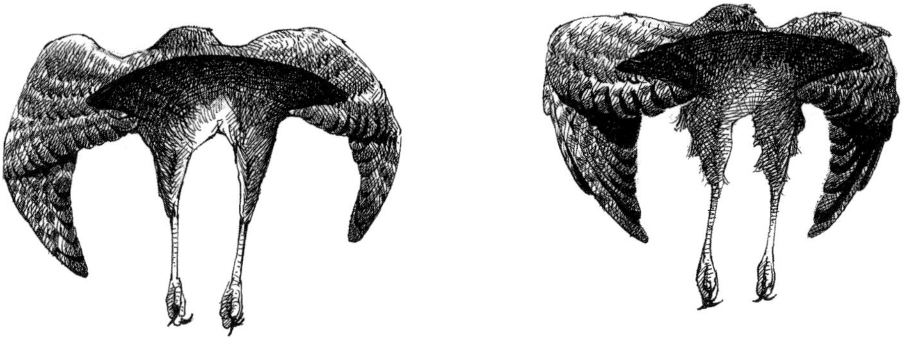

African Marsh Harriers) occurred just before and during breeding initiation and tailed off as young reached independence. Rarely, leg-lowering escalated into physical contact. Male African Marsh Harriers sometimes hit the back of unguarded neighbouring females (once), intruding nomads (five times), and even their own mates when another male was close by (three times). The attacked females reacted by squealing and dropping into the vegetation. Talon-locking, the most aggressive interaction observed, was recorded six times among adult birds, and was typically intra-sexual and arose when an intruder penetrated deep into the territory. Females in two instances flew behind and under such intruders, and pulled them down by locking talons and falling earthwards. These encounters lasted 1–7 min, while the birds were presumed fighting or entwined on the ground. Females were never seen to hit their mates or neighbouring females, but immature birds, particularly naive, recently airborne young, were frequently hit and talon-grappled by neighbours.

Among African Marsh Harriers, aerial interactions in which two birds locked together and fell earthwards were seen six times over 2200 hours of observation in South Africa. Some examples will suffice (Fig. 3.4): in one case two females fighting over a territory locked and fell 20 m, before reappearing 1.5 minutes later to resume their territoriality. A second case involved a male defending his flying young from attack by a neighbouring male; the adult birds fell 3 m locked together at a mutual boundary and remained on the ground for seven minutes. The neighbouring male reappeared first and once again struck the youngster, sending it to the ground. This latter behaviour was seen on four further occasions. A third event occurred when a nomadic harrier flew over an active nest and the breeding female flew behind and below the intruder to pull it earthward from about 10 m. They remained grounded for 7 minutes before the intruder flew swiftly away escorted by the female. The remaining talon-locking involved a neighbouring male and a neighbouring female, both of whom fell to the ground on different occasions, locked together with intruding young of the year. Each event was accompanied by extreme aggression, including one occasion in which the male continued diving at the young harrier for 8 min attempting to strike it, having spent 4 min grappling with it on the ground. Paired (patagially tagged) birds were never seen to cartwheel or even lock together,

Fig. 3.4 Northern Harriers about to clasp talons in a physical display of territorial aggression. Most talon-locking in raptors is territorial in nature, and in the harriers observed, it only occurred in this context.

during mutual soaring or inter-territorial interactions, ruling out a courtship function. Hence all talon-locking in these harriers was aggressive, similar to that recently found in a re-appraisal of cartwheeling in all other birds of prey (Simmons and Mendelsohn 1993).

Among Australasian Swamp Harriers similar incidents involving adult and immature birds have been seen. Baker-Gabb (pers. comm.) observed a marked adult male swoop at an immature harrier which flipped over and locked claws. They tumbled about 10 m before disengaging, whereon the immature fled the territory. In an inter-specific interaction recorded in Tasmania, Nick Mooney (*in litt.*) observed a breeding female Brown Falcon attack a female Marsh Harrier, foraging close to the falcon nest. The two grappled from 30 m to 2 m, where they broke apart and parted company. Cramp and Simmons (1980) report that cartwheeling among European Marsh Harriers *C. aeruginosus* occurs when the polygynous females of a bigamous male both fly towards the prey-carrying male and lock talons and fall into the reeds. In a similar interaction in Canada, two close-nesting polygynous female Northern Harriers locked talons in the early breeding season, following intense leg-lowering interactions, and fell together into a drainage ditch. The polygynous male flew overhead but did not interact. The two females were not resighted for at least 10 min. In the same study in April 1980, a foraging female nearing the nest of another was attacked by the breeding female. The two locked talons over a small copse of trees and fell about 10 m to the ground. They were not re-sighted. Similar aggression is recorded for both Scottish Hen and American Northern Harriers (Watson 1977; Craig *et al.* 1982). Note that most recorded instances (8 of 11) are female–female interactions, a generally overlooked component of sexual selection ideas on Reversed Size Dimorphism treated below. Let us now look at the mate-attraction component of this spectacular flight and ask: are males that display most vigorously attracting mates more often, as predicted if sexual selection was driving the process?

Display vigour and mate attraction

Among Northern Harriers in Canada all displays were recorded over three seasons and ascribed to individual territorial males by observing where displayers landed. Because this species is strongly polygynous, I could then relate the amount of displaying to the number of females in their harem. Since I spent roughly equal periods in focal males' territories to measure food provisioning rates, the displays were recorded with roughly equal effort. I found that those males performing the most displays in the season attracted up to four mates while others displaying less often remained monogamous (Fig. 3.5; Simmons 1988c). A strong and significant correlation was apparent between number of mates and the number of displays recorded. Moreover, the measure of vigour (Us per display) was also strongly correlated with the males' harem size (Fig. 3.5). This is the first such demonstration that display vigour is associated with mate attraction for any raptorial species, and corroborates evidence from other species. For example, Mather and Robertson (1992) using similar methodology found that those polygynous male bobolinks on Canadian grasslands that displayed for longer attracted more females than other males. They also wing-clipped some males and found that display duration and number of females declined, suggesting male displays like those of harriers are energetically expensive and thus honest advertisements of quality. Among lekking birds, whose mate-attracting abilities are easier to measure because all males occur in small discrete groups, similar patterns occur: male Sage Grouse *Centrocerus urophasiansus* spend most time on the lek, strut at a higher rate, and mate more often than those with less vigorous displays both in the field and under laboratory conditions (Gibson and Bradbury 1985; Spurrier *et al.* 1994). However, correlations alone are not conclusive evidence that females are using the display vigour as a choice cue since we have to know whether displays are unfakeable (costly) cues.

If displays are costly one would predict that males in the best body condition perform the most displays. This is an important concept because if displays were easy to fake, then females may choose poor-quality mates and reproduce poorly. Selection would soon act against such females. We predict that for species where

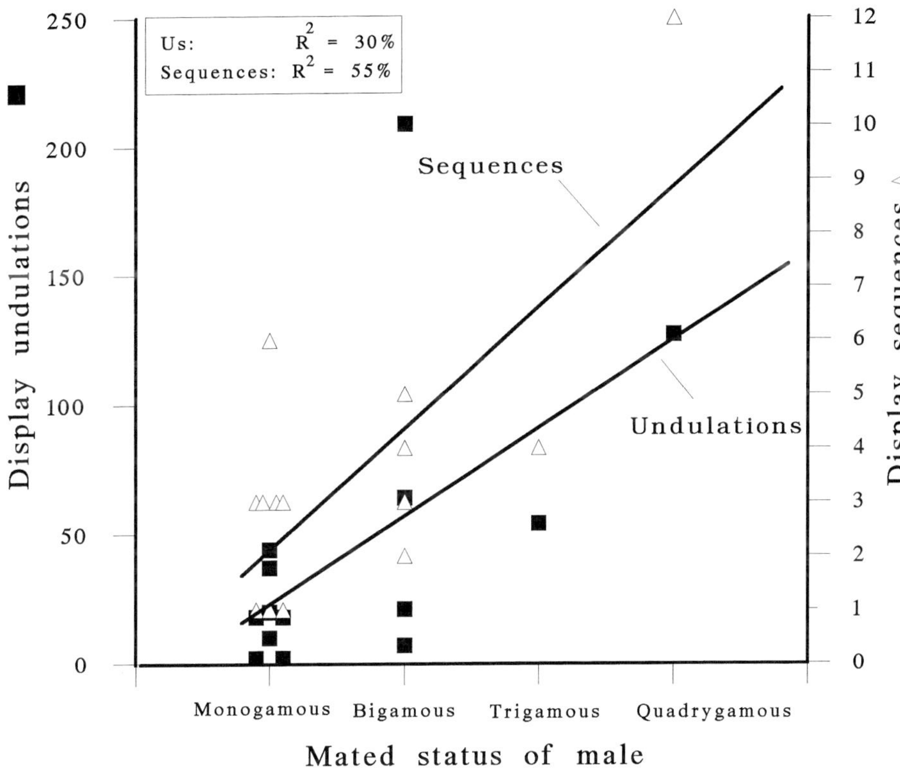

Fig. 3.5 Skydancing displays in relation to the number of mates attracted by polygynous male Northern Harriers in New Brunswick. Both the number of undulations and each complete sequence were significantly correlated with the harem size.

displays are important in mate choice that only the costly (reliably honest) ones remain (Kodric-Brown and Brown 1984; Andersson 1994). This is easy to test in species where colouration can be experimentally changed or tail-length altered, the results of which are impressively supportive of the reliable signalling of quality (Andersson 1982b), material benefits (Hill 1991), or genes for long survival (Alatalo *et al.* 1991). In the case of harriers it is more likely to be a badge of parental ability since, as we shall see in Chapter 5, males who display most are generally better food providers, although they cannot maintain their food provisioning performance to all females; this explains why females choosing mated males fare so poorly in reproduction (Simmons *et al.* 1986a).

So do male harriers that display most attract more mates because they are demonstrating they are in good condition, and are they better food providers? In eastern Canada and southern Africa I measured small mammal populations (voles in Canada, mice in Africa) to look for correlations of food abundance in the environment with display activity. The answer appeared to be yes: in Canada display frequency of adult males was strongly related to food abundance, which varied greatly between years—in food-rich years more display sequences and more Us/sequence were recorded than in food-poor years (r = 0.99, Table 3.1). Second, young inexperienced males, known to be up to five times less efficient at securing prey (Bildstein 1987; Chapter 4) only displayed in the richest food year despite being present in all years. This suggests display frequency and food availability for both adults and youngsters are directly linked. It is unlikely that the higher display rates were simply correlated with the larger number of females available, and that they stimulated males to display regardless of food availability per se, because similar numbers of females were present in each year in Canada and indeed always outnumbered males by about 1.5:1 (Simmons *et al.* 1987).

The strongest evidence that displays are energetically costly and thus reliable cues comes from the supplementary feeding of African birds in the mid-1980s, described in Chapter 1. While natural food supplies were dwindling over the three years (Table 3.1) seven pairs were supplemented with up to 245 g extra food per day for approximately one month before breeding. As expected seven fed pairs ($\bar{x}$ = 10.7 displays/100 h) performed 2.5 times as many displays as seven controls (4.3 displays/100 h). The results were statistically just short of significance because I was unable to record all displays by fed males since they displayed for longer and from greater heights. In 1984, when no birds were fed, the average display rate was similar (3.3 displays/100 h) to that of unfed controls in other years. These results suggest that males in better condition could display more frequently, verifying suspicions (Simmons 1988c) that aerial displays might be a highly reliable choice cue for females seeking good-condition mates.

If males in good condition display more often, and attract females, one would predict that polygyny might increase. While some evidence for this is apparent in supplemented Hen Harriers (S. Thirdgood, S. Redpath unpublished), African Marsh Harriers are monogamous and the increased display frequency did not result in polygynous matings. This is an interesting result for two reasons: first it confirms that these displays are mainly territorial (few females were attracted and none were mated with). Second, and of greater interest, other factors seem to be constraining polygyny in this species—mated males did attract females on two occasions and on both occasions the males acted aggressively towards the females and drove them away. Reasons for this unusual behaviour are discussed further in Chapter 5.

Support for female preference for frequently displaying males comes from a recent study of supplementary fed, polygynous Redwinged Blackbirds *Aegalius phoenicues* on marshes in Washington, USA (Langston *et al.* 1997). Supplemented males not only displayed significantly more often than unfed controls but harem sizes were 50% greater than for unfed males on the same marsh. Feeding females alone had no similar effect. This suggests that female blackbirds, harriers, and probably many other species, are using similar cues in choosing polygynous mates—a costly measure of male vigour. In summary, the evidence seems fairly good that skydancing male Northern Harriers use the display to attract mates, with those polygynous males that display most vigorously attracting most females, and that the display is linked to food resources and body condition in at least two different species. The display is therefore unlikely to be fakeable and is probably a costly, reliable cue for females choosing a quality mate. The ideal test of the entire sequence in harriers is for experimental food supplements to be provided to a polygynous species (particularly to late-settling or younger birds), and their mate acquisition success (including extra-pair copulation success) to be compared with non-fed controls. This has never been done for any raptor.

Reversed Size Dimorphism

Harrier skydancing displays also raise interesting questions about the much debated 'small male, large female' syndrome. Reversed Size Dimorphism (hereafter RSD) is common among raptors, and especially among bird-eating ones (Snyder and Wiley 1976; Newton 1979; Mueller and Meyer 1985). However, RSD is also very common in a vast array of other organisms ranging from hummingbirds and shorebirds (Jehl and Murray 1986), to hyenas, seals, and the world's largest living creatures, the baleen whales (Andersson 1994). There has been a general misconception, that RSD is associated with predatory habits (e.g. Amadon 1959; Cade 1982) but this is far from true: Reversed Dimorphism occurs in tiny nectivores (hummingbirds), invertebrate consumers (shorebirds), large insectivores (the bustards), and largest of all, vegetarian ratites. Indeed, because of the wide variety of non-predatory avian families in which it occurs (24 of the 32: Jehl and Murray 1986), it can be stated that RSD is most frequent in families with a non-predatory lifestyle. Predatory birds, comprising just 25% of the 32 families exhibiting RSD, are therefore somewhat unusual. Along with this interest in explaining RSD in raptors, there has been a plethora of raptor-specific hypotheses which seek to explain RSD only in this group. This one-sidedness may have even hindered the field, rather than advanced it, since universal solutions have not been sought (Jehl and Murray 1986). What *is* special about the raptors is that RSD is more common among the various genera than in any other group (Newton 1979), and it is this that attracted so many different hypotheses among raptorphiles.

Hypotheses and tests

RSD poses three main questions:

1 Have males become smaller or females larger?

2 Does natural or sexual selection account for these trends?

3 Can selection pressures identified in raptors be applied to other families?

Small agile males

For raptors, one side of the question (why males are smaller) was largely settled by the comparative work of Snyder and Wiley (1976), Newton (1979) and follow-up quantification by Temeles (1985). Snyder and Wiley were the first to quantify that the degree of RSD was closely correlated with bird prey in the diet (it was previously thought to be related to vertebrates in the diet). Raptors that existed mainly on easily caught insects (or large mammals) showed little dimorphism compared with those that caught small mammals, which in turn showed less dimorphism than the extreme dimorphism shown by the bird-catching sparrowhawks, in which males can be half the size of females. The greatest dimorphism is shown by European Sparrowhawks in which females are on average 1.9 times heavier than males (Newton 1986), with American Sharp-shinned Hawks close behind (1.8: Snyder and Wiley 1976).

Snyder and Wiley proposed that the link between RSD and bird catching by males was that males who provide the bulk of the food to incubating mates and dependent young must be particularly adept at catching difficult prey. They proposed that bird prey are declining at the most demanding stage of breeding, just as females begin to hunt. Divergence in the prey spectrum would greatly assist the sexes in providing prey for their demanding young, without competing for the same food. Mueller and Meyer (1985), however, pointed out that one would expect greater RSD in those species that share a territory year round (because they would always compete). Yet as found by Kemp (1995), sedentary African Greater Kestrels *Falco rupicoloides* exhibit (low) levels of RSD little different from other kestrels that do not share a foraging territory. Moreover, the reason why it is not the females that are smaller is not intuitively obvious from this hypothesis and Snyder and Wiley argued that females need to be larger because they defend the nest most often. This is now known not to be true for all species (Mueller and Meyer 1985) and is not true for Northern or African Marsh Harriers in which males are easier to catch because they attack stuffed owls more vigorously and more often than females. In testing the idea that female European Marsh Harriers defend against intruders more often than males, Fernandez and Azkona (1994) also found that it was the males that did most defending. In 280 cases when the intruder was another adult, males responded 81% of the time, while females responded only 53% of the time. When both were present and an intruder appeared, males responded significantly more often than their mates. Others have suggested females are large for energetic reasons: to lay down large fat stores for egg laying and the ability to weather short-term declines in food supplies (Mendelsohn 1986; Ydenburg and Forbes 1991). We have to ask in each case, why is this so important for raptors and not for other species, with typical dimorphism? Such species surely also face short-term food shortage and females would benefit by being large? Furthermore in the harriers, at least, it is the *smaller* females that lay the larger clutches and the largest total clutch volumes (Chapter 8), not the largest as this idea would predict.

Ian Newton (1979) emphasized the *agility* of the prey by pointing out that extremes in RSD were not just related to bird prey; some raptors that caught agile mammals (i.e. small ones) were also relatively dimorphic compared with ones that caught insects or ate carrion. Bird prey are simply at the far end of the prey agility spectrum. Dimorphism evolved, argued Newton, for males to be agile to catch the prey for a large part of the breeding cycle extending from pre-egg laying through to mid-nestling stage. The males' efficiency and agility would be enhanced by being smaller (Andersson and Norberg 1981), and this would be most important for the bird eaters. Harriers would thus be expected to show moderate to extreme levels of dimorphism, depending on each species' particular dependence on agile birds at critical times of the breeding season. One of the

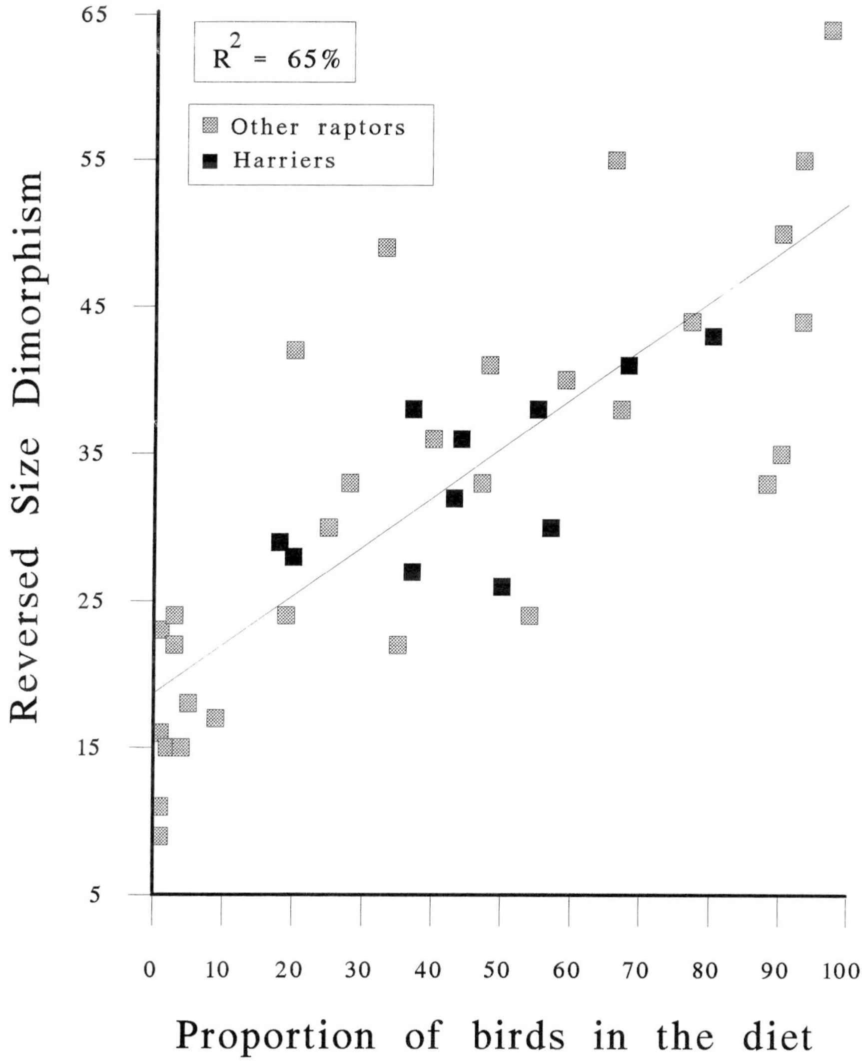

Fig. 3.6 RSD indices (female–male mass, divided by the mean mass of the sexes) for 30 species of raptors (light squares) in relation to the proportion of birds in the diet. The harriers (dark squares) generally fall above the regression line for all raptors, indicating high dimorphism indices. Data derived from Newton (1979), Snyder and Wiley (1976), Steyn (1982), and own data including small and larger raptors (60 to 6000 g).

problems with depicting these RSD relationships with the categories 'mainly mammal' prey or 'mainly bird' prey is that is qualitative, and difficult to test statistically. To quantify RSD in relation to bird prey for harriers I followed Snyder and Wiley's approach in determining the percentage of birds in the diet of each species and correlating it with RSD. Where data exists I have added to the original data base of North American birds with data from European and African species. This gives a sample of 30 species ranging from the smallest sparrowhawks to largest eagles (Fig. 3.6).

For the 30 species the relationship between the degree of RSD and proportion of birds in the diet was very close and highly significant ($r = 0.82$, $n = 30$, $P << 0.001$). While bird diet explained 68% of the variation in RSD, the relationship is not exact. There are curious exceptions that show high dimorphism indices such as the African Goshawk *Accipiter tachiro* (RSD = 55%) but whose diet consists of only 65% birds (Simmons 1986b). Its RSD predicts it should take 80–90% birds. We shall return to these exceptions below.

Harriers exhibit relatively high levels of dimorphism compared to other raptors (Fig. 3.6) since several of them lie above the line relating RSD to diet. This is itself unusual since many harrier species feed on both birds and mammals and in many cases have a higher proportion of small mammals than birds in their diet. The greatest bird diet is about 80% for the Longwinged Harrier, which has correspondingly the highest dimorphism index of 42% (female 1.5 times larger than the male). This is in contrast to the European Sparrowhawk with an index of 64% (female 1.9 times larger). However the relatively high RSD of harrier may arise because of body size trends found by Jehl and Murray (1986): smaller species tend to be more dimorphic than larger species within families. This is also true of large and small raptors (Fig. 3.7). Jehl and Murray (1986) did not provide an causative explanation for this, but discounted simple allometric or statistical bias because the reverse trend is apparent in shorebirds—it is the larger species that are the most dimorphic. One possible explanation for raptors is that males in larger species do not have to be as relatively agile as smaller males in smaller raptors because the number of highly agile and difficult to catch passerines in their diet reduces to zero (in buzzards and eagles for example: Steyn 1982). With a reduction of this prey component in the diet of larger raptors, males need not be so agile and are therefore relatively larger. This is not universally true, however. In *Aquila* eagles, for example, Watson (1997) reports that larger species are more dimorphic than smaller species (bar Wahlberg's Eagle which he omitted from the genus), and he attributes this to relatively agile prey in the diet of the larger species. Clearly other groups should be evaluated and their diet quantified to give us a complete picture.

The greatest dimorphism among birds is shared between nocturnal raptors and species such as Montezuma's Oropendela with non-reversed dimorphism of 2.1:1 in mass (Webster 1997). Australian owls are peculiar in that the *Ninox* (e.g. Boobok and Powerful Owls) are also non-reversed in their dimorphism and feed on large sugar gliders and other marsupials. Sooty Owls *Tyto tenebricosa*, on the other hand, show extreme forms of reversed dimorphism with males

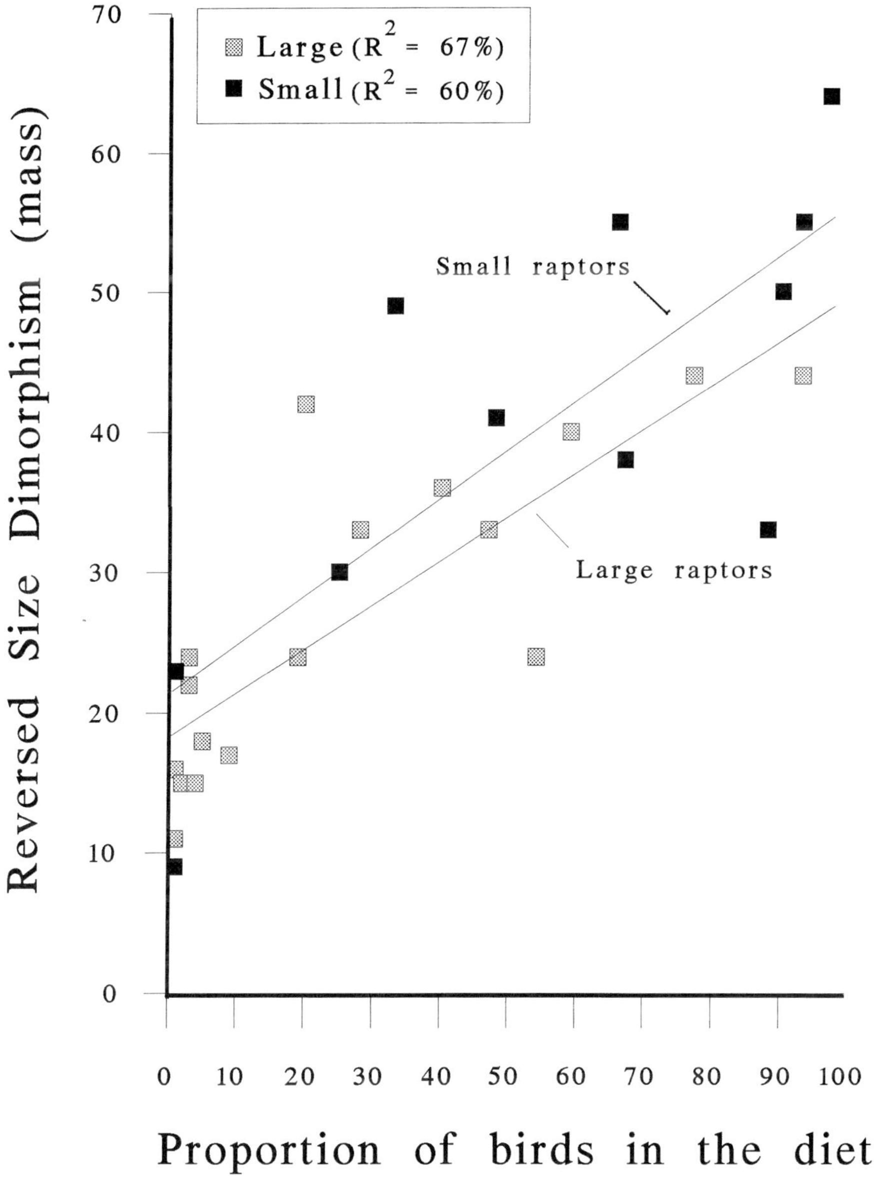

Fig. 3.7 Differences in RSD of small (< 500 g) and large (> 500 g) raptors in relation to the proportion of birds in the diet. Smaller raptors show greater dimorphism for the same bird diet than larger raptors.

averaging about 500 g and females almost twice as much at 900–1000 g (N. Mooney pers. comm.). These Australasian owls could tell us a great deal about the selection pressures operating on males and females but are unstudied in this regard.

Geographic variation in RSD

The greatest dimorphism shown by any harrier is in the Longwinged Harrier of South America. Females are about 42% heavier than males (the average is about 34% for the 13 species of harrier: Table 2.2) and females have legs (tarsi) 9% longer than males. The mean difference for other harriers is about 6% (Table 2.2). True to its highly dimorphic body plan, the diet comprises about 80% birds, of which over 60% are passerines (Bó *et al.* 1996)—more than any other harrier. We saw in Chapter 2 that this species varies geographically in morphology, with southern birds being larger than those in the north. This difference also affects the degree of RSD. In regions furthest from the equator (southern South America) RSD in wing chord measurement is 7.3%, while in equatorial regions it is 4.8% (body mass data are not available from both areas). It is the female harrier that causes this difference because while both female and male are larger in southern than northern regions, it is the female that is proportionately larger than her mate in the south (Table 3.2). The reasons are unknown but because the female's size may influence her competitive ability in intra-sexual competition (below), and such competition should be stronger in polygynous populations, we would predict that Longwinged Harriers are possibly polygynous in their southern range but not in the north. Data are lacking to test this prediction since only six nests have ever been studied (M. S. Bó unpublished).

For Montagu's Harriers we know that birds also differ geographically (Chapter 2) and that RSD indices decrease from 27% in a sample of Russian birds (Nieboer 1973) to 16% in France (Leroux and Bretagnolle 1996). This is the lowest value for any harrier. Contrary to Longwinged Harriers it was the increase in body mass of the *male* harriers in France (265 g to 295 (± 25) g, n = 64) that caused this difference since females were similar (345 + 30 g, n = 62) in French and Russian samples. From the foregoing arguments on the importance of agility in prey capture for males, we would predict that male harriers in western France would capture fewer birds than those in other regions. This is so: Arroyo (1997) found that Montagu's Harrier diets in all areas other than western France (i.e. Hungary, Italy, The Netherlands, Britain, and Spain)

Table 3.2 Geographic variation in the degree of RSD in Longwinged Harriers from northern (5° N) and southern (~ 30° S) South America. Average measurements from Nieboer (1973). The '% increase' refers to the size increase of southern birds over northern birds. RSD refers to the difference in males and females in two different body measurements (wing and leg), with southern birds always more dimorphic.

	Wing chord (mm)			Tarsus (mm)		
	Northern	Southern	(% increase)	Northern	Southern	(% increase)
Male	395.4	413.4	4.6	77.1	77.9	1
Female	415.0	444.8	7.2	82.7	85.6	3.5
RSD	4.8%	7.3%		7.0%	9.4%	

were dominated by birds (23% of 3096 items) or insects (43%), while those in France were dominated by small mammals. So strong is this dependence in western France that the Montagu's Harriers' clutch size, breeding density, and fledging success are each significantly correlated with population fluctuations of their chief prey (*Microtus arvalis*: Butet and Leroux 1993). Their low RSD, therefore, fits well with their relatively atypical diet of small mammals.

Could the larger size of males in western France be a result of birds being weighed at different times or simply being heavier because they are better fed? Both are unlikely because if males were heavier for these reasons females would be expected to follow similar patterns—which they don't. Also the samples were relatively large at over 60 birds in Butet and Leroux's sample, giving a fairly low variance in weight (standard deviations given above) which show no overlap with the weights provided by Nieboer (1973). Unfortunately other body measures (wing chord or tarsus) are not available to confirm the lower RSD values in western France.

Small males and elusive prey

Temeles' contribution to our understanding of RSD and prey agility was to show that prey capture success (successful hunts/all hunts) was positively linked with RSD. In other words, the difficulty of prey capture was closely linked with reversed dimorphism and males may have diverged in size from females to reduce inter-sexual food competition. This differs slightly from Snyder and Wiley's ideas in that the competition should be greatest in species where more time and effort is needed to capture a prey item. While Temeles (1985) stated that none of the prey-agility/vulnerability models can explain why it is the males that are smaller (it would pay for a female also to be small to increase her foraging efficiency), it may be adequate to state that if the selection comes at a time when males are providing females with pre-laying food, they must be as large as possible to utilize that food in order to lay either large numbers of eggs or large size eggs (Mendelsohn 1986)—egg size and body size are allometrically linked (Rahn and Ar 1974; Saether 1985). To decrease female body size may reduce her ability to lay large eggs.

Large competitive females and sexual selection

The foregoing arguments, reviewed by Mueller and Meyer 1985 and Andersson 1994, are based on natural selection thinking—small agile males will capture more prey than larger males, survive better, and leave more offspring. But arguments for normal dimorphism in which males are larger than females are usually based on sexual selection (Darwin 1871; Payne 1984). That is, males are larger than females because males normally compete against each other for the best territories or mates, and size in physical combat is often a good predictor of the victor (Andersson 1994). Moreover, where competition (and sexual selection) is more intense among polygynous species males are relatively larger, leading to greater dimorphism (Payne 1984; Andersson 1994). Where roles are reversed, as in polyandrous species, and females clearly compete against one another, females are larger and more colourful than males (e.g. jacanas, spotted sandpiper, and phalaropes: Jehl and Murray 1986).

Penny and Jerry Olsen (1984, 1987) developed the sexual idea for raptors in the following way: females are larger because it is they who are competing among themselves for a limited and critical resource: a small agile male who is essential for the success of the breeding attempt. While males may compete for territories, it is the females who then compete for males. In few other groups are males so important to breeding. The trend for greater RSD in the bird catchers is explained in this hypothesis by the good quality males in such species being even more 'sought after' by females because their prey catching abilities are even more essential. Hence, female–female competition should be even more intense. Indeed the Olsens do not believe that small male size needs to be explained at all by their prey catching abilities—it is only females that increased in size.

Comparisons with shorebirds

Comparisons with shorebirds are both relevant and fascinating because the span from normal to reversed dimorphism is greater than in any other group, and female–female competition may also explain RSD in this group. In their review of RSD in shorebirds and its relationship with foraging, migration patterns, and mating systems, Jehl and Murray (1986) found that RSD was not related to any particular type of foraging mode, nor to migratory patterns. They did find, however, that RSD was loosely related to mating patterns in the 217 species considered. Shorebird mating systems are particularly diverse and interesting and include monogamy, polygyny, polyandry, and polygynandry (several males, mated to several females). As expected, polygamous species tended to have larger degrees of 'normal' body size dimorphism as predicted by Darwinian selection. In all but one case of polyandry, females are larger than males, as expected if females compete among themselves and males are attracted to females. In the unusual polygynandrous species, the first pair bond breaks up and females join a second mate, and the male another female, each producing their own second clutch. All such species exhibit slight reversal in dimorphism (Jehl and Murray 1986). Thus species with polygamous mating systems exhibit greater degrees of dimorphism than monogamous species and the role-reversed species show greatest reversed dimorphism, with jacanas the extreme.

However, the patterns do not end with mating system; Jehl and Murray (1986) proposed that a third important factor is involved in selecting for degrees of dimorphism—aerial display activity. They explained it as follows:

> Although larger size often confers an advantage in intra sexual contests for dominance, we think that this is so when such contests occur mainly on the ground. When inter-male contests involve acrobatic displays in the air with steep dives and climbs, twists, and turns, we propose that the smaller, more manoeuvrable individuals are able to dominate the larger less manoeuvrable birds. Smaller size of males may also be selected if females choose as their mates males with superior acrobatic courtship displays. (p. 39)

Since little quantitative evidence was then available, they gave qualitative examples of shorebirds for which aerial displays were known and all but four species fitted their expectations of RSD. Aerial display is reported to be important in the mate choice of the Redshank *Tringa totanus* in which males display to females both in the air and on the ground, before the pair bond is fully established (Hale and Ashcroft 1982, 1983). Recent work on RSD in Dunlins, *Calidris alpina* shows that those males that displayed for longer and most vigorously were the smallest in body size (Blomqvist *et al.* 1997). This is the first such demonstration that aerial agility and its attractiveness to females may be selecting for small rather than large male body size, in a reversed size dimorphic wader. Such relationships should be possible to establish for any species with measurable aerial displays which are easy to catch (and are preferably polygynous).

Returning to raptors, could it be that males are smaller because they are selected for aerial agility in display and they prove their prowess through vigorous and prolonged displays (Jehl and Murray 1986; Simmons 1988c; Andersson 1994)? Females may be larger because they must compete for the best males. Second, females should be proportionately larger than their mates (RSD greater) in those species catching agile prey because the selection pressure on getting the best mate for successful breeding is likely to be more intense. The strength of this sexual selection hypothesis lies in the fact that it can explain both the direction and degree of reversal.

Evidence for selection on female body

Empirical evidence that selection for large female body size occurs comes from one of the most dimorphic of raptors, the Eurasian Sparrowhawk. In his 14-year study of the lifetime reproduction of this bird, Newton (1986) found first that lifetime production of young was significantly correlated with the lifespan of 142 females. Lifespan itself was significantly correlated with only one variable—wing length. Larger females lived longer and produced more young than smaller ones. Thus selection would favour larger females. Newton found no correlation with territory quality and the trends did not arise because larger females started breeding earlier. Thus the evidence shows at first hand that large female size is favoured in one of the world's most dimorphic hawks. Could it be that larger females are able to find and remain mated to better food-providing mates than smaller females? Is it that the larger females are better able to survive the severe winters when bird numbers are low? The answer remains unknown for sparrowhawks.

Evidence for female–female competition

While there is little first hand evidence for female–female fighting in raptors, it does exist. Its apparent absence has been used as an argument against the Olsens' ingenious hypothesis, but the evidence appears to have been clouded by the misinterpretation of the behaviour when it has been seen. The behaviour in its most extreme form is the so-called cartwheeling displays recorded in at least 37 species ranging from Pygmy Falcons to Martial Eagles (Simmons and Mendelsohn 1993). In a critical re-appraisal of the subject we found that the behaviour, seen as courtship by many prominent ornithologists, had been misinterpreted and instead represented mainly aggressive interactions. In cases where sexes could be distinguished, virtually all were same-sex interactions, and it seems that some of these would be females fighting over territories. Among harriers, while only 11 incidents of talon-locking and cartwheeling were noted above, 73% of them were recorded between females. In about 200 cartwheeling incidents in African Fish Eagles *Haliaetus vocifer*, induced in Botswana, it was the females that aggressively cartwheeled with each other, while the males were more concerned with securing the prey put on their territory boundary (Tim Liversedge in Simmons and Mendelsohn 1993). Among European Sparrowhawks Newton (1986) also observed more female–female interactions than male–male aggression around the nest, suggesting competition for territories or males in this species too.

Malte Andersson (1994) has argued that if female–female competition were indeed important, then traits that improve success in agonistic interactions, such as body armaments or plumage colour, should be favoured in females. He states that there is no evidence to suggest these occur in females and it is only males which show enhanced colour dimorphism. Andersson (p. 281) doubted his own argument because there is nothing to stop sexual selection operating on the sexes in different ways. Furthermore, among raptors it is only necessary for female body size to increase for greater competitive ability since all raptors have the necessary armaments for fighting (claws and beaks) and bright plumage colour is unnecessary in these birds. Given that females may compete and males may be essential and limiting, how might a male prove himself as good provider and parent?

RSD and male agility in display

As argued for size-reversed shorebirds, small male harriers, by displaying more vigorously and energetically, may attract and mate with more females than less agile, larger birds. This seems more likely than such males physically competing with one another over territories as Jehl and Murray (1986) also proposed. Since agility is related to small size (Andersson and Norberg 1981), selection will favour smaller and smaller males. Sexual selection (via display vigour) rather than natural selection (via enhanced predation ability) may therefore play a role in promoting reversed dimorphism. While the evidence that male harriers displaying most attract larger harems (Fig. 3.5, Simmons 1988c) is not direct evidence that *agility* per se is important, it does suggest that vigour is important. The critical step is a morphological analysis of those males displaying most often to determine if they are smaller than males performing less often.

We must next ask what it is about vigorous displays that are so important to females? One of the most critical aspects already identified is that males provide all the food for the egg-laying and incubating female in many species of raptors (Newton 1979), including harriers. Do vigorous displays indicate that the male is a good food provider? Among polygynous male Northern Harriers which displayed most vigorously, significantly higher provisioning rates than monogamous males were apparent (Simmons 1988c), supporting this idea. This is also apparent in European Marsh Harriers (see Fig. 8.6). Further support is provided by two other lines of evidence. Older harriers are known to be more efficient foragers than younger ones (Chapter 4; Bildstein 1987), and individual Hen Harriers are known to attract larger harems as they grow older (Picozzi 1984b, see Fig. 5.14). They may be doing so because as they get older their foraging return is greater and they can provide females with greater food supplies at critical times. Thus females are choosing them because they make better fathers.

The second line of evidence is less direct: among two species of kestrels, females in laboratory studies appeared to prefer more colourful males over duller coloured birds (Palokangas *et al.* 1994; Wiehn 1997). In the first of these lab studies with European birds, it was established that the brighter coloured birds were indeed more efficient hunters and females mated to them produced more young than those mated to duller males. Hence by choosing brighter males, females were increasing their chances of being supplied a larger amount of food and rearing more young. While in none of these species was the entire intense display—brighter (thus older?) males—better food providers sequence established, older male harriers are generally more brightly coloured, with less brown in the plumage and brighter yellow eyes (Hamerstrom 1969; Watson 1977; Picozzi 1981). Thus body/eye colour may be the first cue to condition and future parenting, the display may be a second (less fakeable) cue to condition and parenting by the male, and each male's actual (early) provisioning may a third and more definitive cue to parenting abilities.

RSD: natural and sexual selection combined

Can we distinguish between these two main hypotheses: (i) natural selection (small males can catch the more agile prey); (ii) sexual selection (females compete for small agile males whose display activity is a guide to their vigour and future parenting). This is possible for our target group because harriers are unique among falconiforms in exhibiting varying degrees of polygyny. We can thus determine RSD for each species in relation to the proportion of birds in the diet and the degree of polygyny found. We would expect that if sexual selection had promoted RSD, those species exhibiting highest degrees of polygyny would show greatest reversal in size dimorphism. We might also see greater dimorphism in plumage colouration since colour and mate-choice preferences are related in some species as described in the Kestrels above.

It should be made clear that sexual selection on females leading to larger size can occur not only because they are competing for a particularly good food provider, but also because the intensity of female–female competition is expected to be greater since interactions continue *after* they have settled in the territory. Females attaining the α position in the harem frequently try to prevent other females from settling and receive the highest provisioning rate, hatch more eggs, and fledge more young than any other female within the harem (Chapter 5). Thus the premiums on winning the α position are great indeed among harriers and should lead to high competitive ability.

We have seen that as a genus harriers are relatively dimorphic on the RSD scale (Fig. 3.6). We also see (Fig. 3.8) that RSD is significantly correlated with

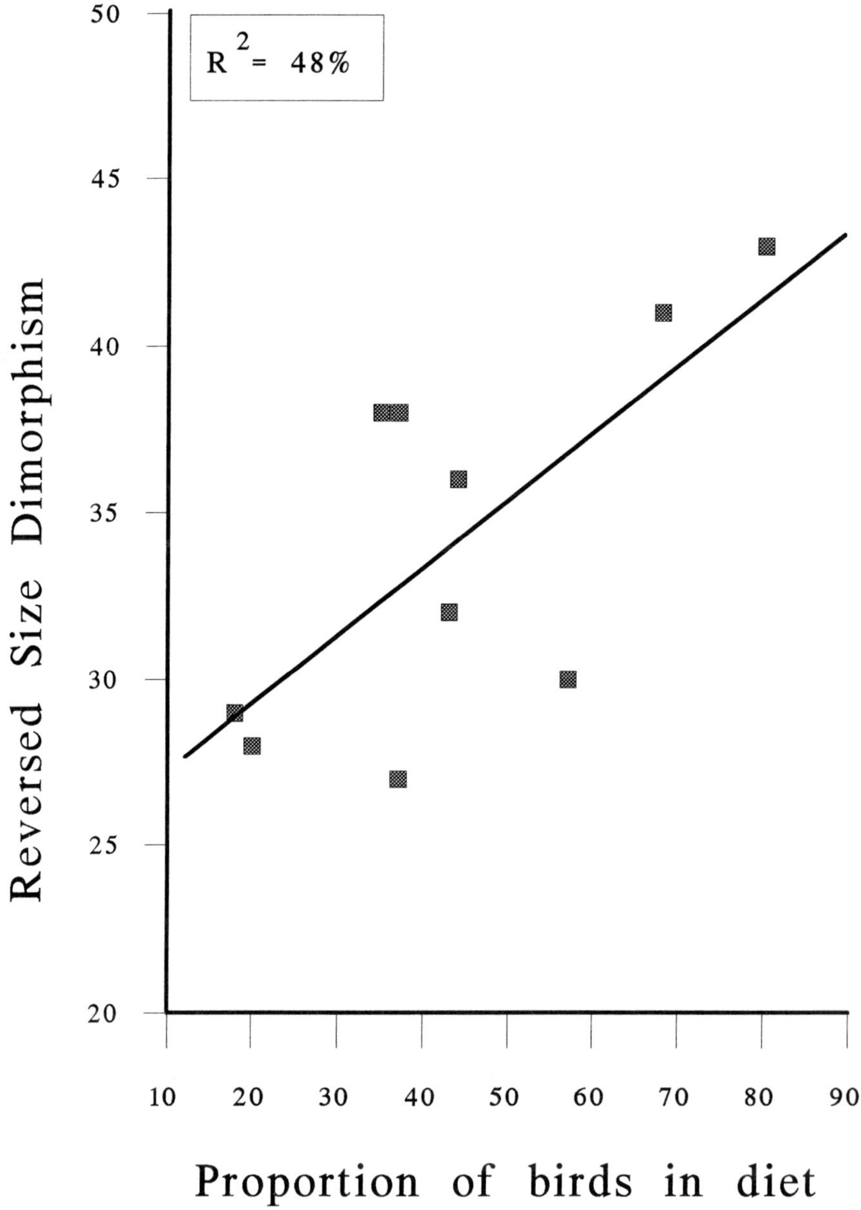

Fig. 3.8 Harrier dimorphism indices based on mass differences in relation to the proportion of birds in the diet.

the proportion of birds in the diet of 11 harriers for which there are data (R^2 = 50%, 10 df, $P < 0.01$), conforming to known RSD–diet trends. If Temeles and Newton are correct and it is *agility* and difficulty of prey capture (rather than just birds per se) that reduces male size, we should see stronger RSD correlations

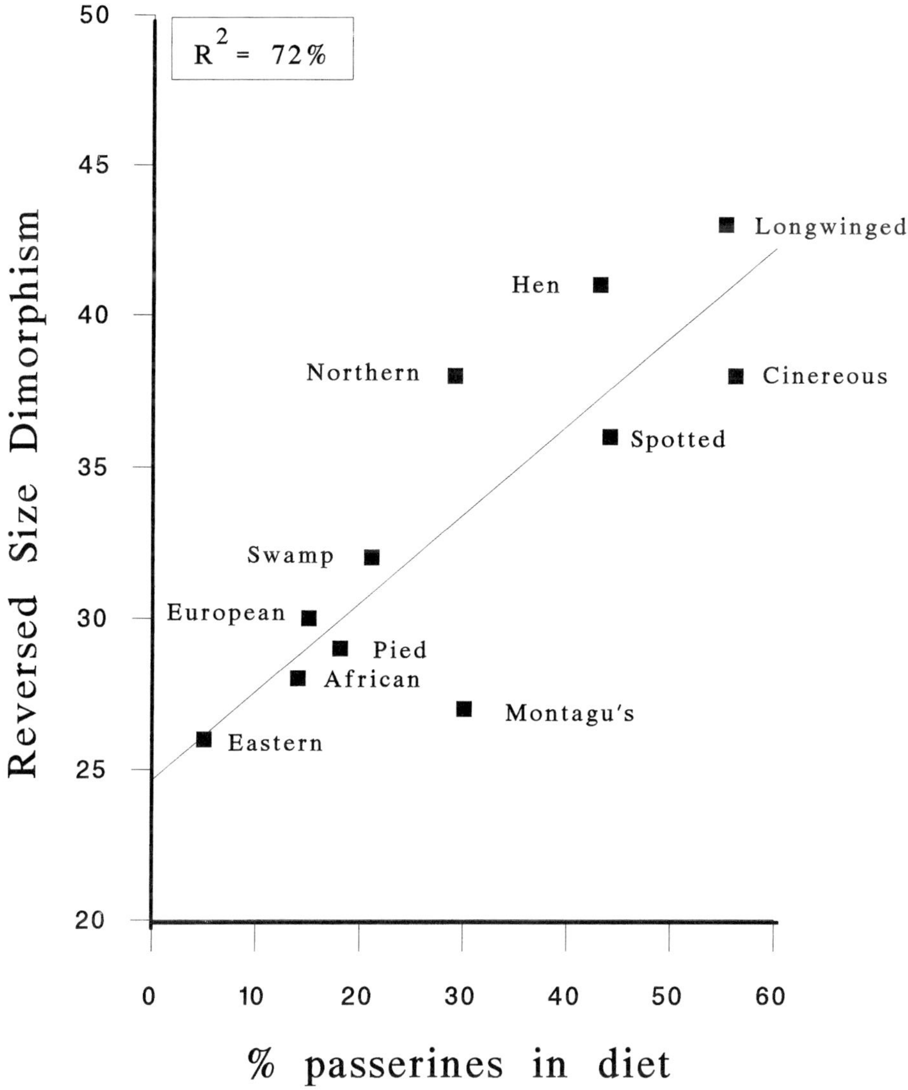

Fig. 3.9 Highly significant correlations between harrier dimorphism and the small, agile component of birds in the diet: passerines.

with just agile prey in the diet. Thus to fine tune this analysis I reassessed harrier diets, using only the proportion of passerines in each species diet (details given in Table 3.3), and correlated it with dimorphism indices. RSD correlations were substantially enhanced (Fig. 3.9). The degree of variation in RSD explained increased by 22% for passerine prey alone, and was highly significant (R^2 = 72%, 10 df, P = 0.005). The relationship between RSD and prey agility was thus confirmed for the harriers. This is surprising because one would expect fairly high levels of RSD anyway for harriers that take small agile mammals.

Table 3.3 RSD, frequency of polygyny, and avian diet of the world's harriers. RSD derived from Table 2.2. Polygyny and dietary data from sources shown.

Species	RSD[a]	Polygyny: % of nests (total nests)	% Birds in diet[b] [% passerines]	Sources
European Marsh	30	Monogamous/polygynous 2% (171), 51% (39), 28% (216), 0% (33), 4% (235), 20% (383), 8% (26), 2% (150), 4% (70), 14% (65) **$\bar{x}$ = 13% (1388)**	60% [27%], 43% [7%], 68% [20%], 32% [14%], 36% [8%] **$\bar{x}$ = 48% [15%]**	Woets 1989; Schipper 1973, 1978; Altenburg *et al.* 1982; Blanco *et al.* al. 1995; Fernandez and Azkona 1994; Witkowski 1989; Underhill-Day 1984, 1985 Bengtson 1967; A. Gamauf unpublished; Bavoux *et al. 1995*
Eastern Marsh	26	Monogamous (90 nests)	50% [5%]	Fefelov 1996 and unpublished
Australasian Swamp	32	Monogamous/polygynous 0% (18), 21% (19) **$\bar{x}$ = 11% (37)**	43% [21%]	Baker-Gabb 1981, 1982
African Marsh	28	Monogamous (24)	20% [14%]	Kemp and Dean 1988; Simmons *et al.* 1991
Reunion	—	No data (monogamous?)	~ 50%	Clouet 1978; Cheke 1985
Madagascar		No data (monogamous?)	~ 20% [**]	R. Thorstrom unpublished
Longwinged	43	Monogamous? (6)	80%, 65% **$\bar{x}$ = 73%** [61%] [49%] **[$\bar{x}$ = 55%]**	Bó *et al.* 1996; Cáceres 1996
Cinereous	38	Monogamous (14), (10)	27%, 94%, 86% **$\bar{x}$ = 69%** [26%] [83%] [58%] **[$\bar{x}$ = 56%]**	Jimenez and Jaksic 1988; Caceres 1996; Bó *et* al. MS
Spotted	36	Monogamous (34)	44% [44%]	Baker-Gabb 1982
Black	—	Monogamous (15), one case of polygyny suspected	13% [8%]	Chadwick 1997; Steyn 1982
Hen	41	Highly polygynous: 23% (43), 35% (275), 88% (345), 93% (214), 4% (111), 0% (49) **$\bar{x}$ = 59% (1037)**	72%, 47%, 86% **$\bar{x}$ = 68%** [48%] [32%] [49%] **[$\bar{x}$ = 43%]**	Balfour and Cadbury 1979; Schipper 1973, 1978; Picozzi 1978, 1980, 1984a; Cormier 1989; B. Arroyo unpubl.

Table 3.3 Continued

Species	RSD[a]	Polygyny: % of nests (total nests)	% Birds in diet[b] [% passerines]	Sources
Northern	38	Highly polygynous 24% (330), 29% (119) **$\bar{x}$ = 25% (449)**	37% [29%] (n = 5)	Hamerstrom *et al.* 1985; Simmons *et al.* 1986a; MacWhirter and Bildstein 1996
Pallid	35	Probably monogamous	No data	Lundevall and Rosenberg 1954; Cramp and Simmons 1980; Davygora and Belik 1994
Pied	29	Monogamous (?)	18% [18%]	Neufeldt 1964; Narayan and Rosalind 1990
Montagu's	27	Monogamous; low level polygyny: 5% (43), 3% (71), 8% (25), 2% (776), 2% (224) **$\bar{x}$ = 2% (1139)**	50%, 23% **$\bar{x}$ = 37%** [43%] [16%] **[$\bar{x}$ = 30%]**	Schipper 1973; Arroyo 1997

[a] RSD mass index = (F – M)/[(F + M)/2].
[b] Proportion of birds in diet was taken from breeding studies, combining pellets, direct observations and remains. Pellet analyses alone were not used due to biases towards micromammals.

This analysis confirms that it is agile bird prey rather than mammals that appears to be selecting for high RSD values.

But what of sexual selection: is RSD further enhanced by the extent of polygyny in each species? Figure 3.10 indicates a strong trend supporting the prediction that polygynous harriers show greater reversal than monogamous ones. Species above the line indicate greater than expected RSD for their bird diet and the Northern and Hen Harriers are indeed the most polygynous of all harriers. Proportion of polygynous nests in these species averages 59% of 1037 nests, and varies from 0% in southern regions to 93% in northern regions (Table 3.3, Fig. 5.9). Some males attract up to 5 (Northern) or even 6 to 7 (Hen) females to their harems (van der Kraan and van Strien 1969; Balfour and Cadbury 1979; MacWhirter 1985). This degree of polygyny is greater than in other species of harrier (and most other birds). South American Cinereous Harriers lie below the line yet have relatively dimorphic plumage suggesting some polygyny. However, their mating system was monogamous in two studies (Jimenez and Jaksic 1988; M. S. Bó unpublished data). Two other species that lie above the line are polygynous as predicted, though less so than the Northern and Hen Harriers. These are the Australasian Swamp Harrier, which is occasionally polygynous (11% of nests: Baker-Gabb 1981), and the well known European Marsh Harrier. From a sample of 1323 nests from 10 populations the average proportion of polygynous nests is 13% (Table 3.3). Polygyny varies widely in this species from 0% to 51% of nests, yet it too conforms to the idea that its level of RSD is greater than that predicted by its (low) ability at capturing passerines. One other species lies above the line—the Longwinged Harrier—yet its mating system is unknown because a large sample of nests has never been studied. Its position here again suggests it should be polygynous in some parts of it range; this is strengthened by the finding (above) that RSD indices increased southwards, and polygyny is known to increase with latitude in two species (Fig. 5.9). Thus I would predict that polygyny will be found in the southern part of its range (i.e. southern Argentina and Chile).

Species below the line, as predicted, are essentially monogamous and include the least dimorphic of all the harriers—Montagu's. This species tends towards semi-colonial monogamous breeding with very low levels of polygyny averaging 2% of 1139 nests in five studies (Table 3.3). Spotted, African, and Pied Harriers have only ever been recorded as monogamous (Neufeldt 1964; Baker-Gabb 1982; Simmons 1989; Table 3.3) and thus fit the trend. We can combine these two measures statistically using multiple regression. The results confirm that polygyny is important: together diet and the degree of polygyny explained a very high 85% of the variation in RSD, and remained highly significant ($n = 10$, $P < 0.001$) with the Longwinged Harrier removed. (If we include this species as monogamous, $R^2 = 83\%$, $n = 11$, $P = 0.0008$.) Since the explanatory power (R^2) rose from 72% to 85%, the degree of polygyny explained a further 13% of the variation of RSD in harriers.

The beauty of this combination of hypotheses is that it overcomes a major difficulty that virtually all other single hypotheses have not answered: why it is that males are smaller and females are larger. Individually none of the hypothe-

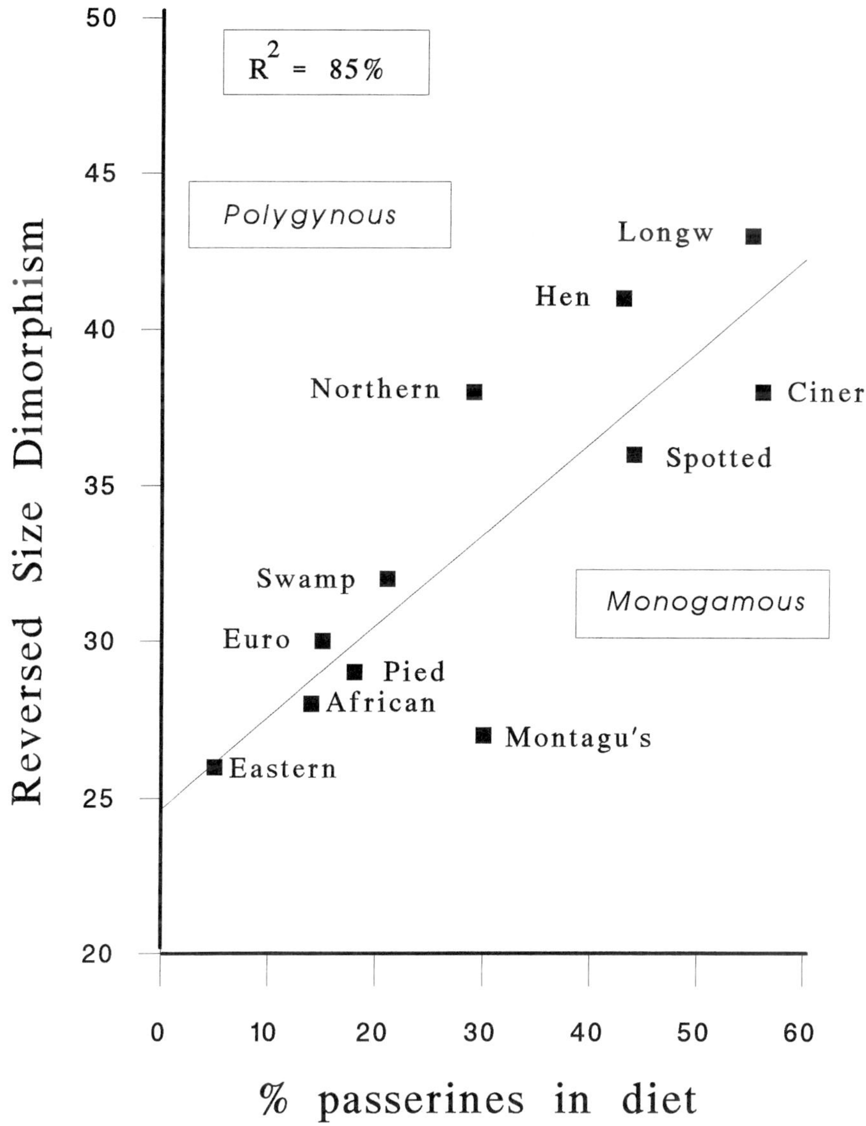

Fig. 3.10 RSD of the harriers in relation to passerines in the diet and mating system. Polygynous harriers all fall above the line, and are expected to shower higher dimorphism indices because sexual selection should be great on females to be larger (intra-sexual competition). The R^2 value reflects the variation explained by a multiple regression correlation between RSD, and mating system and per cent passerines in the diet.

ses can answer this and rely either on additional explanations such as intersexual food competition (Snyder and Wiley 1976; Newton 1979; Temeles 1985) female defence of nests (Cade 1982) or large female body size to maintain body reserves (Mendelsohn 1986).

Evidence from other species

Recent support for the sexual selection view has come from skuas and jaegars studied in Scotland. Catry *et al.* (1999) found that, as above, females rely completely on their mates for provisioning, the females are not the main nest defenders, and female dominance within pairs did not affect any aspect of breeding output. They also found no support for the idea that RSD arose from the specialized roles during breeding. However, they did find that female size and male age (indicative of high quality) were positively correlated in Great Skuas *Catharacta skua*. This suggested that large female skuas were winning contests for quality males; indeed females often fought and like raptors (Simmons and Mendelsohn 1993) fatalities are sometimes recorded (Catry *et al.* 1999). The authors concluded that sexual selection was probably driving increased female size as proposed by Olsen and Olsen 1984, 1987). They did not, however, find any benefits for males to be smaller such as enhanced foraging efficiency, enhanced reproductive output, or increased aerial display ability (skuas rarely undertake such displays). It would seem in the skuas that it is sufficient for the larger females who have won female–female contests to then choose the best available mates, and directional selection on females is sufficient to maintain the RSD in these species. This evidence follows closely the hypothesis of Olsen (*in litt.*). However, in raptors it seems from the above data on passerines in the diet that there is directional selection for *both* large female size and small male size, but it is not a prerequisite for RSD that both sexes come under directional selection pressures.

It is tempting to predict that in species like skuas, in which directional selection appears to occur in only one sex, that the degree of RSD would not be as great as in those species where directional selection is identified in both sexes. The body mass RSD values of the skuas at 10.3 (Great Skua) and 14.2 (Parasitic Jaeger *Stercorarius parasiticus*) are very low when compared with harrier RSD levels (average 34), but this assumes everything else is equal. Knowing the selection pressures on both males and females in other taxa would be useful in evaluating this idea.

Unsolved problems: agile males and aerial display

The results in Fig. 3.10 also support the hypothesis of Jehl and Murray (1986) and Widén (1984) that aerial display agility per se (whatever its purpose) is selecting for deviations from the RSD–passerine correlation: above the line the Hen and Northern Harrier are both noted for their intense skydancing while below the line are two species whose displays are less acrobatic and devoid of the vigorous U-shaped elements seen in other harriers (Montagu's: Robinson 1950; Arroyo 1995), or in which skydancing is not performed at all (Spotted Harrier: Baker-Gabb 1982). Aerial agility trends in harriers therefore also appear to correspond with the Jehl and Murray hypothesis on RSD. Are the two results (male display agility and female–female competition selecting for high RSD indices in polygynous species) mutually exclusive? I suspect not. Since aerial display vigour and polygyny are strongly associated in the mating behaviour of the Northern Harrier (Fig. 3.5; Simmons 1988c), and monogamy and a lack of display are associated in the Montagu's and Spotted Harriers, at least for harriers the agility of aerial display is linked with polygyny, and one (agility) is simply a measure of the other (degree of polygyny). Thus deviations of RSD above and below the RSD–passerine line (Fig. 3.10) may be indicative of aerial agility, because aerial agility is a guide to the mating system in the harriers. This may be so with shorebirds (Figuerola 1999) and other aerially displaying birds.

What is required is quantitative analysis, of display vigour across populations with and without polygyny, and within those, of the body size of displaying males. The prediction is that RSD will be greater in polygynous populations and within populations the smallest males will display most vigorously and attract the largest number of females—or sooner than other males (see Blomqvist *et al.* 1997). Ideal study areas for testing this would be European Marsh Harrier populations in The Netherlands, where polygyny is high (Schipper 1978; Altenburg *et al.* 1982, 1987) in contrast to populations in Spain, where polygyny is very rare (Gonzalez 1991; Fernandez and Azkona 1994). Spanish populations would be predicted to exhibit lower levels of RSD for the same bird diet relative to Netherlands' populations.

To conclude, skydancing displays of harriers appear to be chiefly territorial in nature but may also attract females. Skyspiralling displays by contrast are courtship orientated, leading females to likely nesting stations. Males in good condition in rich-food years, or those that are food-supplemented, display more often and vigorously than other males. In Northern Harriers, high display frequency is associated with larger harem size. Reversed Size Dimorphism is relatively high in the harriers and species with high RSD have a high proportion of birds in the diet. In support of the idea that it is the agility of bird prey and their difficulty in prey capture that is important, the correlations with RSD increased substantially when only passerine prey were included. Of most importance, because harriers are polygynous, it allowed us to determine if sexual selection (through females competing for small agile males) was operating. This was supported: greater RSD than predicted by a passerine diet alone was evident in those species known to be polygynous. Combining polygyny and the proportion of passerines increased the variance explained. This is the first evidence that supports the idea that both sexual and natural selection are operating in determining the degree of RSD in raptors, and it therefore explains both the extent and degree of the dimorphism.

4 *Of mice and harriers: who wins the arms race?*

One of the frustrations of watching harriers ironically arises from one of their most interesting attributes—their long distance foraging behaviour. Unlike other raptorial birds which hunt by perching, soaring, hovering, or ambushing prey from short flights, harriers hunt by active flying—quartering over grassland and marshland for hours on end. Invariably birds eventually pass into the haze, crossing a lake to a distant and favoured marsh, or become a distant speck, quartering over reed beds and the lake shore fringes of their wetland habitat. Sometimes birds check in flight, wheeling around and momentarily hovering on sighting a fleeing mouse or bird, only to resume their foraging, eventually becoming so distant that the arms tire from holding binoculars to the eyes. As one watches, patterns start to emerge—birds only begin hunting at certain times; in certain weathers all birds are active, at others times nothing happens for hours and the temptation to take a nap can be overwhelming. At midday breeding birds begin to soar and display, at other times they hunt favoured patches, quartering and re-quartering the same area time after time. Eventually the hunting bird will flash into the vegetation, hitting the ground hard, flapping excitedly, and eventually settling to eat their hard-won prey. After eating some of the head (for reasons unexplained, the favoured morsel), they may then efficiently transfer it to their dependent mates in an aerial food pass.

But consider the consequences of this hunting, which may encompass fully seven hours of a summer's day. On the plus side harriers, like many other birds of prey, cover every square inch of ground, day-in, day-out, month after month, sometimes for years at a time, often exploiting cover and surprise in securing their dinner. Clearly, this searching and researching gives the predator intimate knowledge of both the habitats and the areas where prey are most vulnerable: where the young birds can be found, or where mice perhaps cross an open patch once too often. That learning is involved is obvious by watching young (1–2-year-old) harriers foraging. These birds often take on inappropriately large prey, chase passerines they have no hope of catching, and make strike after strike into long grass after mice they can probably hear but not see. We will see later how immature birds and adults compare in predatory efficiency. For adults, however, there can be no more efficient predator than an aerial raptor whose livelihood depends on knowing where and when to catch the naive and unwary.

The downside is that the prey, hunted day after day, especially those, like mice, that can move swiftly and hide in impenetrable habitats, also become aware of this constant searching behaviour by their main enemies. The result is predictable: prey become more and more wary, seeking out the densest habitats, moving around perhaps when harriers and other predators are not hunting. The successful ones pass on these behaviours to their offspring, the unsuccessful ones don't. Natural selection at its simplest. However, harriers face an added problem: mice don't need to be caught to learn anti-predator behaviour. Despite their exquisite adaptations, harriers miss their prey about 50% of the time. And those are the adults. Mice then can learn from unsuccessful predation attempts—near misses—when and where they are most vulnerable (Vermeij 1982). The prey trying to save its life generally outwits the finely tuned predator which merely misses dinner—this is the Life–Dinner Principle of Dawkins and Krebs (1979). Because the selection pressures on mice to save their lives are greater than those on predators to catch their dinner, one expects mice to be ahead in the escalating arms race between predator and prey. They also have shorter generation times in which an evolutionary advantage can be expressed. Thus there is good reasons to expect that mice may be ahead in this arms race. The point of this Principle is that predator–prey studies must examine both sides of the coin—ecologists must carefully examine the behaviour of the prey in order to understand the predation tactics of their main predators. As important, the predation tactics themselves should heavily influence where prey live and when they are active.

Communication between prey and predator?

Harriers often try to outwit their prey by suddenly appearing over high vegetation or hunting between habitat types (Schipper 1977; Underhill-Day MS), but less obviously may receive information directly from their prey as to its suitability. This is not as altruistic as it seems since such signals are useful for both the predator and the prey. For the prey it may mean evasive action is unnecessary and the predator may accept that it has been spotted and therefore not waste time in needless pursuit. The mammalian equivalent is the tail flagging displays of deer, or, on the African plains, the stiff-legged stotting behaviour of gazelles in the presence of pursuit predators like cheetah and wild dog (Caro 1986).

There are two known avian examples in which prey signal to the predator. Among gallinaceous waterbirds such as coots and gallinules, birds constantly flick their tails to expose white under tail feathers—a highly visible flag for any birdwatcher—and predator. On investigating tail flicking in response to European Marsh Harrier presence, Alvarez (1993) found that birds were not signalling to their neighbours since they tail flicked both at and away from their neighbours and while solitary. However, they significantly increased the rate when harriers approached within 200 m, and no subsequent predation events were reported. Similar conclusions—that gallinules are signalling to the predator that they have been seen—were reached by Sue McRae for moorhens (unpublished data). Will Cresswell (1994) also noted that singing skylarks, vulnerable to attack by aerial predators such as Merlins *Falco columbarius*, were much *less* likely to be attacked than non-singing larks. Singing may therefore be a form of pursuit deterrent—good condition birds are signalling that they are difficult to catch. Clearly such systems can only persist if the predators occasionally test that signalling prey really are uncatchable, otherwise prey may constantly signal and fool the predator. That young raptors often chase everything (Sherrod 1983; Simmons 1984b) suggests that prey are tested, and that the raptors eventually learn what is inappropriate prey.

There is also good evidence that prey show fine tuned recognition of their predators and in some extraordinary contexts may even fight back. For example, Seyfarth and Cheney (1986) showed that alarm calling by African Vervet Monkeys in response to an overflying predator was more frequent when Martial Eagles *Polemaetus bellicosus*, one of their most dangerous adversaries, approach than when Blackbreasted Snake Eagles *Circaetus gallicus* approach. Snake eagles are virtually identical to Martials in colouration, yet never pose a threat to monkeys. Ground squirrels studied by Paul Sherman (1985) give different calls to dogs and weasels than they do to aerial predators. Neighbouring squirrels react appropriately to the different calls by scurrying to the nearest vantage point and searching when the terrestrial predator call is given, and fleeing to their burrows instantly when the aerial predator call is given. Similar calls are given by dwarf mongooses in the presence of different predators (Ann Raza pers. comm.). Among birds Grey Louries *Corythaixoides concolor* give an 'urgent' predator call, very different to the more familiar 'go-way' call in response to approaching raptors (P. Frost pers. comm.). That predatory birds are aware of the context of such calls is evidenced by species such as Wahlberg's Eagle *Aquila wahlbergi*, Gabar Goshawks *Melierax gabar*, and African Hawk Eagles *Hieraetus spilogaster* instantly scanning the sky on hearing such calls (P. Barnard and own observations).

Prey are not entirely defenceless when caught and in certain cases show remarkable anti-predator behaviour. Jon Rood (1983), observing mongoose families, witnessed three such events in Kenya. An immature Martial Eagle caught an adult mongoose from a Banded Mongoose family and flew a short distance to perch in an Acacia tree. The pack members ran up the tree, and the eldest male lunged at the young eagle causing it to drop the still living mongoose. The pack continued to bait the eagle for a further seven minutes before making off. In the two other incidents, Rood reported one successful rescue and one unsuccessful, while several rescues, with a 38% success rate, have been reported for the even smaller dwarf mongoose by Rasa (1986). Group baiting by monkeys of large eagles is well known (Struhsaker and Leakey 1990) and includes lunging at, or even attacking perched or incubating birds (Brown 1971). In these cases it may reduce further attacks by reminding the predator that the prey is not an easy proposition. Female ungulates can also successfully defend their young against eagle attacks (e.g. Darwin 1871; Locati 1990) although these are more even contests where the adults stand little risk of self injury or death.

Predictions of the Life-Dinner Principle

While it is established that predators can make optimal decisions about where and how to hunt their prey (Pyke *et al.* 1977; Krebs and Kacelnik 1991), many studies of non-sessile prey neglect anti-predator behaviours because they complicate simple predictions of optimal foraging by predators. For example, many predator–prey studies, particularly those of aerial predators, assume that prey are most available where they are most abundant (e.g. Linden and Wikman 1983; Temeles 1987). Field biologists subsequently assess where predators forage in relation to prey abundance. This is misleading because, as predicted by the Life–Dinner Principle, prey such as mice are expected to avoid areas where they are vulnerable to predators and congregate in areas where they are least vulnerable. Availability and abundance then become inversely related.

Several testable predictions arise from the Principle. First, within suitable habitat prey should be more abundant, that is live at higher densities, where predators are least likely to catch them. From a predator's viewpoint one expects, therefore, that harriers should hunt mice where they are least abundant. This is a powerful prediction of the Life–Dinner Principle since it reverses usual expectations of predator–prey studies which assume that predators, particularly aerial ones able to move swiftly from one habitat to another, should concentrate their predation on areas where prey are most abundant. Clearly, predation pressure alone may not account for the habitat choice of small mammals, and food, shelter, and competition may play a role. However, if predation is the main determinant of spatial habitat use by prey, we should find inverse correlations between predation and mouse density; if not, other factors are likely.

The second main prediction is that prey should tune their activity patterns to avoid the main activity periods of their principal predators. Most of the work in this areas has focussed on adaptive foraging patterns of the predator (Endler 1991), and shown, in apparent contradiction of the Life–Dinner Principle, that aerial predators tune their hunting activity to coincide with the main activity periods of their vole prey (Rijnsdorp *et al.* 1981; Daan *et al.* 1982). However, support for the Life–Dinner Principle comes from studies which show that some small mammals become rapidly more active during heavy rain (Bider 1968; Getz 1968) which is precisely the period when many aerial predators cease hunting (Belanger 1981; Rijnsdorp *et al.* 1981).

Predation tactics

To evaluate the predation tactics of harriers we will first look at generalizations that can be made from previous studies, notably that of Wibe Schipper and associates who spent several years studying three species of European harrier in several study sites in western Europe. We will then focus on an example from southern Africa where predator and prey were simultaneously assessed and compare it with other studies.

Sympatric European harriers

Two species of harrier winter in the polders (reclaimed marshland) of north-western Holland, and Schipper's study assessed the degree of overlap between the two species: Hen Harriers were found to take predominantly voles from low, less structured vegetation while the European Marsh Harriers occurred in the more structured reed habitat where they chased more birds, especially ducks. Hen Harriers flew faster than the marsh harriers in an attempt to surprise small passerine birds, a technique rarely used by the heavier marsh harrier. As

expected, hovering was more pronounced when head winds allowed it and this was particularly evident among the males. The same hunting techniques were found in a summer breeding study in which all three species—Hen, European, and Montagu's—were compared. Schipper (1977) found that marsh harriers hunted closer to their nests and searched their foraging habitat more scrupulously than either Hen or Montagu's Harriers. The latter species, the lightest of all harriers, was found to hunt furthest from the nest (up to 12 km), a range that increased as the season progressed. The heaviest, the European Marsh Harrier, by contrast ventured the shortest distance from the nest to a maximum of only 2.5 km. This trend was confirmed by Salomolard (1997) who also found Montagu's Harriers foraged further from their nests than the heavier Hen Harrier. This is contrary to the typical trend found in raptors in which larger birds have larger home ranges (Newton 1979). Foraging distances, however, are expected to vary according to prey abundance and also the level of accuracy of the sightings of individual birds. For example when French researchers Christian Bavoux and co-workers (1994) attached transmitters to wintering European Marsh Harriers they found birds foraged up to 4.5 km from their roost sites, further than breeding marsh harriers, probably because of lower food resources. However, it is equally likely that some bias was introduced by the method of observation. In a test of the idea that observations of known birds may underrepresent the true range of foraging birds, Marc Salomolard radio-tagged five Montagu's Harriers in western France and compared their observed range with that derived from tagged birds in the same terrain. The maximum distance from the nest was 3.0 km ± 1.0 km, for wing-tagged birds, while that for radio-monitored birds was significantly larger at 5.0 km ± 1.0 km. The area covered rose even more dramatically from just over 310 ha to 1420 ha. Thus we must treat with caution hunting distances and areas derived only from direct observations.

Nevertheless, when the three species are compared, the winter observations held true for summer behaviour: Hen Harriers more often tried to surprise their prey by flying transects, fast and low over ditches and hedges, and indeed were found to take the most agile prey (Schipper 1973). This is in accordance with their high degree of RSD (Chapter 3). European Marsh Harriers preferred the higher vegetation and often took young mammals, while Montagu's hunted over shorter grass areas and chased birds. Schipper's (1978) study was unable to confirm any competition between the species, but he did note that the narrow prey range of the Montagu's and Hen Harriers, and their unusually large foraging area relative to the larger marsh harriers' was consistent with the idea that the two smaller species were avoiding competition with their congener. What these comparative studies don't tell us is whether these harriers were foraging in areas where their prey were most abundant or most available.

African Marsh Harriers

To answer this question I undertook an intensive study of African Marsh Harriers and their main prey, the Striped Mouse *Rhabdomys pumilio*. These prey animals are widespread and relatively common throughout central and southern Africa, inhabiting damp to dry grasslands from sea level to altitudes of over 2000 m in East Africa (Taylor and Green 1976; Smithers 1983). They are relatively well studied and breeding is known to be influenced by rainfall (Perrin 1986), although large year to year cyclical fluctuations are not a feature of this species as they are in vole populations taken by temperate zone raptors (Chapter 5). Numerically, they are the main prey of African Marsh Harriers and together with another damp grassland species, the Vlei Rat *Otomys irroratus*, they constitute 74% of the prey in total (Kemp and Dean 1988; Simmons *et al.* 1991). In any one month they could form up to 84% of the prey.

To separate the effects of small mammal abundance and availability (to harriers) mouse availability was determined independently of abundance, using measures of vegetation density and mouse-runnel density. Abundance was determined by monthly trapping programmes before and during the harrier breeding season in the four major habitats searched by harriers (Simmons 1989). Effective trapping (i.e. allowing for false closures) totalled 3116 trap days, and was spread evenly over six months (July–Dec) from 1984 to 1986. Trapping indices gave the number of previously untrapped small mammals per 100 trap days. Each of the 881 animals caught was identified, weighed, sexed, assessed for reproductive condition, toe clipped (to recognize recaptures) and released. Of these, 89% were either *Rhabdomys pumilio* (687 animals) or *Otomys irroratus* (99 animals). *Otomys* is well known as trap-shy (e.g. Bond *et al.* 1980) and I certainly under-sampled it. Years were combined because trap success did not differ significantly (ANOVA $F_{2,13} = 1.00$, $P = 0.394$). Similarly, neither the mean body mass of *Rhabdomys* ($\bar{x} = 47.2$–49.5 g), nor *Otomys* ($\bar{x} = 125.3$–131.5 g) differed greatly between years, so trap success and biomass were comparable.

The following analysis attempts (possibly for the first time) to combine spatial and temporal analysis of prey availability to predatory birds, even though spatial (Wakeley 1978; Bechard 1982; Erickson 1985; Barnard 1987; Preston 1990) and temporal prey availability (Rijnsdorp *et al.* 1981, Daan *et al.* 1982, Greene 1990) have been independently determined several times.

Harrier foraging data

I wished to answer three questions with respect to foraging:

(1) in which habitats were harriers most successful (selection pressure presumed to be greatest);

(2) when during the day did harriers forage most commonly;

(3) when did they catch most prey?

Habitat-specific foraging data were collected over three years, during 2200 h of nest observations throughout the day, from 20 breeding birds (12 males, 8 females). The sex and age of each uniquely colour-marked bird were recorded, as were the number of attempted prey captures by habitat. This was easily assessed because birds landed in the habitat they were attempting to capture prey from.

Predation success is defined as the number of successful prey strikes per attempt, and is differentiated from foraging yield, the number of prey caught and delivered per hour of foraging. Harriers attempted to capture prey by (i) striking the ground hard from low level flight; (ii) plucking prey from water; or (iii) flying fast towards perched or flying birds. Harriers stalling in flight but making no ground strike were not classed as making a predation attempt (cf. Schipper *et al.* 1975). Hard strikes on the ground were classified as attempts on small mammals because each time birds were successful mice were caught. The only exception was on musk shrews (*Crocidura* spp.), which do not attempt to evade aerial predators; harriers caught them by parachute descent. Their strong musk prevented the harriers from eating them. Bird prey other than nestlings were always caught in a short chase. Identifying prey caught in the field was relatively difficult, so sample sizes are lower, but sufficient to gauge differences in prey-specific capture success. Since those prey that were distinguished in the field (n = 82) differed insignificantly (X^2 = 0.83, 1 df, n.s.), in proportions of mammals to birds, from prey types delivered to nests (n = 374), samples were not biased to easily identified prey.

Foraging methods

Like other harriers (Schipper 1977), African Marsh Harriers employed long distance quartering flights to locate prey. Foraging speed, recorded three times on a speedometer over 800+ m, varied with wind speed. One male hunted at average speeds of 25 and 29 km/h in calm conditions over *Typha* reeds, and another male was recorded at 20 km/h in strong breeze (wind rank 2). Northern Harriers of the same body mass exhibit slightly faster flight speeds up to 38 km/h, but quarter over other habitats at about 20 km/h. Males always fly faster than females in this species whatever the flight type, despite their lower wing loading (MacWhirter and Bildstein 1996). Breeding African Marsh Harriers actively fly between 38% and 50% of the day (Simmons 1989) while European Marsh Harriers do so for between 18% and 50% of the day (Altenburg *et al.* 1982; Bavoux *et al.* 1994). Combining hours of flight activity with foraging speed reveals that male African Marsh Harriers cover a minimum 91–120 km per day in the austral spring (September) in moderate breeze, and slightly more in calm conditions (132 and 174 km per day) when they fly faster. During 14-hour days in December, maximum distance would average over 200 km. Hence male harriers should be under considerable selection pressure to optimize both the habitats and the times at which they hunt their prey.

Predatory success: age and sex differences

Young (1–2-year-old) and adult harriers differed significantly (X^2 = 21.47, $P < 0.01$) in their prey capture success. Adults were five times more efficient at

Table 4.1 Predation success (captures/pounce) of young and adult African Marsh Harriers by year and sex, recorded in the Wilderness Lakes, 1984–86 [*n* = sample size.]

Year	1984	1985	1986	Totals
Overall success (age and sex combined)	41% [70]	51% [55]	29% [45]	41% [170]
Adult vs. Imm. success (sexes combined)	54.9% 0% [50]	58.1% 25.0% [55]	50.0% 4.8% [45]	55.1% 11.1% [150]
Adult female vs. adult male success	55.6% 54.5% [51]	61.1% 56.0% [43]	40.0% 52.6% [24]	56.1% 54.5% [118]

Table 4.2 Foraging time and yield for five male harriers observed in September 1984.

Total time observed (h)	304.8
Total time seen (h)	204.4
Time lost (%)	33
Hours observed foraging	49.1
Total prey returned to nest	84
Prey strikes/h	3.1
Net foraging yield (prey/h)	1.7

catching prey than young birds (Table 4.1) and made relatively few predatory attempts per hour (c. one every 20 min). No sex-related differences between adults were apparent, however: adult males and females averaged about one successful prey capture for every two attempts (55% vs. 56%: Table 4.1). There was little difference in adult predatory success between years except for 1986. Harriers then chased birds more often, females being less successful and making fewer attempts than their more agile mates. Foraging yield was estimated only for males during 304.8 h of observation (Table 4.2). During that time males were observed foraging for 24% of the time, and returned 84 prey to their nests. Foraging yield therefore averaged less than 2 prey per hour.

Predatory success and prey type

Like other raptors (Temeles 1985), harriers found birds significantly harder to catch than small mammals. Adults and immatures caught birds (excluding nestlings) in only 13% of 24 attempts, while rodents were caught on 41% of 58 attempts. For sub-adult and adult harriers (> 3 yr) the disparity was even greater: birds were caught on 17% of 18 occasions, but small mammals were caught on 55% of 77 attempts, a highly significant difference ($X^2 = 6.95$, 1 df, $P < 0.01$). Northern Harriers show similar predation efficiency with amphibians and reptiles caught in 74% of attempts, small mammals in 34% of attempts, and birds only 14% of the time (Toland 1986).

Predatory success and habitat type: mouse abundance vs. availability

Vegetation was categorized into one of the four previously defined categories used in the trapping program. I was unable to consistently determine the habitats over which each male foraged because harriers often hunt on the border between habitats (Schipper 1977; Underhill-Day 1993). Therefore, I could only analyse predation successes by habitat, not by where harriers spent most time hunting and their foraging yield there. However, foraging yield (i.e. the rate of predation), which may be the most important selective pressure affecting where mice live, is very closely associated with predation success by habitat in other raptors eating small mammals ($r \sim 0.92$, $P < 0.001$; Fig. 4, Masman *et al.* 1988).

Spatial small mammal availability and harrier success

Availability of small mammals is probably related to vegetation depth and density for two reasons: small mammals should be less visible in thick vegetation, and even if detected by the acute auditory abilities of harriers (Rice 1982), they should be less catchable. The three main foraging habitats were sampled by walking 200-m transects in areas where harriers were known to forage. Every 10 m a thin calibrated dowel was inserted vertically into the vegetation and the height of the highest blade or leaf that touched the dowel measured. Twenty measurements were taken per transect, and 22 transects were walked in total. While walking the intervening 10 m, all the visible mouse runways within an approximately 1-m band were counted, simulating the view that a harrier foraging at 1.5–2.0 m might have. Thus, runnels are expressed per 10 m^2, and were assumed to represent an alternate index of mouse availability, because they exposed the mice using them to aerial predators.

So where did most mice occur? Dense habitats such as *Stenotaphrum* grass supported almost twice the number of rodents (22% trap success) that mixed grass and bushes offered (11.5%). Comparing trap success between all four habitats revealed a significant difference in mouse abundance (Table 4.3). Predation success of adult harriers, however, was lowest in the areas of high mouse abundance (Table 4.3). This inverse relationship was confirmed for all

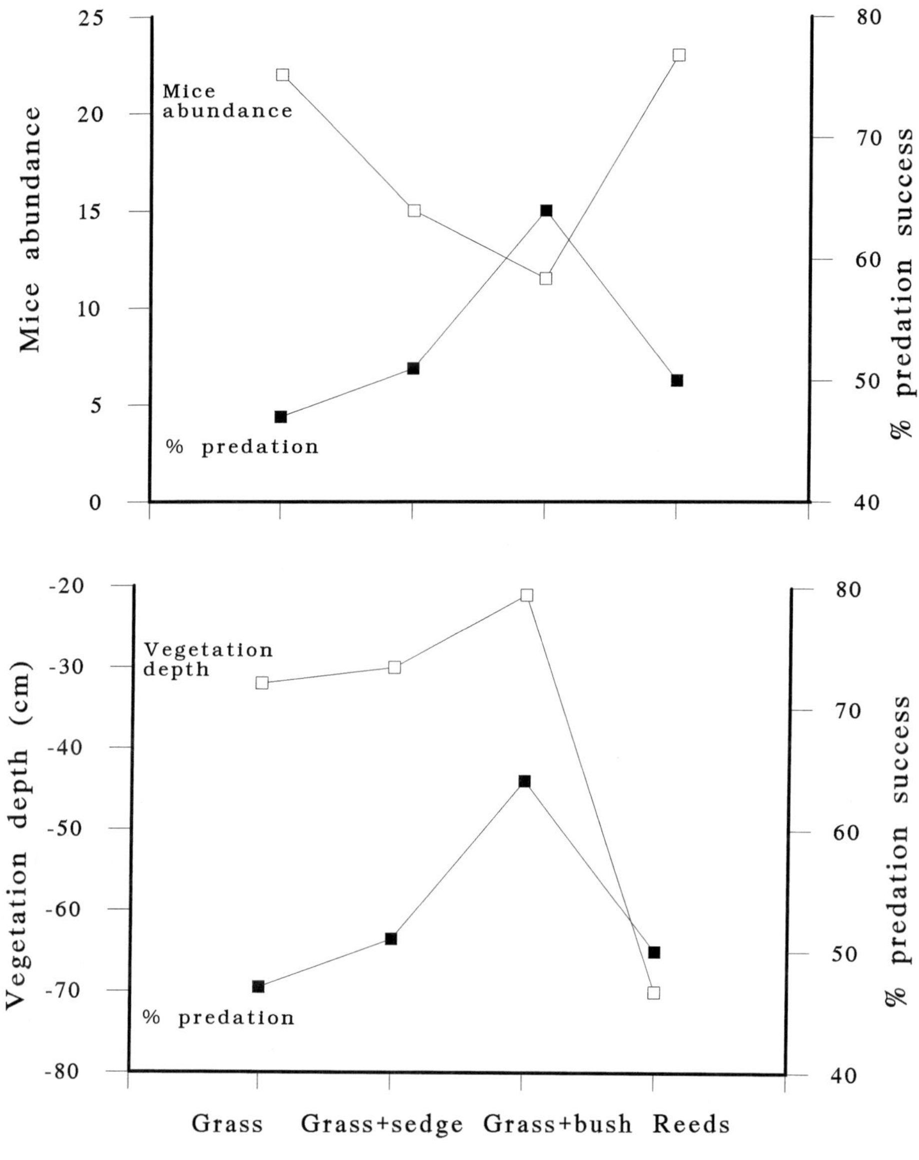

Fig. 4.1 African Marsh Harrier predation success in relation to mouse abundance (top) and mouse availability (bottom) in different habitats. Availability, which closely mirrors predation success, is given as vegetation depth (cm) where 0 corresponds with the vegetation surface and the negative figures, the average depth below the surface.

Table 4.3 Predatory success of adult African Marsh Harriers in different habitats in the Wilderness Lakes. Success is expressed as a percentage of all attempts on mice and all prey (except nestlings), and is compared with small mammal abundance. Number of new captures (mice per 100 trap nights) and runnel density (runnels/10 m^2) is given for each area; 1984–86.

Habitat type	% Successful predation attempts [*n*]		Abundance of small mammals ($\bar{x} \pm 1$ SD)	Runnel density ($\bar{x} \pm 1$ SD)
	Mice	All prey		
Mixed grass/bush	63 [8]	64 [11]	11.5 ± 5	2.2 ± 0.8
Steno/juncus/phrag	59 [27]	51 [37]	15 ± 8	3.3 ± 1.6
Phragmites/Typha	53 [19]	50 [35]	23 ± 6	–
Thick Stenotaphrum	48 [23]	47 [34]	22 ± 11	4.4 ± 1.2

[a] Habitats differed significantly in trap success: Kruskal–Wallis $H = 7.97$, $P = 0.047$.

four habitats sampled—harriers caught fewer mice per strike where mice were more abundant (Fig. 4.1). As expected, this unusual situation was resolved when the relative availability of small mammals in each habitat was examined. Predatory success then mirrored the relative vegetation depth of the habitats (Fig. 4.1); harriers were less successful in habitats with high prey abundance because mice were least vulnerable there. The small number of habitat types precludes statistical analysis, but the trend held for small mammals alone and mammal and bird prey combined (Table 4.3).

Since presumed prey availability was measured in two ways (vegetation depth and runnel density), the relative importance of harrier prey detection by sight and sound could be assessed. Harriers were most successful in habitats with low vegetative cover (Fig. 4.1), but were least successful in those with high mouse runnel density (Table 4.3). Small-mammal availability is therefore best judged by vegetation depth, and by implication sound not sight was used most by harriers to catch prey. Runnel density merely reflected the abundance of mammals in each habitat (Table 4.3) and was a poor measure of availability.

Viewed from the predators' standpoint, the fact that harriers were least successful where mice were most abundant is paradoxical. For the prey, however, it is not—mice should occur most commonly in areas where they are least vulnerable—as predicted by the Life–Dinner Principle. The main measure of prey availability (vegetation depth) further indicated that harriers were hunting in areas of lower mouse abundance because mice were more available there. We cannot, however, assume that high predation success is the only reason for fewer small mammals in low density habitats, and other reasons are explored below.

Evidence from other studies

The African Marsh Harrier's propensity to be most successful in habitats offering the lowest cover supports findings in several previous studies (Wakeley 1978; Baker and Brooks 1981; Bechard 1982; Barnard 1987). These studies show that raptors spend more time hunting over low cover habitats. Preston (1990) has shown further that frequency of use of certain habitats was related to a combination of small-mammal biomass and vegetative cover. None of these studies, however, showed that prey were less abundant in the areas where most successful foraging occurred, the main prediction of the Life–Dinner Principle.

Does predation alone explain mouse habitat preferences?

For these spatial results to be both generally applicable to other raptor–mouse situations and support the premise that predation per se is the main reason for small-mammal habitat preferences, one would expect favoured small-mammal prey to be consistently found in areas of high cover. In support of this expectation, *Rhabdomys* and *Otomys*, the two commonest diurnal small mammals in southern Africa (Smithers 1983), always occur more commonly in areas of high cover (Brooks 1974; Rautenbach 1976; Smithers 1983), and do so in many different habitat types and altitude zones (Bond *et al.* 1980). This was consistently true in this study even for wet versus dry habitats. Similarly, *Microtus* voles, a common prey item of numerous aerial predators in Europe and North America (e.g. Erlinge *et al.* 1984; Korpimaki 1984; Simmons *et al.* 1986b; Masman *et al.* 1988), commonly occur in areas of densest cover (reviewed in Birney *et al.* 1976). Clearly there are several other selection pressures that may cause mice and voles to prefer dense cover. Food and microclimate are two obvious contenders. However, on investigating the preference of *M. pennsylvanicus* for dense cover, Getz (1961, 1971) concluded that the lower relative humidity and higher temperatures in sparse cover would not place them under significant physiological stress and was unlikely to play a significant role in the avoidance of such areas by voles. Getz (1971) suggested that predation pressure and behavioural preferences were the principal reasons for the main habitat choices of voles. Similar conclusions are apparent in an increasing number of small-mammal studies (Desey *et al.* 1990; Dickman *et al.* 1991; Kotler *et al.* 1991; Longland and Price 1991; Bowers and Dooley 1993; Norrdahl and Korpimaki 1993). Thus, like well-studied *Microtus* species, I believe that Striped-mice habitat preferences are mainly governed by predation, but we cannot rule out food or competition without experimental manipulations.

Direct selection or learned behaviour?

In this study I measured prey vulnerability only in terms of direct selection, that is whether mice were more likely to be caught in a certain habitat. I had poorer measures of two alternative selection pressures. First, many unsuccessful predation attempts ('near misses') may have influenced habitat selection by small mammals. Since this is incorporated into the measure of predation success (i.e. 1—success) it has already been assessed. The other more likely influence is the rate at which mice are caught from each habitat, that is the predation pressure per unit time. For example, if mammals were exposed to twice as much predation per day in habitats where they were typically less vulnerable (because harriers hunted more often there), this may affect their micro-habitat choice, even though the habitat per se may offer poor rewards for hunting harriers. It is unlikely that mice could simply move away from vulnerable habitats because of the territorial behaviour of neighbouring mice (cf. Choate 1972) and only micro-habitat adjustments are likely. I was unable to evaluate this important component, but note that if predation rate was much higher in dense habitats, the results would have shown no support for the predictions of the Life–Dinner Principle. Moreover, strike success and foraging yield are known to be closely related from month to month in other small-mammal specialists (cf. Masman *et al.* 1988). There are as yet no studies of aerial predators which have determined habitat-specific prey yields and related them to prey abundance.

Two other possible scenarios may explain the low abundance of mice in habitats where harriers were most successful. First, mice may simply be removed more frequently by predators from these habitats. This is not unlikely since Brooks (1974) noted that predation explained most of the mortality of a closely monitored *Rhabdomys* population near Pretoria. This clearly reinforces the likelihood that natural selection will play a role in habitat selection by later generations, and is not a mutually exclusive explanation of low abundance. Second, factors other than food, microclimate, or predation may be operating (i.e. competition between *Rhabdomys* and *Otomys*). This is unlikely because the species are often found side by side, use the same runnels, and show minimal dietary overlap (Curtis and Perrin 1979). Competition is therefore unlikely to play anything but a minor role in habitat segregation and their co-occurrence in the densest cover (Bond *et al.* 1980) is best explained by direct or indirect predation on both species.

The question arises why any small mammals occur in areas where they are vulnerable to predation. The simplest answer is that it is due to the dispersal of certain individuals from optimal habitats because of high density. Other studies show that dispersing individuals are more likely to suffer mortality than residents as they move into marginal habitats (Errington 1946; Liddicker 1975), and these individuals are often smaller or younger than the population at large (Beacham 1979; Perrin 1986; Dickman *et al.* 1991).

Predatory success and seasonal changes in availability

Vegetation depth, the best predictor of availability, might be expected to be low during winter and increase uniformly thereafter, making mice less and less available relative to the number present. Monthly changes in vegetation profiles were, however, more complex than this. Availability in the frequently foraged *Stenotaphrum* habitats (where mice were most abundant) in fact increased to a peak in spring (as did mixed *Stenotaphrum/Phragmites*), and thereafter decreased (Fig. 4.2). Since abundance also peaked in spring (October) it is not surprising that harriers made more successful strikes on mice (80%) in October than in any other month (Fig. 4.2). Success was lowest in August (28%) as they began breeding, and at median levels (50%) in December when large young were taking wing. In mixed grass and bush, however, the opposite was true: vegetation profiles became deeper (decreased availability) with the approach of the austral spring. Predation success on mice paralleled the increase in availability in the *Stenotaphrum* habitats (where abundance was high), and simple

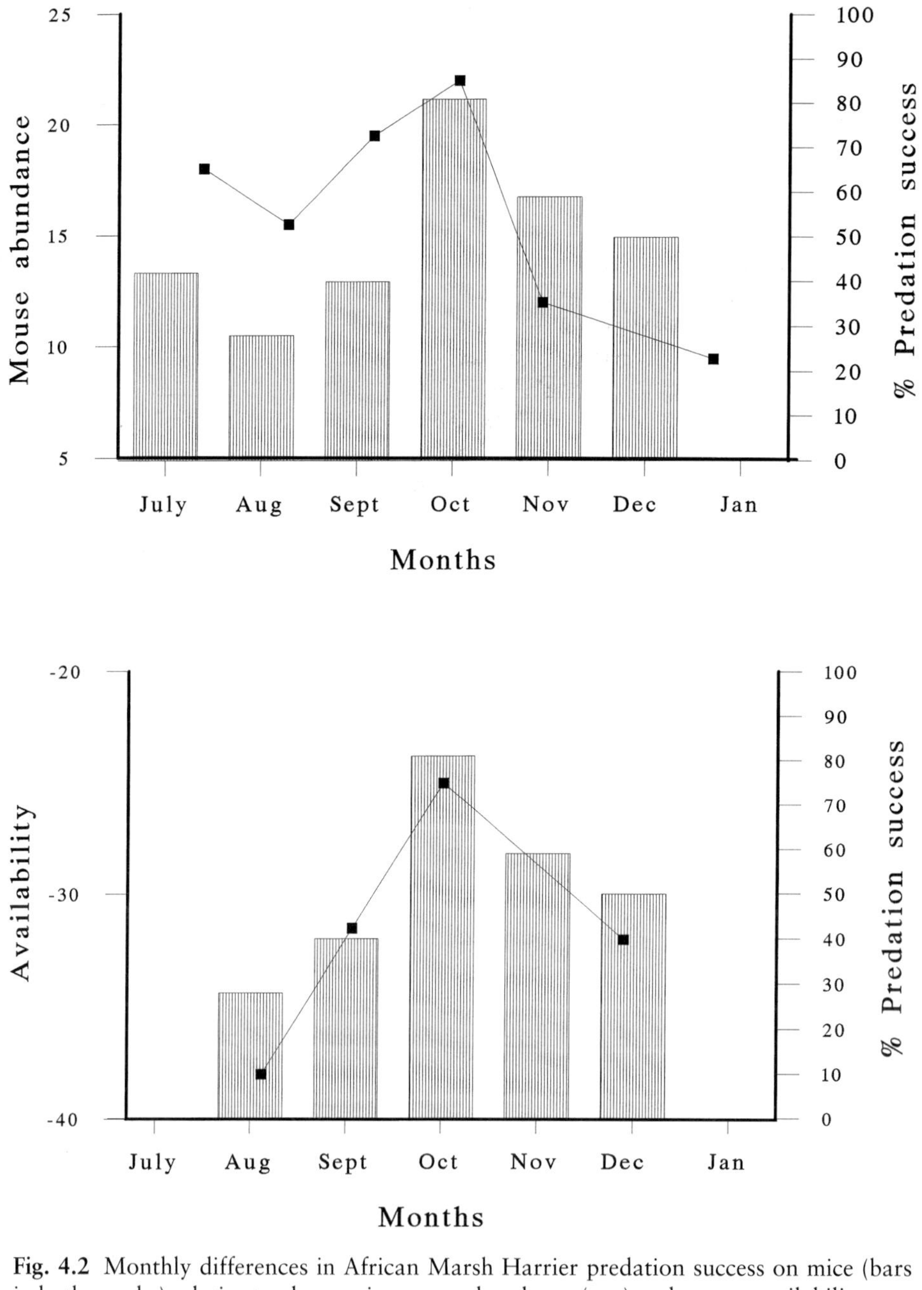

Fig. 4.2 Monthly differences in African Marsh Harrier predation success on mice (bars in both graphs) relative to changes in mouse abundance (top) and mouse availability vegetation depth (bottom). The trend for highest predation success (October) was reflected by both mouse abundance (top) and the increasing availability of mice in *Stenotaphrum*, the mouse-rich habitat. Closer correlations were apparent between availability in the latter habitat and predation success than any other combination (see text). Mouse availability was not assessed in July and abundance was combined for December/January.

correlation suggested that availability in *Stenotaphrum* (r = 0.94, n = 4) was a more important influence on predation success than availability in the mixed grass habitats (where mice were half as common: r = –0.57, n = 5). Mouse abundance alone was poorly correlated with predation success per month (r = 0.25, n = 6).

Was predatory success influenced by young naive animals entering the population? *Rhabdomys* below 30 g (Henschel *et al.* 1982), and *Otomys* below 50 g (Davis 1973) were classed as juveniles, and their seasonal appearance was plotted and correlated with predatory success. Juvenile mice played no role in enhancing success at these times. Juveniles comprised up to 25% of the trapped-mouse population (Fig. 4.3), but the highest levels were recorded in July in two of the three years, and in December in 1984. Newly available juvenile mice could not, therefore, explain the October high in harrier predatory success.

Predation success depends not only on increasing and decreasing availability of mice but also other prey types after September (these are not included in Fig. 4.2). For example, after September nestling passerines and waterbirds enter the diet. In August and September, 5% or less of the 12 birds attacked were juveniles; by October and November/December 29% and 30% of 40 attacked birds were juveniles. These easily caught prey increase predatory efficiency even though mouse availability is then decreasing in concert with mouse abundance.

To summarize, relatively simple rules appear to govern where mice should live and where predators should forage. Seasonal analyses show that harriers face con-

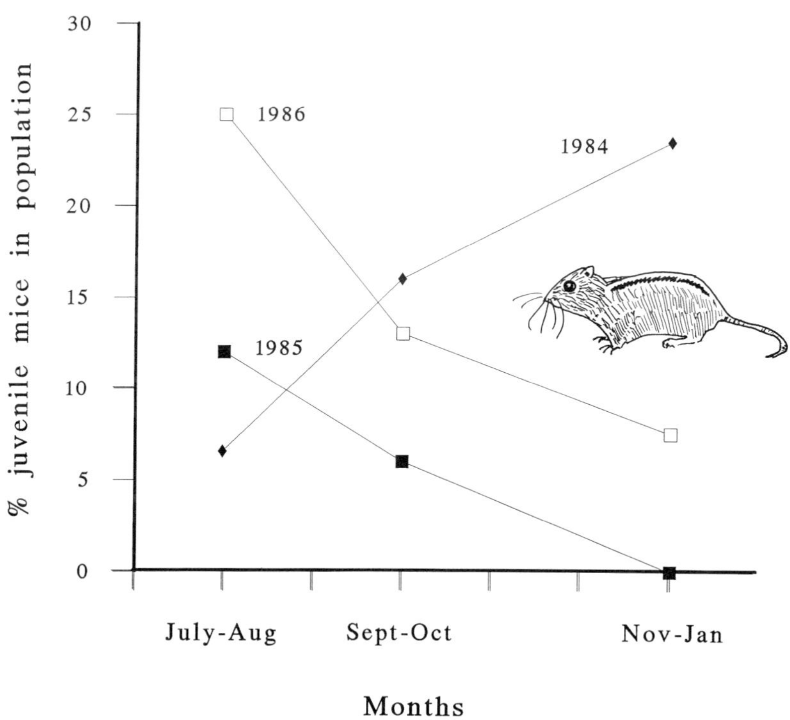

Fig. 4.3 The proportion of juvenile *Rhabdomys pumilio* in the trapped population per year.

stantly changing indices of availability and abundance in different habitats. Like other harriers (Barnard *et al.* 1987; Redpath 1992; Arroyo 1997), alternative prey types become available later in the season, enhancing predatory success as mice decline. Juvenile mice appeared to play no role in differential predation success, possibly because they did not spend extensive periods above ground or occurred in denser habitats. While predation success is a complex interaction of fluctuating mouse numbers, increasing then decreasing availability, and seasonal changes in newly available bird prey, predation trends mirrored changes in availability in the mouse-rich *Stenotaphrum* habitat. The results support the idea that prey availability, not abundance, is the key variable influencing predation.

Hourly mouse availability and timing of hunting

Mouse availability can vary not only seasonally but diurnally. The diel activity patterns of *Rhabdomys* were calculated from five days (2860 trap-hours) of live-trapping in 20-minute periods in September 1984 from dawn to dusk. *Otomys irroratus* activity could not be similarly assessed because of the difficulty in capturing them, but they are known to be mainly crepuscular (Davis 1973). The number of *Rhabdomys* (n = 193) captured per hour was plotted for all five days, and used as a guide to mouse activity. Mice were again assumed to be more available or vulnerable to detection when they were most active.

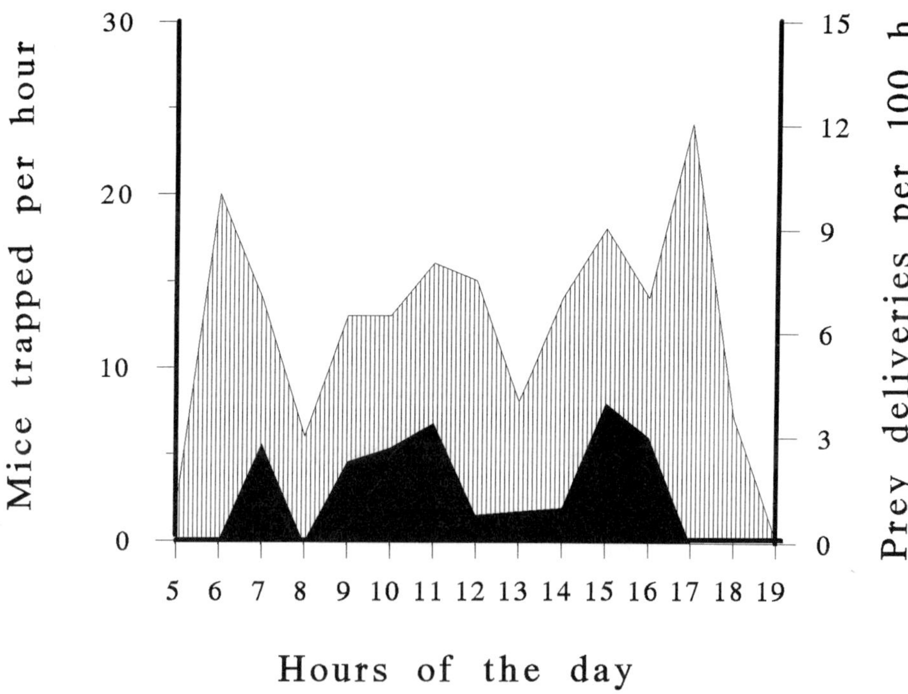

Fig. 4.4 *Rhabdomys* activity patterns (from September 1984) in relation to their capture and delivery (black graph) mby harriers throughout 1984.

Two pronounced activity peaks were identified in the early morning and at dusk with two minor peaks in mid-morning and mid-afternoon (Fig. 4.4). During the lulls between these peaks, *Rhabdomys* were three times less trappable. Assuming these patterns reflected the availability of *Rhabdomys* to foraging harriers, did harriers show mammal capture peaks that coincide with their prey? Using the hourly return rate of adults with prey to their nests, harriers did exploit the mid-morning peak, rested at midday, and again brought more *Rhabdomys* to nests in mid-afternoon. In September 1984, when mouse activity was assessed, there was a significant correlation between activity and prey deliveries ($r = 0.76$, $n = 10$, $P = 0.011$).

While this method may underestimate the total number of captures, I have no reason to believe that it biases the timing of prey captures because most observed captures were either delivered within about 10 minutes of capture, or, in very few cases, were not delivered at all. There was also no trend for foraging birds to eat all their captured prey more often at certain times of day. Instead, the strategy of male harriers is to eat some of the prey following most (82% of 38) prey captures, then deliver the remainder to their dependants.

This method does not, however, directly test the idea that harriers are timing their foraging bouts to increased mouse activity because they might merely forage all day and catch (and deliver) mice at times when mice are most active.

Thus I additionally undertook a month-long (24 day) 2-minute analysis (Altmann 1974) of male harrier foraging behaviour in September 1984. For each of five closely observed males, I recorded their activity as either foraging, or non foraging (i.e. flap-sailing, soaring, perching, eating). I did so for 20-minute blocks throughout the day and summed all males across identical 20-minute blocks to gain a proportionate measure of male activity per hour. A total 122 hours of data were collected for up to 3 birds at a time, giving 304.8 bird hours of observation. Birds were lost for 33% of the time giving 204.4 hours of data on activity profiles. Proportion of time foraging was calculated as the number of 2-minute intervals when birds were observed foraging divided by all 2-minute intervals when birds were in view. (The number of prey delivered in this time is taken as a rough estimate of foraging yield.)

Harriers did exhibit broad hunting synchronization with mouse activity during the day ($r = 0.84$, $n = 5$, for 2-hourly intervals, $P < 0.05$), although less so on an hour by hour basis ($r = 0.55$, 9 df, $P = 0.1$, Fig. 4.5). Most importantly, however, in 1984 (and 1985, data not shown) harriers *failed* to exploit the early morning or late evening peaks in prey activity (Fig. 4.4) because they were completely inactive before 07h00 or after 17h30 (Fig. 4.5). Harriers typically began foraging only after early-morning sunning and preening, and roosted or loafed close to their territories about one hour before sunset, except during peak nestling demands (October–November).

Other selection pressures on mouse activity

Since *Rhabdomys* mice were most active when harriers were not hunting them at all, this supports the second prediction of the Life–Dinner Principle. More striking examples of this phenomenon are reported by Belanger (1981), who showed

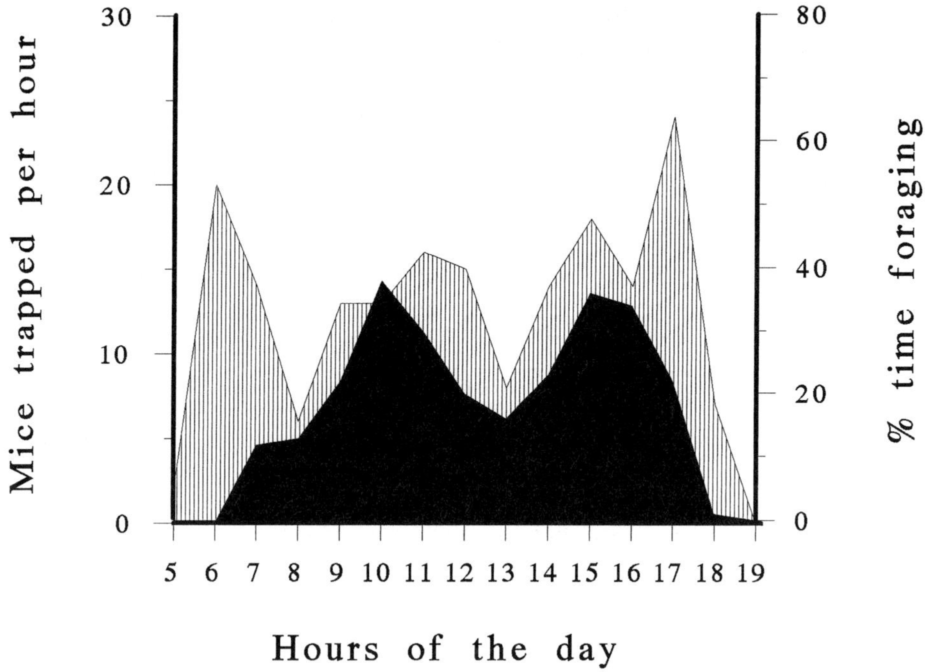

Fig. 4.5 *Rhabdomys* activity patterns (from September 1984) in relation to harrier foraging patterns (black graph) in the same month. Note the lack of foraging during both main mouse peaks. Sunrise occurred at 06h20 and sunset at 18h20.

that abrupt increases in activity of small-mammal prey occurred during rain when their main avian predators stopped hunting. For nocturnal predators Clarke (1983) and Kotler *et al.* (1991) found decreased foraging behaviour of *Peromyscus* mice and *Gerbillus* gerbils under full moon conditions when risk of predation by owls increases. However, Halle (1993), reviewing several predator–prey studies, found that there were no periods when predation risk was significantly reduced for *Microtus* voles, and he argued that crepuscular activity peaks could not be explained as an avoidance of predators. So, before increased *Rhabdomys* activity can be attributed to a lack of predation pressure we must assess the activity of other small-mammal specialists present on the study area (Blackshouldered Kites and Grass Owls *Tyto capensis*). Both in this study and as found in other areas (Tarboton 1978; Mendelsohn 1982; pers. obs.) kites are generally (but not always) inactive at dawn and dusk. Indeed on this study area distinct predator-free windows were apparent when harriers had stopped foraging and the nocturnal owls had yet to start hunting. These coincided precisely with the activity peaks of *Rhabdomys pumilio*. No canids or cats were present on the study area although small numbers of Egyptian Mongooses (*Herpestes ichneumon*), which take rodents (Estes 1991), and one Large Spotted Genet (*Genetta tigrina*) were seen and trapped. Their activity patterns were not studied.

While mouse activity supports the idea of avoidance of predators, others have argued (for *Microtus* species) that activity patterns may instead be associated with natural internal rhythms (Daan and Slopsema 1978; Rijnsdorp *et al.* 1981) or with feeding requirements (Perrin 1981). Feeding requirements seem less likely to explain activity rhythms for *Rhabdomys*, because one would expect consistent activity patterns from different studies if this was so. However, in five studies which have assessed *Rhabdomys* activity patterns considerable variation was the rule and included some nocturnal behaviour. The only consistent finding was for early morning activity (Smithers 1971; Choate 1972; Brooks 1974; Christian 1977; Perrin 1981). This is unlikely to be explained solely by hunger-related rhythms. Perrin (1981) further suggested that foraging for termites, a principal food item with crepuscular activity, may explain *Rhabdomys* activity. However, these termites did not occur in my study area, hence it cannot explain the crepuscular peaks there. It is possible that if variable predation intensities occurred in these studies they might well explain the different patterns found. However, little work has been done on this aspect by African small-mammal ecologists, so it remains untested, and largely overlooked. On present evidence,

aerial predation patterns are likely to strongly influence the activity of certain small mammals that face consistently strong predation pressure (Lima and Dill 1990; Sih 1980; Kotler *et al.* 1991), and this study provides some of the first evidence for this in African species.

Why do mouse activity peaks occur at all?

If predators are able to cue in on, and time, their hunting activity to short term prey activity peaks, as found in this and other studies (e.g. Rijnsdorp *et al.* 1981; Daan *et al.* 1982), why do small mammals exhibit such peaks at all if it apparently increases chances of predation? One plausible answer is that individuals are less likely to suffer predation by foraging together because they dilute the probability of their own demise. Hence, just like schooling or herding behaviour may dilute predation in a spatial sense (Hamilton 1971), simultaneous activity in a temporal sense may have the same effect. Two studies offer both support (Daan and Slopsema 1978) and contradictory (Daan *et al.* 1982) evidence for the risk associated with communal foraging by voles. The difference depended on whether the predators could adjust their hunting to prey activity or not. In this study the crepuscular peaks decreased risk, since no aerial predators were foraging then, but predatory synchronization occurred at other times, possibly increasing risk.

Hunting patterns and weather variables

The effect of weather variables on foraging return was gauged by ranking wind from 0 (calm) to 6 (gale) for each hour of the day, and cloud cover from 0/8 (clear) to 8/8 (complete overcast or mist). Temperature was recorded with a maximum–minimum thermometer on site. In 1984, these variables were correlated with prey deliveries to nests. The object was to determine if weather variables constrained foraging return and thus might explain the lack of foraging when mice were very active. According to the delivery patterns of 263 items in 930 nest hours, temperature had no effect on delivery rate. Assuming an equal temperature-dependent rate across the range of temperatures (10° C – 35+° C), no significant variations were apparent ($X^2 = 6.04$, 6 df, $P > 0.1$). Marginal differences occurred only at temperatures of 30–34° C when 35% fewer items than expected were delivered. All other differences were less than 17%.

Differing wind speeds, however, had a significant influence on both the overall delivery rate ($X^2 = 21.6$, 5 df, $P < 0.0001$) and the number of definitely identified *Rhabdomys* delivered ($X^2 = 12.5$, 5 df, $P < 0.05$). Prey deliveries during calm conditions and light breeze were 23% less than expected from an equal wind-dependent rate (Fig. 4.6). This is consistent with the finding that some African small mammals are more active during windy weather (Bowland 1987). Similar findings were apparent for Northern Harriers in New Brunswick; males and females combined delivered about 30% fewer prey per hour in calm conditions (1.04 prey/h) relative to breezy conditions (1.33 prey/h). Assessing small-mammal prey alone MacWhirter (1985) found that male and female Northern Harriers delivered about 20% more prey per hour in windy conditions

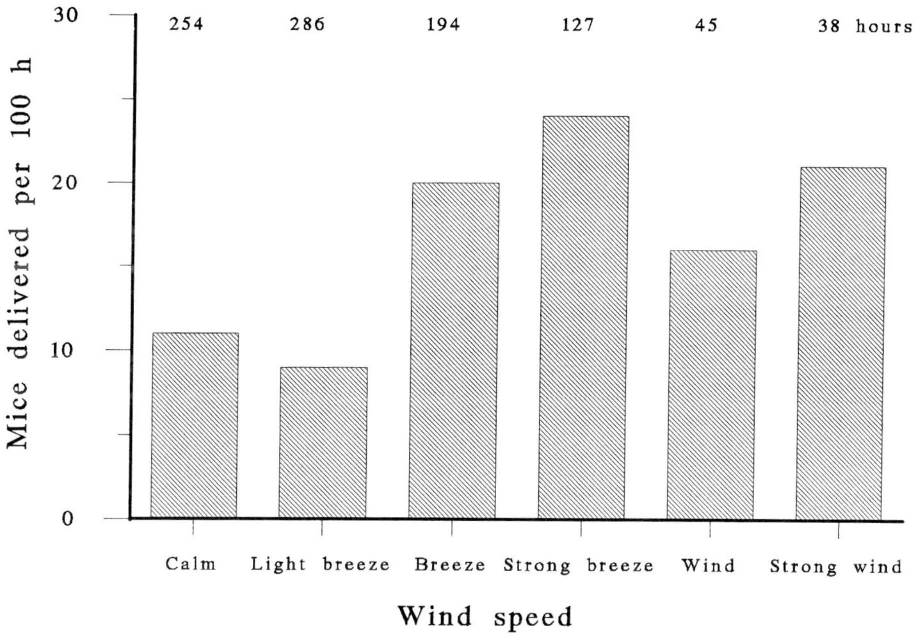

Fig. 4.6 Delivery rate patterns of all small mammals in relation to differing wind speeds. Significant decreases were apparent at lower wind ranks. Hours of observation in each wind rank are given over each bar.

than in calm conditions. That similar significant correlations have recently been found for Hen Harriers in northern Britain (Redpath and Thirgood 1993) suggests that harriers are greatly aided in their foraging return by medium to strong winds. Surprisingly cloud cover also significantly influenced small mammal prey deliveries ($X^2 = 56.3$, 9 df, $P < 0.0001$). Delivery rates under clear skies (0–1) were 33% lower than the expected 15 mammals/100 h, while peak deliveries under partially clouded skies were 76% higher than expected (Fig. 4.7). Cloud-influenced prey delivery patterns are also known for African sparrowhawks hunting birds but in the opposite direction: the first prey deliveries of the day were delayed by over an hour on overcast mornings, relative to completely clear mornings (Simmons 1986a). Possibly mouse predators (and their shadows) are more visible on bright days, alerting prey to the presence of the predator.

Can environmental conditions explain the lack of harrier exploitation of the dawn and dusk *Rhabdomys* activity peaks? For them to do so it is necessary to show that either clear skies or still conditions were typical of dawn and dusk periods in the study area. I analysed wind speeds in the hours 7h00–11h00, for morning watches in August and September 1984, and found, as expected, that the hours 07h00 and 08h00, when few harriers hunted, experienced average wind speed ranks significantly lower (0.3–0.5) than the hours 9h00 and 10h00 (1.3–2.0: Kruskall-Wallis $H = 10.1$, $P < 0.02$). This was not so for the late

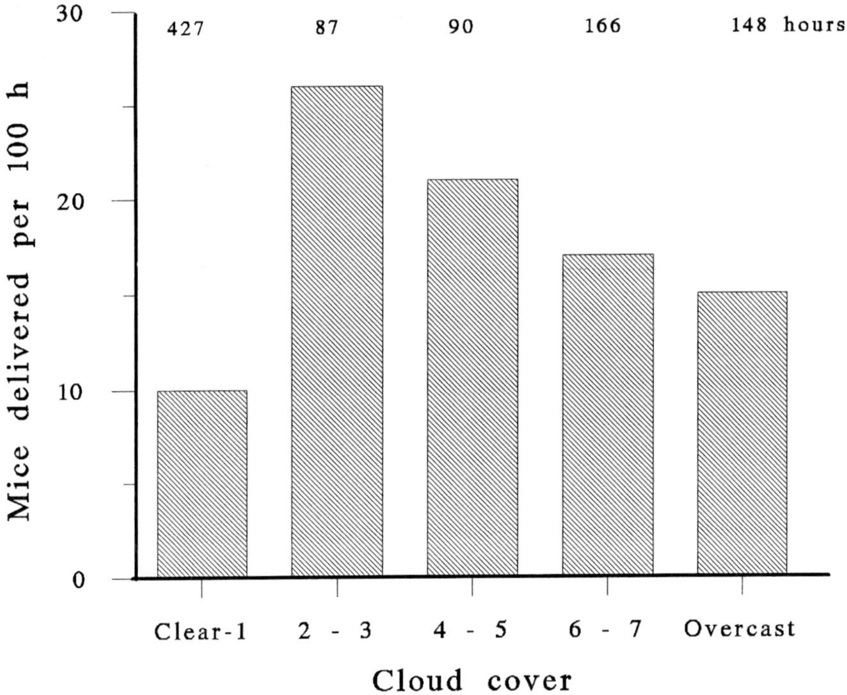

Fig. 4.7 Delivery rates of small mammals in relation to cloud cover. Significant decreases were apparent under clearer skies. Hours of observation in each cloud rank are given over each bar.

afternoon lull however ($H = 0.68$, $P = 0.7$). Thus calm conditions may have accounted for the lack of provisioning in the early mornings.

Two other factors may explain the lack of exploitation of active mice by harriers. First, predators such as raptors, which sample areas only rarely, may face constraints on fine-grained information about small-mammal availability (Rijnsdorp *et al.* 1981). This seems unlikely since other peaks were exploited, indicating intimate knowledge of local *Rhabdomys* activity. On the other hand, the loafing behaviour of harriers during the early morning peak suggested that harriers were more concerned with maintenance (sunning and preening) and territorial behaviours (calling and taking short flights over the nesting territory), than hunting for mice. This was also the main period of copulation activity early in the season (Chapter 6; Simmons 1990), and one re-nesting male spent twice as much time perched in his territory when his mate was egg laying, as did two others whose mates were incubating (Chapter 6). Thus, it is likely that males were mate guarding at these times to prevent extra-pair copulations (Birkhead and Møller 1992). This could explain early season loafing, but it could not explain missed micro-mammal activity later on. However, when weather variables are introduced it seems probable that the harrier's apparent dislike of foraging under clear skies and at low wind speeds (Figs 4.6 and 4.7) inhibited them from foraging because of energetic constraints. It is unlikely that low light levels themselves were responsible for this lack of hunting because harriers are auditory and visual predators (cf. Rice 1982) and some successful prey captures were observed after sunset. The midday foraging lull was associated with significantly increased display activity by harriers when soaring conditions were good (Chapter 3). The evening lull was more difficult to explain but may have arisen because of fatigue, lack of motivation, or even satiation (Rijnsdorp *et al.* 1981). These remain untested possibilities but late afternoon retrieval of cached prey could be ruled out because African Marsh Harriers rarely cached prey.

Comparisons with temperate harriers

In common with foraging studies of relatively monomorphic temperate raptors, African Marsh Harriers were found to differ with respect to age, but not sex, in predatory success. Also, like other harriers (Temeles 1985; Toland 1986), *C. ranivorus* found avian prey more difficult to capture than mammalian prey and predation success varied with habitat (Collopy and Bildstein 1987; Redpath 1992). In these respects, African birds differ little from their temperate congeners. One striking difference, however, was apparent in the foraging yield (1.7 prey/h) which was less than that reported in other temperate zone studies of similar-sized harriers hunting mainly small mammals. For example in The Netherlands, Schipper (1977) recorded between 3.6 and 1.7 prey per hour of foraging, Daan *et al.* (1982) recorded 3.6 prey per hour in a winter study in the same area, while Redpath (1992) recorded 4.7 prey per hour on Scottish moorland. Similarly low return rates are apparent for other subtropical small-mammal specialists (Mendelsohn and Jaksic 1989) in comparison with their temperate counterparts and the reasons are explored in Chapter 9.

I conclude, contrary to Halle (1993), that sustained predation pressure on small mammals by a relatively small predator guild plays a major role in where mice live and subsequently why harriers are least successful at catching mice where they are most abundant: abundance and availability are inversely related. Overlaid on this spatial variation is a temporal variability in which harriers both hunt and capture mice more often when they are active, yet miss out on peak activity times. It is these predation-free windows that mice appear to exploit, and they appear to be more active because harriers are least likely to catch them. The behaviour of the Striped Mouse in this study therefore supported both predictions of the Life–Dinner Principle.

Effect of predators on prey populations: who regulates who?

To end this chapter we will consider whether predators such as harriers are able to substantially decrease or even control population numbers of their main prey.

One school of thought proposes that predators fluctuate in response to fluctuations in prey numbers, and prey are governed by other intrinsic or extrinsic factors, such as parasites or forage quality (Errington 1946). The alternative school of thought championed by Keith *et al.* (1977) is that predators interact strongly with their prey and cause the cyclic changes in populations seen in some small mammals, or the stable equilibria seen in others. Swedish researchers studying a large guild of predators including five species of mammalian predators (e.g. foxes and weasels) and four species of raptor (e.g. buzzards owls, and kestrels), concluded that small-mammal populations in southern Sweden do not cycle as they do in other areas of Scandinavia because they are regulated by their predators (Erlinge *et al.* 1984). The predators did so by removing the entire production of the voles in summer, while in the winter the decrease in vole numbers could be explained entirely by the consumption of the remaining predators. The key element to emerge from this comprehensive study of predator–prey dynamics was the presence of generalist predators. These species (four mammals, two birds) did not move away when voles became scarce, but simply switched to other prey (rabbits). Thus when voles became abundant again they simply switched back, giving the voles no time to increase to high population levels. Predation by the generalists accounted for 80% of the voles taken and they were thus a major influence on regulating the cycles which are common to these voles in other parts of Sweden.

An African example of predators at least changing the behaviour of their prey is given by Davies and Ferguson (1998) for Black Eagles *Aquila verreauxii* and their main prey the Rock Dassie. In areas where Black Eagles have been shot out in South Africa's Great Karoo, dassies forage much further from their rocky refuges, exposing themselves to increased predation. Where Black Eagles occur, dassies rarely forage beyond about 15 m. This has implications not only for dassies, but for the Karoo farmers whose sheep suffer from grazing competition with dassies in the absence of their eagle predators.

Examples of harriers responding to fluctuations in their prey are well documented from Hagen's (1969) work in Norway and Hamerstrom's (1979) and our own work (Simmons *et al.* 1986b) in North America. This is discussed more fully in Chapter 5 with respect to mating patterns and vole populations. But are there any examples of harriers regulating their prey? I will discuss three studies, one in which a generalist predator appeared to make a major impact on vole populations; one in which rabbits were the chief source of food, and last, a controversial but well documented example which caused a storm among conservationists when it was recently published.

North America: voles, coyotes and harriers

The Northern Harrier population on the Tantramar Marsh of New Brunswick has been studied for a total of ten years: initially from 1980 to 1985 and then by Bruce MacWhirter from 1990 to 1993. In all years small mammals were snap-trapped and in the first six years populations underwent marked fluctuations every three years. At population highs, the number of voles trapped averaged over 6 animals per 100 trap nights and individual trap lines revealed up to

12.5 voles/100 trap nights (see Fig. 5.5). At populations lows numbers averaged about 0.5 voles/100 traps (Table 5.3). Following this field work, the next phase of work on harriers began in 1990 and from previous peak vole numbers in 1980 and 1983, vole populations were predicted to be at intermediate levels in 1990, low in 1991, and high again by 1992. However, on snap-trapping in similar areas on the study area, MacWhirter (1994) found that vole cycles were no longer evident and vole populations were 'extremely low'. As usual, harrier populations followed these vole population levels and few birds nested; harrier populations remained unusually stable and non-fluctuating in the three years of study. Land use had not changed and the area of the marsh had not been reduced. The most obvious change was the appearance of one predator not seen during the early years of the study. The coyote *Canis latrans* had been extending its range in the mid-1980s northwards through the eastern United States and up into eastern Canada. By 1990 it was reported on the study area for the first time. As a generalist predator it stayed throughout the year whereas the specialist such as the harriers and Short-eared Owls *Asio flammeus* migrate south. If the principles derived from Erlinge *et al.*'s (1984) work are universally applicable, this generalist predator is the most likely cause of the major change in population dynamics of voles, smoothing out the peaks, by switching to voles as they became common again. Because harrier ecology is so dominated by voles in this population (Chapter 5), the presence of the coyote appeared to affect the population dynamics, mating system, and productivity of Northern Harriers by dampening fluctuations in its main prey. Here, therefore, is a system in which a generalist predator dampens both a prey population and that of a specialist predator. Ideally the vole–coyote–harrier system should be re-evaluated when coyotes are either controlled or naturally disappear from eastern Canada once again.

Australia: rabbits and harriers

To evaluate the impact of harrier predation and that of eight other raptors in south-eastern Australia on rabbit populations, Baker-Gabb (1982) first assessed the consumption of rabbits from the diet of each raptor. He then assessed the daily food requirement of each species and thus how many immature rabbits they would eat in a four-month period, and so the total number of rabbits that would be killed by all the raptors in his study area. Myxomatosis had reduced the rabbit population from sometimes plague proportions of 3500 rabbits/km^2 to an average of 52/km^2. At these lower densities rabbits are highly fecund producing 16.8 young per annum, many of which die from disease, food shortage, or predation. He calculated that 58 500 rabbits were produced on his study area over a four-month period and only 14% of these would be consumed by the large raptor guild in his area. Thus it is unlikely that his raptor population was exerting a significant influence on the population dynamics of the rabbits. Baker-Gabb did suggest, however, that those sedentary raptors that stayed the winter in his Mildura study area would exert a greater influence on rabbit numbers because they would then be taking only adults that might later breed.

Europe: grouse and harriers

The third study was a comprehensive five-year attempt at answering a particularly emotive question that has occupied the minds and trigger fingers of many scientists and gamekeepers in the United Kingdom. Steve Redpath and Simon Thirgood were set the task of determining if Hen Harriers breeding and wintering on grouse moors in northern England and Scotland were reducing grouse numbers through their predation on chicks or adults. For decades, gamekeepers on grouse moors have routinely shot or poisoned predators such as foxes, crows, and harriers, in an attempt to reduce predation on grouse nests and increase the bags available for wealthy hunters in the shooting season. The question is then: do harriers reduce grouse numbers or are they simply taking a doomed surplus that would die from other causes anyway? A review by Newton (1993) found that in up to 86% of 28 studies that experimentally removed predators from an area, nest success of the birds monitored was improved and 62% of 16 studies showed that the removal resulted in an increase in breeding numbers.

Redpath and Thirgood (1997) first ascertained from observations of 26 nests covering an impressive 2614 hours on three different moors, that Meadow Pipits *Anthuspratensis* were by far the most frequent prey brought to nests. They comprised 44% of 2324 delivered prey, while grouse, 96% of which were chicks, comprised 15% of all identified species. They also found that the delivery of grouse was proportional to their abundance in the environment, and at densities of 50–70 grouse chicks/km^2, prey deliveries of chicks increased sharply. By assessing the number of harrier pairs, the rates of delivery per pair, and the proportion of prey that were grouse chicks, they concluded that between 8.6 and 28 grouse chicks/km^2 were removed by the 4–7 males and 4–12 females present on Langholm Moor over four years.

In winter, when predation may have had a greater effect on subsequent breeding populations, 30% of the adult grouse present on the moor were killed by raptors. The same proportion of adult grouse were killed in the breeding months (April–June) and the authors suggested that this was additive—in other words, these birds would not have been lost to other causes had the raptors not taken them. The combined effect of this heavy predation on adults and chicks by the harriers and Peregrine Falcons *Falco peregrinus* was to reduce grouse chick density by about 37%. Because this was seen as adding to predation rather than just removing birds that would have perished anyway, the overall effect was estimated to reduce the post-breeding density by a very large 50%. Thus this comprehensive study of raptor–bird prey was similar to that of Erlinge *et al.*'s, of raptor–small-mammal prey study (1984) in concluding that predators at these densities could regulate their prey. By modelling the dynamics of their prey–predator interactions, Redpath and Thirgood (1997) predicted that with no harriers on Langholm Moor the grouse population would increase to 2.5 times its present level within two years. They cautioned, however, that this level of increase is only likely on a grouse moor exhibiting such high raptor:grouse ratios and is likely to be lower where predators are less numerous.

Given that grouse are less susceptible to predation in habitats with good heather coverage, one wonders what the impact of man is on this whole dynamic. Under natural circumstances one would expect virtually complete heather cover because of fewer sheep and less artificial burning. Indeed in the four decades up to 1988, almost half of the heather-dominated vegetation was lost from Langholm Moor through heavy sheep grazing (Redpath and Thirgood 1997). This may reduce not only the vulnerability of grouse to harrier predation, but also decrease the availability of other prey such as voles and pipits that are crucial to harriers at other times of their breeding. If this paucity of alternative prey in turn reduced harrier densities, then the intense predation levels found on grouse may well be reduced. Thus it could be argued (although Redpath and Thirgood wisely did not do so!), that the high predation levels are artificially maintained by the actions of the landowners and their land-use practices.

What is lacking in many such studies is an experiment in which the raptors are removed and prey levels gauged and compared with a control area nearby. Many 'natural' experiments are already underway because of the illegal activities of landowners against predators on their land, so biologists have a poor excuse when they claim it is unethical or unwise to perform such experiments. Some imaginative planning, once land changes hands and predator removal is curtailed, may be sufficient to provide some non-correlative answers to this fascinating challenge.

The Redpath–Thirgood study, while probably giving grist to gamekeepers and landowners to remove raptors from their grouse moors, also allows a chance to determine if the Life–Dinner Principle can explain the patterns of predation on grouse. The first prediction would be that grouse should be found more often in areas where they are least susceptible to predation. From measures of vegetation height, vegetation density and heather cover, Redpath and Thirgood determined that grouse density was explained best by heather cover (the amount of heather in a 2×2 m quadrat), and neither vegetation height nor density had any effect of grouse density. True to expectation, grouse losses were lower in areas with greater heather cover, and the authors concluded that grouse were moving into such areas with more heather cover. A similar analysis for losses in summer found that areas with high heather cover and tall vegetation were those areas in which grouse losses were lowest.

The second prediction is that the predator should enjoy greater predation success in areas where grouse are least common because it is in these areas that grouse should be most susceptible to predation. Weaker support was gained for this prediction since the trends were not significant. Nevertheless, harriers were most successful in a mixture of heather and grass (50% of 46 strikes) and least successful where grouse where densest in mature heather (35% of 107 strikes). Part of this greater success arose from greater detectability of broods in heather/grass in which 36% of all interactions were observed despite only 6% of all broods occurring there. Thus Redpath and Thirgood's study not only provides an important assessment of the impact of predation and its regulatory role on a species of game bird, it also illustrates that the Life–Dinner Principle is just as applicable to predator–bird interactions as it is to predator–mammal situations.

Who wins the arms race?

We began this chapter by asking: who wins the arm race between predator and prey? The answer seems to be neither! In most situations involving natural systems, the predators have evolved sufficient skills to outwit and consume a certain proportion of their prey. That proportion appears to vary according to the predator's population size relative to that of its prey and most importantly whether it is a generalist or a specialist. On the other hand the prey have evolved sufficient anti-predator adaptations to either live in impenetrable vegetation, be most active when their predators are constrained from hunting (early morning, moonless nights, during rain), or, in certain circumstances, live in groups large enough to fight back. In certain sophisticated systems, prey appear to signal to predators that they have been detected or are of sufficient quality that further pursuit will be unprofitable. In situations where predators do win the arms race, the evolutionary race was begun elsewhere (i.e. another landmass) and the predator has moved into a new habitat or ecosystem by natural (i.e. landbridge) or unnatural (introduced) means and may drive its prey to extinction. Many examples are found on islands on which introduced rats and cats have eliminated ground nesting birds (Qarmen 1994). These are atypical cases, however, and many predator–prey systems enjoy some sort of equilibrium.

5 *Mating patterns: polygyny and deception*

When Charles Darwin wrote in 1871 'Males that conquer their rivals do not win females independent of the choice of the latter' he raised a question that has probably occupied more research in behavioural ecology circles than any other. How do females choose their mates and what cues do they use? Raptors, like all other sexually reproducing species, face particular problems when they choose their mates. Do they choose, for example, to breed with one good quality mate who offers the potential of good genes for their offspring? If so, how can they assess whether a particular bird possesses such qualities? Or do females instead choose a male who will provide them with critical resources, upon which they and their young will depend for a major fraction of the breeding period? If male, do they breed with one good quality female or several? Should they breed simultaneously or in rapid succession to spread the energetic load?

Most raptors exhibit apparently monogamous mating systems. I say *apparently* because numerous studies are now showing that individuals in many species adopt 'mixed reproductive strategies'. That is, monogamous birds (one male, one female) copulate with other, often higher quality, birds outside the pair bond. Indeed so common is this trait that it is becoming increasingly difficult to find any species that can be truly called monogamous (Birkhead and Møller 1992). We will return to this subject in more depth in Chapter 6.

Here we will explore one of the most interesting aspects of harrier ecology: their polygynous mating system. Of all raptorial species, harriers are the only genus that regular exhibits polygyny, where several females breed with one male. The mating pattern is clearly a product of what a female chooses (or is constrained) to do, since, unlike primates, a male cannot force a female either to join him or, in most cases, copulate with him. In the case of harriers, several females appear to choose to breed with a single male. Hence to understand the mating pattern we must understand female choice and what shapes it. However, few studies of raptor mate choice have been undertaken, possibly because the many different aspects of female choice can only be teased apart by experimental manipulation. Mate choice cues range from the exaggerated ornaments of males that add little to the breeding effort per se (widow birds and whydahs in Africa and swallows in Europe (Andersson 1982a; Møller 1988; Barnard 1990). In others, males contribute directly to breeding by feeding their mates. Female terns, for example, choose males with high provisioning rates (Nisbet 1973; Wiggins and Morris 1986) because males with high feeding rates not only give rise to larger clutches, but 'courtship-feeding' is a good guide to feeding during the critical chick-rearing stage. Some insights into what females choose are offered by the order in which females choose their mates, based on the assumption that a female will always choose the best remaining option open to her (Lenington 1980). One can then assess various mate-quality aspects based on the order in which they were chosen. We will look at choice in Northern Harriers, compare it with experimental work undertaken on kestrel mating preferences, and ask why, among those polygynous birds, do polygynous female harriers do as badly as any studied to date?

Reasons for rarity of polygyny

Interest in polygyny among raptors arises partly because of its rarity. It has been recorded in only 4.4% species of 292 diurnal raptors (Newton 1979; Lendrum 1979; Steyn 1982; Baker-Gabb 1983), of which five are harriers. Four of these species have been well studied: the European Marsh, Hen, Northern, and Montagu's (Chapter 3, Table 3.2). In the first three species polygyny occurs regularly, while it is rare in the latter. Polygyny is also suspected in the Black Harrier based on close nesting and provisioning by a single male in one area (Steyn 1982), while it is confirmed in Australasian Swamp Harriers. Baker-Gabb (1983) found 11% of 38 swamp harrier nests were provisioned by a shared male. Curiously, the same proportion (4.3% of 162 species) of diurnal owls are known to be polygynous (Korpimaki 1988). This contrasts with passerines, in which polygyny has been recorded in 6–20% of species (Lack 1968; Møller 1986).

For most of the raptors, however, the incidence of polygyny within species is very low; most species are only included because they have been thoroughly studied and a few cases have come to light (e.g. 1% of over 1000 European Sparrowhawks nests studied by Newton 1986, 6% of 560 Tengmalm's Owl nests studied by Korpimaki 1992). Only among the harriers can polygyny be considered regular. For the behavioural ecologist the main interest arises because females who choose polygyny over monogamy reproduce especially poorly. They do so because they have to share limited resources with other females in the harem. We must ask why these species make apparently maladaptive choices, and most important in an ultimate sense, how can polygyny persist in present day populations if females practising it leave fewer copies of themselves than birds choosing monogamy? For harriers, polygyny is simultaneous i.e. females in the harem concurrently have dependent young. This raises special problems for females within the harem, since they share critical resources, especially food. This may not be problematic for species with precocial young (e.g. pheasants and grouse) who self-feed, but it may be especially difficult for females who not only have to feed their altricial young and keep them warm (and therefore cannot forage for them) but also need to defend their young against aerial and terrestrial predators when foraging becomes necessary. We would expect, therefore, to find that these females suffer severe reproductive losses in these relationships, and their decisions to join already-mated males become puzzling. Throughout I shall distinguish between those females who make the decision to join an already-paired male (= polygynous females), those that choose an unpaired male (monogamous females), and those that also choose unpaired males but whose mate later becomes polygynous (α females). The latter females are in a special category because they did not choose to be polygynous, but become part of a harem by default.

Those females that do join paired males (β = second, γ = third, and δ = fourth) among Northern Harriers breeding in New Brunswick, reared only 1.22 ($\pm$ 1.7, n = 18) young; less than half the 2.8 ($\pm$ 1.7, n = 11) young that α (first-settling) or monogamous females (2.74 $\pm$ 1.8, n = 54) reared. This was not due to them laying smaller clutches (Fig. 5.1), since clutch size varied non-significantly from 4.57 ($\pm$ 1.3, n = 44) in monogamous females to 3.77 ($\pm$ 1.4, n = 13) for the polygynous females.

Because polygynous females begin breeding on average eight days later than the first-settling female in their harems, we must take care that the lower fledging success of these polygynous females is not due to general seasonal

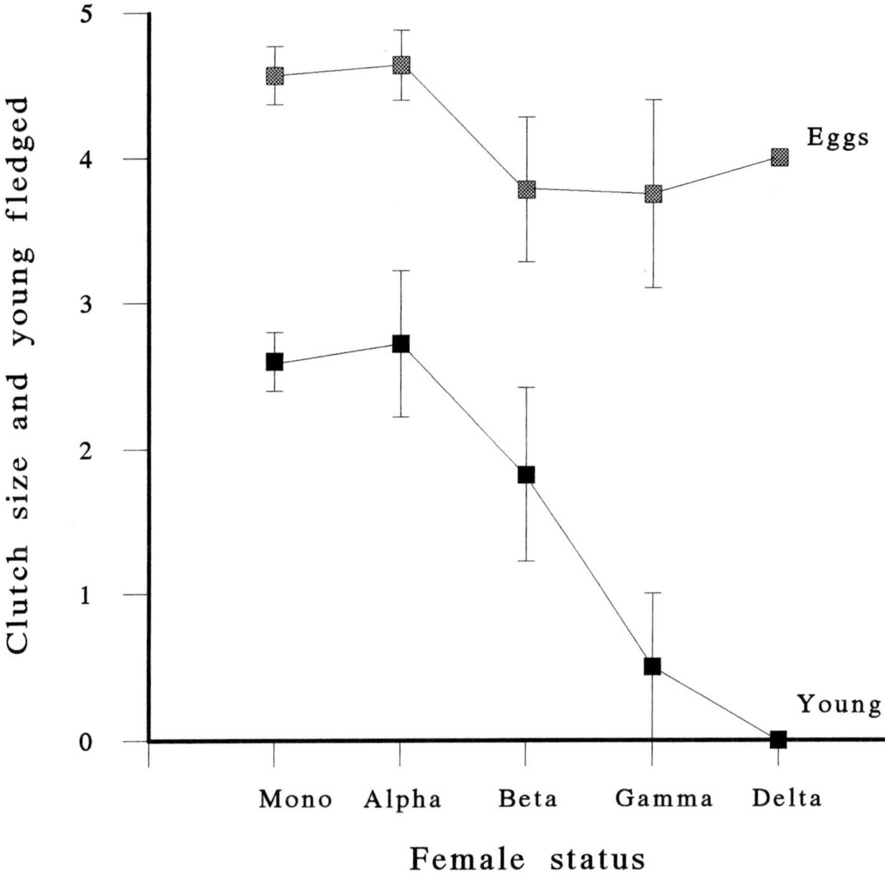

Fig. 5.1 The influence of harem position on mean clutch size (top) and fledging success (bottom) for female Northern Harriers within harems. Note the rapid decline in success of later-settling harem females independent of clutch size. Sample sizes: eggs n = 44, 9, 4, 2; young n = 52, 11, 11, 4, 3 nests respectively. Error bars ± s.e.

declines in success, discussed in Chapter 9. This, too, cannot explain the differences. The fledging success of β, γ, and δ females over four years ($\bar{x}$ = 1.2 young, above) was about half that reared by monogamous females laying on the same day ($\bar{x}$ = 2.67 ± 0.82), derived from equations relating fledging success with season: Simmons *et al.* 1986a). The ratio of polygynous to monogamous young (0.44) is very low, and represents a 56% reduction in what they might expect to produce had they been monogamous. Indeed it is one of the lowest recorded for any species of polygynous bird (Table 5.1). In real terms this means that for every offspring reared by a polygynous female, monogamous females, enjoying the full services of a male, are rearing more than twice (2.3) as many young.

In some bird species, polygynous females do not suffer a reduction in reproductive success (Table 5.1). However, this is only true of passerines, and it may

Table 5.1 Summary of relative fledging success of polygynous raptors and passerines in relation to monogamous pairs settling on the same date. Polygynous females are those choosing mated males, and they are compared to simultaneously settling monogamous birds to circumvent problems associated with seasonal declines in breeding success of later-settling polygynous females. Adapted from Slagsvold and Lifjeld (1994).

Species	Ratio of polygynous to monogamous fledglings	Sample size (poly:mono)	Source
Raptors			
Circus hudsonius	0.44	19:52	Simmons *et al.* 1986a
Circus cyaneus	0.50[a]	52:51	Picozzi 1984a
Circus aeruginosus	0.77[b]	30:156	Altenburg *et al.* 1982
Falco tinnunculus	0.52	9:9	Korpimaki 1988
Aegolius funereus	0.70	23:213	Carlsson *et al.* 1987
	0.55	37:74	Korpimaki 1991
Passerines			
Tachycineta bicolor	0.92	5:21	Dunn and Hannon 1991
Cistothorus palustris	1.10	9:48	Verner 1965
	1.17	9:9	Leonard 1990
Acrocephalus arundinaceus	0.37	10:22	Catchpole *et al.* 1985
	0.83	41:56	Urano 1985
	0.58	24:108	Dyrcz 1986
Phylloscopus sibilatrix	1.56	18:50	Temrin 1989
Ficedula hypoleuca	0.83	45:88	Stenmark *et al.* 1988
	0.80	9:9	Lundberg and Alatalo 1992
Protonotaria citrea	1.05	9:36	Petit 1991

Mean values: **Raptors:** ratio = 0.59; **Passerines:** ratio = 0.92

[a] Laying dates unknown so this estimate not used in mean values.
[b] Estimated only; but polygynous females settled on average only two and seven days after monogamous females in two study areas in The Netherlands, suggesting seasonal effects would be minimal.

occur because males may not be critical for breeding success and females can feed the nestlings unaided. Note also that in raptors in general, polygynous females reproduce more poorly than polygynous female passerines (mean ratio values of polygynously to monogamously produced fledglings, 0.59 vs. 0.92: Table 5.1). Again this may arise from the well-established fact that female raptors are highly reliant on their mates to feed them and their young (Newton 1979) and may not be able to compensate from sharing a mate with other females. To understand this we need to look for trends that distinguish polygynous species from monogamous ones.

Trends among polygynous birds

In 1966, Verner and Willson reviewed known polygynous species in North America and showed that most lived in open, two-dimensional (2-D) habitats such as marshes. This was confirmed 20 years later by Møller (1986) for polygynous European passerines. No other traits such as nest dispersion patterns, cavity nesting, and wintering areas were found to characterize polygynous species. In this respect, harriers fit perfectly into the 2-D habitat trend, since all except the Australian Spotted Harrier nest on the ground in open marshes or grassland, and with one exception (Reunion Harrier), all forage over open ground. Both Verner and Willson and Møller related open habitat, especially highly productive marshlands, to clumped food resources that could be monopolized by certain males who can therefore attract additional females to their high quality territories. Monopolization of the clumped resource is important because one would merely find high density monogamy if males were unable to monopolize resources within high quality patches. This is an attractive idea because the whole concept of polygyny suggests that females are choosing males that differ in some way, either in their attractiveness, ability to provide critical resources for the female, or control of a quality territory that directly provides a female with critical resources. The idea that territories differ by some threshold quality, just enough for a female to share resources with another female in preference to not sharing resources on a poorer territory, is the essence of the polygyny threshold model of Orians (1969). The 'threshold' here refers to that difference between territory qualities (in this case good and bad) that is just sufficient to persuade a female she will do best by joining an already-mated male on a good quality territory.

Importance of open habitats

What may have been overlooked for polygynous species breeding in open habitats is their ability to monopolize any clumping that does occur because of the high visibility that is characteristic of open habitats. Thus while a male harrier may quickly detect intruding males in his territory in open marshland, raptors breeding in woodland, with reduced visibility, are likely to spend increased time searching for intruders. If closed-habitat species spend more time defending their mate and territory then one would expect a lower incidence of polygynous matings. Indeed this *economic* monopolization of resources was stressed by Stephen Emlen and Lew Oring in their classic 1977 paper as a factor promoting polygyny. They did not directly relate open habitats to easily monopolized resources but the correlation between monopolization, open habitat, and polygyny is unlikely to be coincidental. Some support for the idea comes from Nick Davies' (1992) study of Dunnocks (*Prunella modularis*) in Cambridge Botanical Gardens. He found that males were unable to control all resources in the females' territories (which were established first). As a result they suffered from extra-pair copulations from neighbouring males because their mates often sneaked into cover, despite mate-guarding behaviour.

The main point is that correlations of polygyny with 2-D habitats are consistent with harrier mating systems. It is for this reason that I believe that no other group of raptors is as consistently polygynous as the harriers. If open habitat is an essential prerequisite for polygyny, then one might expect Strigidae owls (Short-eared Owl *Asio flammeus* in Europe and North America, Marsh Owl *Asio capensis* in Africa) that use similar habitat, and prey, alongside the harriers also to be polygynous. No such mating patterns have been reported for these grassland owls (Lockie 1955; Korpimaki 1988), possibly because it is harder to detect polygyny by observation alone at night. This is borne out by the fact that most of the owls in which polygyny has been proven use nest boxes (Korpimaki 1988), making it easy to capture the attendant male for identification. This would repay further study. However, not all species of harriers are polygynous either. For some this arises from a lack of appropriate studies. For others, however, including the well studied Montagu's Harriers, this is not so. So are there other trends that may help us explain this unusual mating pattern in harriers?

Polyterritoriality

A trend to emerge in the last decade and argued persuasively by Scandinavian behavioural ecologists working on Pied Flycatchers (*Ficedula hypoleuca*) is that males practise poly-territoriality (i.e. hold multiple territories). That is, a female appears to be fooled into mating with an already-mated male because she believes he is monogamous. The male is able to 'fool' a female about his mated status because he sets up another nesting territory (centred on a cavity nest hole), some distance away from his first or primary female (Alatalo *et al.* 1981; Lundberg and Alatalo 1992). The 'some distance' is almost always a full territory away making it difficult for a female to assess the mated status of her mate within her own territory. Males that sing too close to their original nest hole tend to fail to attract another mate, adding credibility to this idea. This strategy not only occurs in other parts of the flycatcher's range but also on a regular basis in six other species, including the Kestrel (*Falco tinnunculus*) and the Tengmalm's Owl (*Aegolius funereus*), and irregularly in 13 further passerines in Europe (Møller 1986). Intriguingly, this cannot explain polygyny in harriers. Among the best studied, Northern Harrier females not only nest within a male's territory, but nest in two-dimensional habitats (whereas most polyterritorial species breed in three-dimensional habitats: Møller 1986), and in close enough proximity ($\bar{x}$ = 458 m for harem females) to be aware of aerial food transfers from their mates to extra females. Such females also nest closer to the α female's nest than they do to any other nest in the harem, possibly to draw food away from this male-favoured nest (Simmons *et al.* 1986a). This is contrary to reports by Korpimaki (1988) who claimed that all three harriers for which polygyny had been reported (Hen/Northern, European Marsh, and Montagu's) exhibited both polyterritorial and monoterritorial behaviour. There is no evidence for any of these harriers that females nest outside their males' normal territory or that males attempt to fool late-settling females of their mated status by defending a separate territory. Especially important for the polyterritorial idea, female harriers settling second, third and fourth in the harem (β, γ, and δ females) are often attacked by the α female in an apparent attempt to drive them off. These are not merely displays by the α female but involve mobbing the newly arrived female and in some cases physically hitting them. Many leg-lowering displays occur on female-territory borders that become fixed later in the season. Alpha females cannot spend large amounts of time chasing future-settling females, since they are usually incubating. Males also play no role in this activity, and simply look

European Kestrel hovering.

on as these aggressive interactions are underway. The reason for this aggression is that subsequent mates demand food from the male, and thereby potentially reduce his supply to the α female, or his energetic ability to provide food to her later in the critical chick-rearing period.

At least four other reasons for female aggression have been proposed (Slagsvold and Lifjeld 1994); namely: (i) to prevent intruders from stealing nest material or destroying eggs; (ii) to prevent the α female's own displacement; (iii) to prevent egg dumping; and (iv) to space out nests to reduce density-dependent parasitism. No evidence for these ideas has been found in harriers, although in exceptional cases two Hen Harrier females in Scotland will share a nest (Picozzi 1983) which reduces nest success. Nevertheless, reduction in male investment is the most likely cause of aggression and there can be little doubt that later-settling female are aware of additional females, and probably watch their mates feed these females as they wait for males to 'courtship' feed them during the egg-laying stage. This is also likely for highly polygynous Hen Harriers in Orkney where Balfour reported 'It would appear that every Hen Harrier on the breeding ground knows intimately the position of all the harrier nests over quite a large area. Sometimes a female dives at another on her nest and the two have what appears to be a friendly skirmish together' (1962/63, p. 150). There is thus little chance that a female choosing an already-mated male is either unaware of his status or the presence of another female. Harrier polygyny cannot therefore be explained by deception through polyterritorial behaviour. What of the possibility, favoured by European ecologists, that an excess of females drives polygyny in Hen Harriers?

Skewed adult sex ratios

Within a population with equal numbers of adult males and females, given adequate food resources and nesting places, most birds should be able to breed.

However, some populations of harriers appear to be skewed towards an excess of females. One must take care that this really is a skew, and not simply due to the ability of some males to monopolize all the territories and thereby exclude other adults from nesting grounds where adult sex ratios are often determined by researchers. However, given that some populations may be skewed, some females may be forced to accept mating with already-mated males or face the prospect of not breeding at all. This is clearly one possible explanation of polygyny.

The idea of polygyny in harriers as a result of an excess of females arose from studies of the highly skewed population in Orkney, Scotland. There Balfour and Cadbury (1979) found 1.8 adult females for every male and Nick Picozzi (1984a) noted that females were displaying in order to attract males to breed with them. There is also an indication from two studies that polygyny increased as population size increased—possibly forcing some females into polygynous situations. Eddie Balfour documented an increase in polygyny in Orkney from its first record in 1931 (Balfour and Cadbury 1975) as Hen Harriers became more common in the 1970s. Polygyny was also reported more often among European Marsh Harriers since it was first documented by photographer Eric Hosking (1943) as they colonized England over the same period (Underhill-Day 1984). Both studies hint that some females may have accepted polygyny rather than not breed at all, as populations became denser and denser. It should be noted that James Cadbury did not completely support the idea of a skewed sex ratio promoting polygyny. He wrote in 1975: 'Though more females are reared it is possible that polygyny is largely a product of social behaviour by which a proportion of males are excluded from the breeding population.'

However, the idea was given further prominence by Serge Daan's group in The Netherlands working on European Marsh Harriers (Altenburg *et al.* 1982). This excellent study showed clearly that females were attracted to males that had particularly high provisioning rates, and that in general polygynous males reared more young than monogamous males. This arose not because they hunted for longer, but because polygynous males had higher foraging yields—they caught more prey for every hour hunted than monogamous birds (Chapter 9). Despite these differences in male parenting quality, and the possibility that females had something to choose between, the authors concluded that skewed sex ratios forced some females to accept polygyny. They justified this unexpected conclusion by pointing out that females outnumber males in several populations of Hen and European Marsh Harriers, and that females mature before males, further unbalancing the adult sex ratio. Females would thus be constrained by the lack of available unmated males. They proposed this as the ultimate reason for polygyny in harriers.

These conclusions are unusual because sex ratio imbalance was one of the earliest proposed reasons for polygyny in birds (Armstrong 1947; Crook 1964), yet skewed adult sex ratios were found to be exceptional in polygynous birds, and 1:1 adult ratios were more common (Armstrong 1955; Willson and Pianka 1963). Equal sex ratios are also apparent among Northern Harrier populations in America (Burke 1979) in which polygyny is common.

Table 5.2 Sex ratio, polygyny, and population parameters of the New Brunswick Northern Harrier population from 1980–83. Letters in brackets indicate variables important in testing the sexy son hypothesis, used later.

	1980	1981	1982	1983	Total or mean
Total population size					
all nests	30	22	12	37	**101**
males	21	18	10	26	**75**
females	28	19	12	35	**94**
Estimated adult sex ratio [♂:♀]	0.67	0.67	?	?	**0.67**
Breeding sex ratio[s]	0.75	0.94	0.83	0.74	**0.79**
Proportion of polygynous males [f]	0.19	0.06	0.20	0.15	**0.15**
Average harem size [*k*]	2.75	2.00	2.00	3.25	**2.72**
Average productivity of polygynous females (*n*)[a]	1.00 (7)	0.00 (1)	1.00 (2)	1.63 (8)	**1.22 (18)**
Average productivity of monogamous females (*n*)	2.94 (18)	1.71 (14)	2.50 (8)	2.93 (14)	**2.74 (54)**
Ratio [*x*] of polygynous to monogamous offspring	0.34	—	0.40	0.56	**0.44**

[a] β, γ and δ females.

Thus, in essence it appears too simplistic to suggest that polygyny is the result of a female not being able to find a suitable unmated male, and being forced into breeding with a mated male. While an excess of females may enhance the probability of a female becoming polygynous, it cannot be seen as the ultimate reason. In the event that sex ratios within a population were skewed towards females, it is likely that strong selection would occur for pairs to produce sons to benefit from the excess of females. Indeed harrier population in Orkney and western France do just this: (Picozzi 1984b; Leroux and Bretagnolle 1996; Chapter 7). Sex ratios should thus quickly return to parity, and polygynous populations should decay back to monogamous ones.

Nevertheless, it was the case in the New Brunswick population of harriers, with the highest recorded level of polygyny in any North American population, that adult sex ratios were skewed to an excess of females; for every adult male there were an estimated 1.5 females (Table 5.2). This is insufficient evidence to conclude, however, that imbalances in sex ratio lead to polygyny. Instead we can test these ideas by looking at the settling order of females and asking whether all available bachelor males are paired before the first polygynous ones arise. One would expect that bachelor males would all be chosen first, and once 'full-up' that polygynous matings would then occur. I tested this idea using the settling patterns of Northern Harriers in eastern Canada and found, to the contrary, that polygynous matings occurred throughout the breeding season (Fig. 5.2) and were not clumped towards the end of the season as predicted. Indeed in one year one male had attracted two females before other bachelor males had attracted any, and in 1983 one male had attracted five females before any other male had become polygynous.

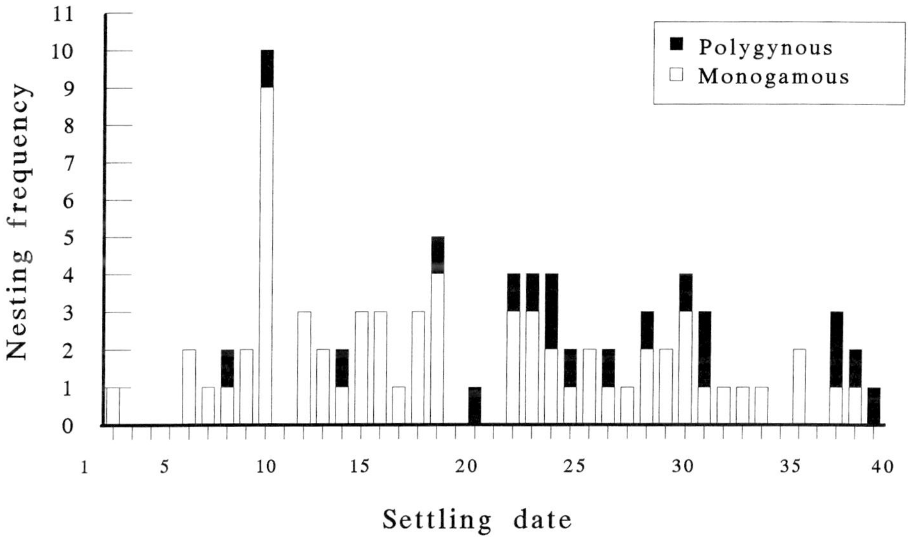

Fig. 5.2 Settling dates of monogamous (unfilled) and polygynous female Northern Harriers (filled bars) in 1980, 1981, and 1983. 1 = 1 May. Note the lack of clumping towards the end of the season.

There is the further possibility that females may have restricted searching time in which to find appropriate mates, and rather than not breed at all they may make a speedy choice of mate that proves poor later. In other words, females in a natural setting may not have all the information on territory quality, male quality, and male mated status at hand to make a choice of the best remaining situation. They may then simply pair with the first male who will accept them rather than not breed at all. In harriers and Great Reed Warblers (*Acrocephalus arundinaceus*) this scenario is unlikely. For the polygynous Northern Harriers studied intensively in Canada, female mate assessment did occur. By closely observing pairs during the early formation period, I found 2 of 19 females were provisioned so poorly by prospective (and intensively nest-building) mates that they deserted them within the day. Since these sites were later used, female desertion was unlikely to have arisen from the inferiority of the site. Poor provisioning was also noted in desertions among harriers breeding in central Wisconsin (Bildstein 1979). Female harriers do not, therefore, simply accept any male that is unmated and territory holding. Furthermore, since the study area in which 30+ males built nests was only about 60 km^2, this is an area that a prospecting female harrier, able to fly at least 200 km in a day could cover many times in a day.

Confirmation of active (rather than passive or random) choice in another polygynous species comes from a clever experiment undertaken by Staffan Bensch and Dennis Hasselquist (1992). They followed radio-tagged female Great Reed Warblers transplanted to a new area in which they had no previous experience and where they had to make new mate choices. They showed that females sampled about six males (range 3–11 males) before making their choice. Most interestingly the females did not simply choose the last male to be sampled, but they often went back to previously sampled males, and, in 7 of 13 cases, they finally chose an already mated male, fully aware of his mated status from their extended sampling. They also did all this sampling in a very short space of time (9–73 hours). These warblers, therefore, showed extensive sampling, active choice, and preferential choice of mated over unmated males in over 50% of choices. Since some harriers on New Brunswick's Tantramar Marsh had attracted and begun to breed with five females before others had attracted any females, these two species show that females are actively choosing some aspect of the territory or mate; they are neither constrained in their choices nor choosing mated males only when no bachelor males remain.

Geographic variation in polygyny

Not all species of harrier are polygynous, and not all populations of the polygynous species are either. Of the 16 species of harrier worldwide, the European and North American species are the best studied, while the Asian and South American birds are least known. To ascertain polygyny, field workers must at least observe copulations in colour-marked birds or distinguish individual males by unique characters. Ideally DNA sampling of adults and their putative offspring is advisable, but this remains a gap in our understanding of harrier polygyny. We must therefore be careful not to conclude prematurely that polygyny does not occur in less well-known populations because of the lack of detailed research.

This is a necessary proviso because differences are obvious between northern and southern hemisphere harriers in their degree of polygyny: European and North American birds exhibit a high incidence of polygyny within and between species. Of four species found in Europe (Pallid, Montagu's, Hen, and European Marsh) three have been recorded in polygynous mating systems (Table 3.2). Amalgamating all major studies of polygyny in Hen Harriers gives a mean value of 59% polygyny from 1037 nests. European Marsh Harriers are second with 13% of 1338 nests polygynous, Montagu's Harriers exhibit occasional polygyny (only 2% of 1139), while Pallid Harriers are apparently not polygynous (Cramp and Simmons 1980, Davygora and Belik 1994), but require further study (Table 3.2).

The degree of polygyny also varies from population to population within a species with the highest number of shared mates recorded in the Hen Harrier on islands. In Orkney, Eddie Balfour and Nick Picozzi recorded up to six females in a harem, while van der Kraan and van Strien (1969) reported a single male on the Dutch island of Ameland with seven females nesting. The male was naturally marked by a damaged flight feather and they observed him provision seven females, one of which disappeared shortly after laying eggs. However, the other six females managed to rear a total of 18 young! This exceptional example is given because it hints that polygyny may be associated with islands.

Importance of voles and latitude to polygyny

In the seven-female example the authors reported an abundant supply of food and further research there has shown that polygyny is the predominant mating system with 92% of 38 nests belonging to polygynous males (Schipper 1978). In Schipper's extensive work on this and adjacent North Sea islands, he noted that large and small mammals were numerous on Ameland (Schipper 1973), and polygyny was common. However, on Terschelling, the very next island in this Dutch archipelago of dune-dominated islands, where voles do not occur, passerines form 66% of the diet and polygyny has never been recorded (Schipper 1978). On mainland sites polygyny was also much rarer and the proportion of voles lower. This is not an isolated example for a parallel situation occurs across the North Sea on the Scottish islands.

In Orkney, where Eddie Balfour's 20 years' of studies were continued and augmented by Nick Picozzi, polygyny reaches extremes. In that island population, nests of polygynous males comprise 88% of all (345) nests investigated. Some of the males also breed with six females simultaneously. Large Orkney voles (*Microtus arvalis orcadensis*) and young rabbits form 48% of all prey and almost 60% of the biomass. However, on the vegetatively similar Shetland Islands just north of Orkney, suitable harrier nest sites occur but no voles are found (Picozzi 1980). Hen Harriers do not occur there. Nor are they common on the island of Hoy, which also lacks voles. Where harriers do occur on mainland Scotland, much reduced levels of polygyny are found (23% of 43 nests) in conjunction with a switch from mammals (14% of the diet) to birds that form 86% of the diet (Picozzi 1980). No year to year fluctuations in voles were measured in these populations to determine if polygyny varied in parallel, as their overall population numbers and breeding success are known to in Norway and Finland (Hagen 1969; Korpimaki 1985). Where small-mammal populations have been studied, striking correlations were discovered between both harrier population numbers and voles and their mating system and voles.

In her ground-breaking study in central Wisconsin, Frances Hamerstrom snap-trapped voles on the Beuna Vista marsh for 20 years. She found a very close correlations between the trap index, which varied from 0 to 8.8 voles/100 trap nights over a 16-year period, and the number of birds nesting on the marsh ($r = 0.91$, $P < 0.01$; Fig. 5.3). For four years (1965 to 1968) DDT was sprayed on the marsh and while this had no effect on the vole population dynamics, it reduced the breeding population of harriers to less than six nests, from an average of 15 (Fig. 5.3). Correlations reported here disregarded this DDT era. She also found that male polygyny varied in concert with fluctuations of the harrier's chief prey ($r = 0.63$, $n = 16$, $P < 0.0005$), as did the number of females within harems (Hamerstrom *et al.* 1985). Again DDT reduced polygyny to zero, despite a vole peak in the middle of this period (Fig. 5.3). Note also from Hamerstrom's work that the adult sex ratio of females to males on the Buena Vista Marsh did not fluctuate in synchrony with polygyny in females ($r = 0.18$, $n = 16$; Fig. 5.4). The relationship between voles and polygyny was not confined just to Wisconsin

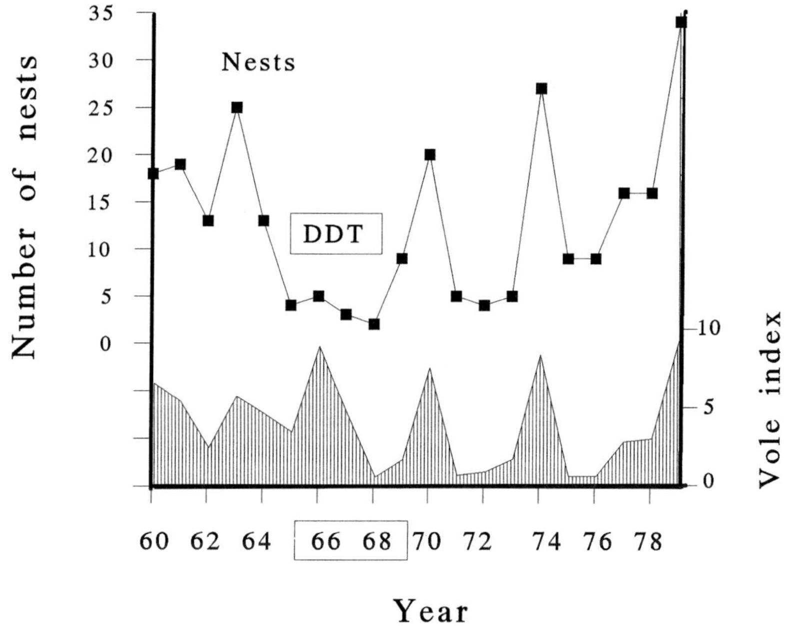

Fig. 5.3 Synchronous fluctuations over 20 years between breeding Northern Harriers and their microtine prey, recorded from Wisconsin, by Hamerstrom and colleagues (1960–79). DDT years interrupted breeding, without influencing the vole population fluctuations. Correlations excluding these years were highly significant. Vole index = voles caught per 100 trap days.

Table 5.3 Vole population indices in May and August in relation to polygyny, and fledging statistics of Northern Harriers in New Brunswick, 1980–84.

	Vole indices[a]		Polygynous			Young fledged by	
Year	May	August	Nests[b]	♂♂	♀♀	All ♀♀[c]	Successful ♀♀
1980	6.6	5.6	30	4	11	2.59 (27)[d]	3.33 (21)
1981	1.9	1.1	22	1	2	1.75 (16)	2.80 (10)
1982	0.5	3.9	12	2	4	2.00 (12)	2.67 (9)
1983	5.5	6.6	37	4	13	2.54 (26)	3.88 (17)
1984	1.8	4.2	18	2	4	1.38 (13)	2.57 (7)
Total	—	—	119	13	34	2.19 (94)	3.22 (64)

[a] Number of voles caught per 100 trap days.
[b] All nests and repeat nests in which one or more eggs were laid.
[c] Excludes nests adversely affected by research activity.
[d] Sample size in parentheses.

birds. In five years' work we found precisely the same patterns in New Brunswick (Simmons *et al.* 1986b). Breeding birds not only tracked vole populations with great precision (nests varied from 12 to 37 nests, $r = 0.90$, $n = 5$; Table 5.3) but so did polygyny in both males ($r = 0.89$) and females ($r = 0.90$) (Fig. 5.5). Each of these relationships was derived from May vole indices—when harriers are mating

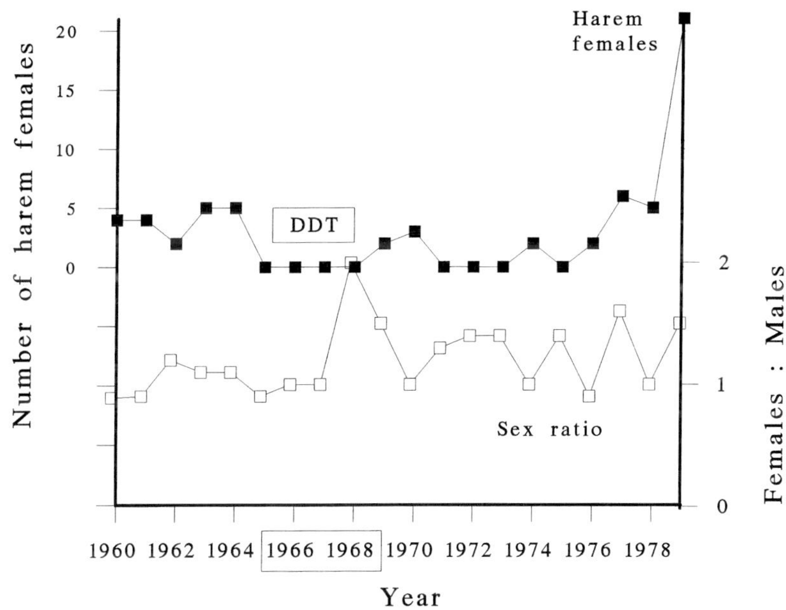

Fig. 5.4 Numbers of polygynous females in relation to the adult sex ratio on the Buena Vista Marsh, Wisconsin by Hamerstrom and colleagues (drawn from data in Burke 1979). No significant relationship was apparent.

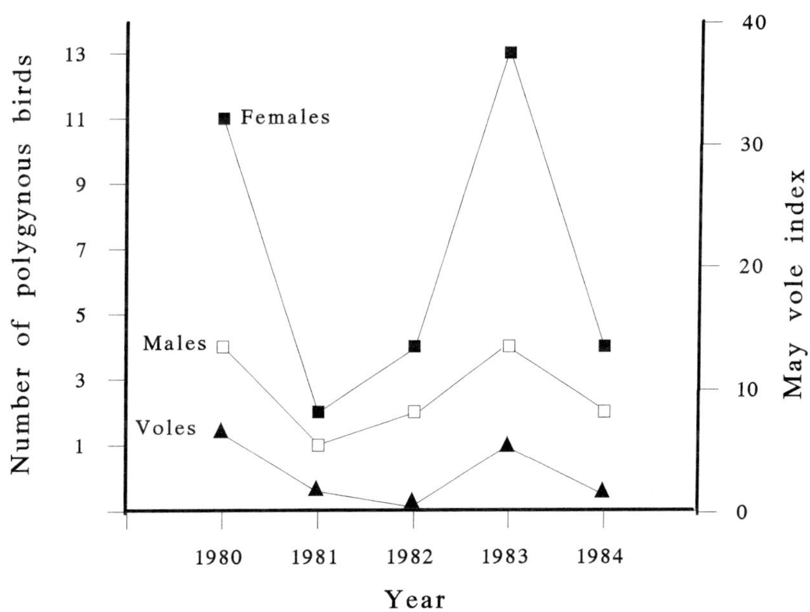

Fig. 5.5 Synchronous fluctuations between the number of polygynous male Northern Harriers, the number of harem females, and vole numbers on the Tantramar Marsh, New Brunswick. Correlations were highly significant.

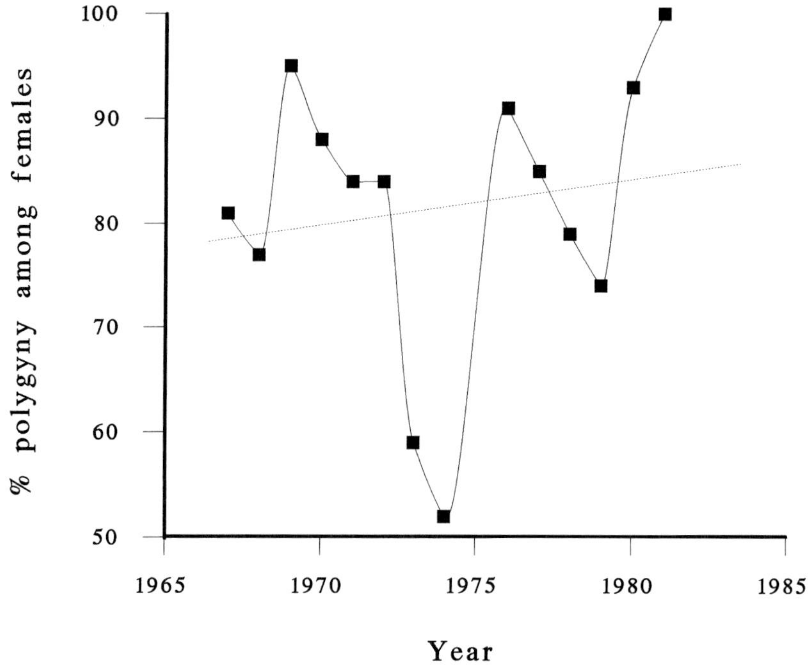

Fig. 5.6 Fluctuations in the proportion of female Hen Harriers in polygynous associations in Orkney, 1967 to 1981. Data collated from Balfour and Cadbury (1979) and Picozzi (1984b). Data from 1975 are missing (not collected). Fluctuations suggest a similar food–polygyny relationship to that found in North America but vole populations were not measured.

and becoming polygynous. However, in 1982, a late, low vole year, polygyny arose in June, and by using the June vole index for this year (when vole populations were rising sharply) the overall correlation rose from $r = 0.89$ to $r = 0.93$ for polygynous males and from $r = 0.90$ to $r = 0.93$ for harem females.

Despite the considerable long-term work undertaken in Orkney, prey populations do not appear to have been monitored to help us understand if polygyny likewise fluctuates with the Orkney vole, which forms a substantial part of the harriers' diet (Balfour and Macdonald 1970). However, if we plot the proportion of females involved in harems over the 15 years it was recorded (by Balfour and Cadbury 1979, then by Picozzi 1984b) we find a similar fluctuation to that found in North America (Fig. 5.6). From one trough to the next is 5–6 years, rather longer than the 3–4 years typical of vole populations elsewhere. Further research would clarify if this fluctuation in polygyny is food-related or not.

Voles and other measures of reproduction

In New Brunswick we also measured clutch sizes in relation to vole indices in May and found that clutches varied from means of 3.6 to 4.9 across years and were also correlated closely with vole populations ($r = 0.94$, $P < 0.01$; Fig. 5.7).

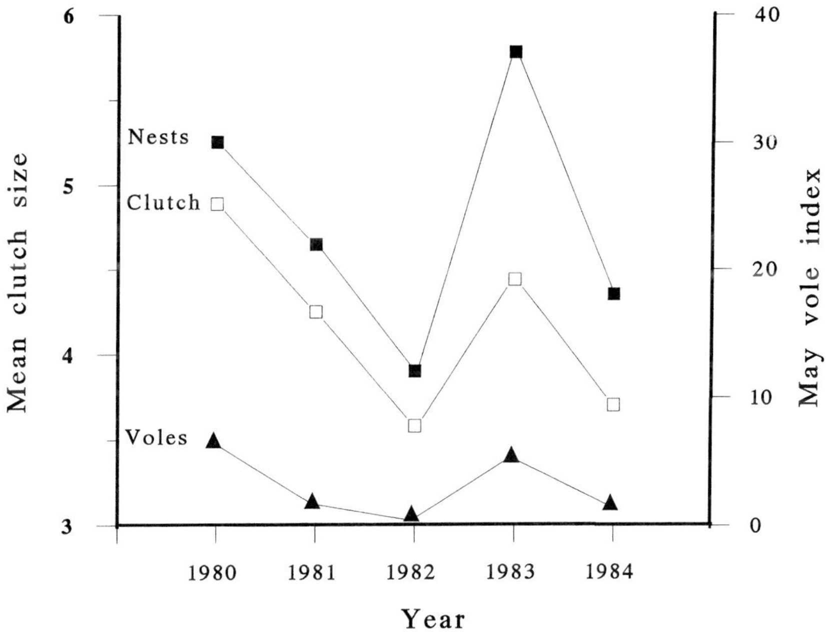

Fig. 5.7 Northern Harrier nesting numbers and mean clutch size in relation to vole population cycles on the Tantramar Marsh, New Brunswick. Both measures were significantly correlated with vole indices from 1980 to 1984.

But were voles per se important or do harriers respond to general fluctuations in the total mouse population? To show that voles, in this case Meadow Voles (*M. pennsylvanicus*) per se were important, I repeated some of the correlations using all small mammals caught in our snap-trapping program. While correlations

Table 5.4 Comparisons of correlations between vole population indices and all small-mammal indices relative to Northern Harrier breeding statistics on the Tantramar Marsh, New Brunswick, 1980–84. An asterisk denotes significance at $P < 0.05$.

	Correlation with		
	Voles	All small mammals	Month
Number of nests	0.90*	0.89	May
Clutch size	0.89*	0.97*	May
Polygynous males	0.89*	0.75 n.s.	May
Harem females	0.90*	0.78 n.s.	May
Productivity: All females	0.68 n.s.	0.34 n.s.	Aug
Successful females	0.72 n.s.	0.43 n.s.	Aug
Breeding yearlings: Males*	0.87*	0.76 n.s.	May
Females[b]	0.66 n.s.	0.66 n.s.	May

[a] Brown plumaged birds.
[b] Birds with predominantly brown eyes (1–2 years old).

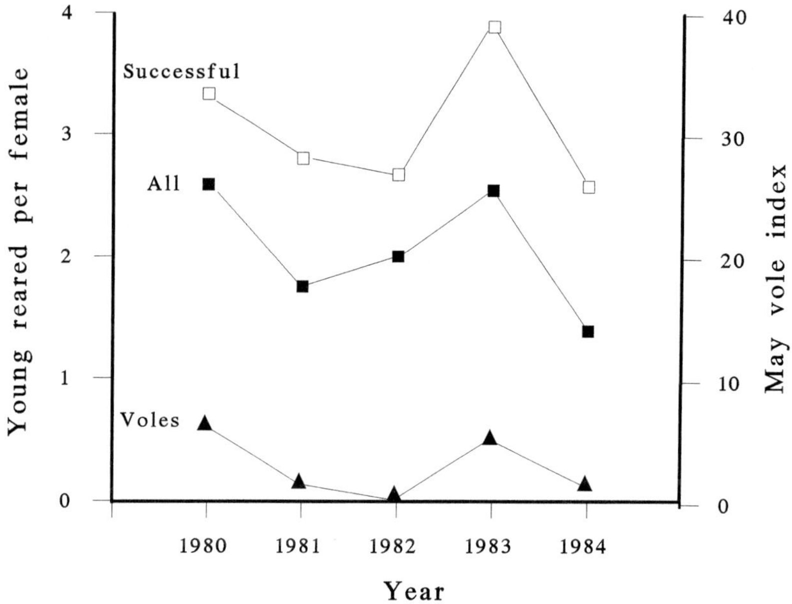

Fig. 5.8 The fledging success of all females and only those females that were successful in relation to August vole dynamics in New Brunswick. Neither relationship was significant over the period 1980 to 1984. Breeding harriers switched to nestling passerines in July–August, reducing their dependence on voles.

with clutch size and number of nests remained significant (Table 5.4), correlations between mouse indices and male and female polygyny both became non-significant. Voles, therefore, appeared to play an important and direct role in harrier polygyny. This is not unique to harriers in New Brunswick since Korpimaki (1984, 1985) found similar correlations with voles alone in Finland, even though shrews and other micro-mammal populations also cycled over a four-year period. If we look further and assess fledging success (number of young reaching the flight stage) of Northern Harriers and concurrent vole indices, we see that parallel fluctuations were also apparent (Fig. 5.8) even though the correlations for all females ($r = 0.68$, $n = 5$) or successful females ($r = 0.72$, $n = 5$) were not quite significant. This arises because harrier diet is almost completely vole-dominated early in the season but switches to fledgling passerines when they first become available in July and August (Barnard *et al.* 1987). The lack of significance may also arise from the low sample size (five years in this case), which is partly confirmed by data from a long-term French study on Montagu's Harriers. Alain Butet and Alain Leroux found that populations in the Saintongeais marshes of western France showed similar populations cycles as the Wisconsin and New Brunswick Northern Harrier populations. Over seven years, harrier numbers varied in perfect synchrony with the common vole ($r_s = 0.88$), although clutch size did not do so significantly ($r_s = 0.67$). Most importantly, they did find that success of all females and the successful females were both significantly correlated with concurrent vole indices ($r_s = 0.85$).

In Europe it is also apparent that vole population numbers are highly cyclical in northern regions. Fluctuations and population densities become lower as one moves southwards (Hansson and Henttonen 1985). In his studies of Tengmalm's Owls, which exist primarily on these *Microtus* species, Korpimaki (1985) found that population fluctuations of the owls in northern areas were also greater, matching that of the voles on which they were living. Because of the dependence of this species on voles, he indirectly confirmed this latitude–vole relationship. This has two implications for harriers: first, if voles are important in controlling polygyny in harriers, as the Wisconsin and New Brunswick data suggest, and if their fluctuations are greater in northern regions, we can predict that polygyny should occur more often in *species* of harriers in more northern regions which exploit these micro-mammals. Second, within those species, we might expect that more northerly *populations* will exhibit greater proportions of polygynous matings (because vole densities reach higher levels). Are these predictions borne out?

Among all species of harrier, the Hen Harrier is well known for its polygyny and exhibits higher levels than any other species, peaking at 94% of all nests in Orkney (Table 3.2). According to distribution maps (Nieboer 1973; del Hoyo *et al.* 1994) it occurs further north in Europe and Asia than any other species of harrier. In North America, Northern Harriers occur in Alaska, and usually breed just below the tundra. European Marsh Harriers exhibit lower levels of polygyny (Table 3.2) and do not reach as far north as the Hen Harrier. The mating systems of other northern-zone harriers are poorly known but further study of the highly dichromatic (suggesting strong sexual selection) Pied Harrier in Northern Russia would be fruitful. In the south, Australasian and southern African birds at about 35° S exhibit almost no polygyny, consistent with this trend (Table 3.2).

Do populations further north exhibit higher levels of polygyny? This too seems to be borne out judging by levels of polygyny in populations studied for three or more years (to allow for fluctuations due to vole cycles). While only five studies satisfied this criterion (polygyny was not reported from Hagen's (1969) or Barth's (1964) Norwegian study sites) there is a significant correlation (Fig. 5.9) between latitude and the proportion of polygyny for Orkney (Picozzi 1984a), mainland Scotland (Picozzi 1980), The Netherlands (Schipper 1978), France (Cormier 1983, 1989), and Spain (Arroyo unpublished in litt.). According to the shape of this relationship (Fig. 5.9), almost no polygyny is expected below about 45° North. While there are insufficient studies of 4+ years in North America to undertake a similar analysis, the two studies detailed above follow the same trend. In New Brunswick (45.5° N), polygynous associations (up to 5 females) accounted for 34% of 89 nests (Simmons *et al.* 1986b), while further south in Wisconsin (44° N) fewer nests (24% of 330) were polygynous and maximum harem size was three females (Hamerstrom 1986). Thus available data support both suppositions: polygyny appears to be influenced by voles,

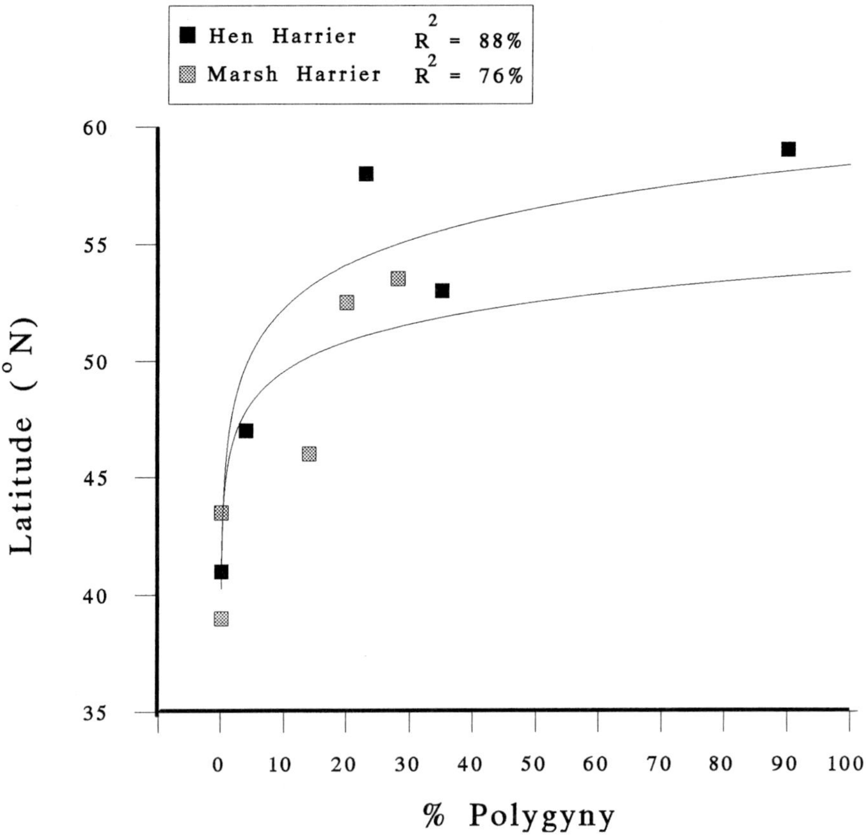

Fig. 5.9 Polygyny in relation to latitude among Hen Harriers and European Marsh Harriers. The significant relationship uses only studies of more than three years to avoid complications involved with variations in polygyny with voles. The study areas (each square) are in order (bottom to top): Hen: central Spain, western France, Scottish lowlands, Scottish Highlands, Orkney Islands; Marsh: Spain, Western France (×2), Netherlands (×2).

because they are often superabundant and relatively easy to catch, and also by latitude. Latitudinal variation may arise because voles occur at greater densities in more northerly areas and reach higher densities at peaks in the more cyclical population in the north, than those farther south.

Polygyny and voles: evidence from other species

Evidence from other species also supports the link between polygyny and voles. European Kestrels, not usually associated with polygyny, only exhibited polygyny in years of high vole abundance, when up to 20% of the males bred with two mates in western Finland (Korpimaki *et al.* 1996). On the same study site, Tengmalm's Owls also only exhibited polygyny (10–20%) in the high vole years (Korpimaki 1991) and not in the vole-poor years.

Other models for the evolution of polygyny

Three additional evolutionary explanations have been proposed for why female birds join harems:

1. Females *cooperate* within harems to reduce predation and enhance success (Elliot 1975; Altmann *et al.* 1977).
2. Hidden genetic benefits are apparent in the form of high quality genes for the *sexy sons* of polygynous males.
3. Females who join harems and reproduce poorly one year may inherit the α female position in future years.

Cooperative harems?

Just as Montagu's Harriers in loose colonies detect predators more quickly than solitary birds (Arroyo 1995), so harem female Northern Harriers who nest relatively close to each other may benefit from (i) faster predator detection and (ii) cooperative mobbing of predators. This may enhance adult survival and longevity (Elliot 1975; Wittenberger 1980). Thus, despite possible reduced annual fitness due to resource competition, harem females may live longer and/or produce more lifetime young. Alternatively, enhanced predator warning may directly increase annual reproduction for harem females (Altmann *et al.* 1977). The latter is verified if female reproductive success increases with harem size.

Without long term data we cannot test Elliot's (1975) 'lifetime' model adequately. There are two simple ways around this. We can test two pre-conditions that must be satisfied if the model is to work at all. These are:

Precondition 1 Harem females should cooperatively mob, or warn of, predators within harems, and such defence should not benefit adjacent monogamous or extra-harem females. Benefits to outlying females would nullify advantages of harem membership and spatially clumped monogamy should prevail (Wittenberger 1980), as found for Montagu's Harriers.

Precondition 2 Harem females should allow other prospective females to settle up to a level at which no further survival or reproductive benefit to individuals is possible. At this point aggression between females is expected (Wittenberger 1980).

To test precondition 1, I gauged mobbing and predator warning responses of Northern Harriers by presenting both a natural predator (a mounted horned owl *Bubo virginianus*) and myself at nests. In both 1980 and 1981, I scored harrier responses as 'mobbing' if harriers dived within 30 m of either predator, and recorded a 'co-mobbing' if additional birds investigated and mobbed. Of 172 nest visits and 23 owl presentations at 45 nests over two years, additional harriers were attracted on only three occasions. Only one of these birds helped mob the owl—a frequency of 0.5%. The co-mobbing also occurred at a monogamous nest, violating the first precondition that mobbing should not benefit females outside the harem. Whether mobbing or not, nesting harriers invariably alarm-called at predators, a call audible from at least 300 m, and to harriers, probably much further (I1'chev *et al.* 1974; Rice 1982). This probably means that females outside the harem may fortuitously benefit from early warning, again violating the first precondition (i.e. don't join a harem if you can enjoy the benefits of early warning by nesting close by with a bachelor male).

The second precondition for Elliot's lifetime model was that first-settling harem members should allow other females to settle in the harem. This is testable on the condition that α females (which have most to lose) suffer little or no fitness costs in the harem or at least can recoup all losses by gains in fitness through cooperation. A comparison of their productivity with that of females which suffer no apparent competition (monogamists) showed that α females not only reared slightly more young ($\bar{x} = 4.64$, $n = 11$) than monogamists ($\bar{x} = 4.57$, $n = 44$), but also reared a slightly greater proportion of their eggs (61% vs. 57%; not significant: Simmons *et al.* 1986a). Precisely the same results occurred in two other sets of data. In the same Northern Harrier population in 1983–84: α females hatched 68% of their eggs ($n = 5$ females), relative to monogamous females ($\bar{x} = 55\%$, $n = 15$: MacWhirter 1985). Among European Marsh Harrier females in different positions within the harem, α females hatched more of their eggs (82%, $n = 133$), than monogamous females (75%, $n = 599$), both of whom

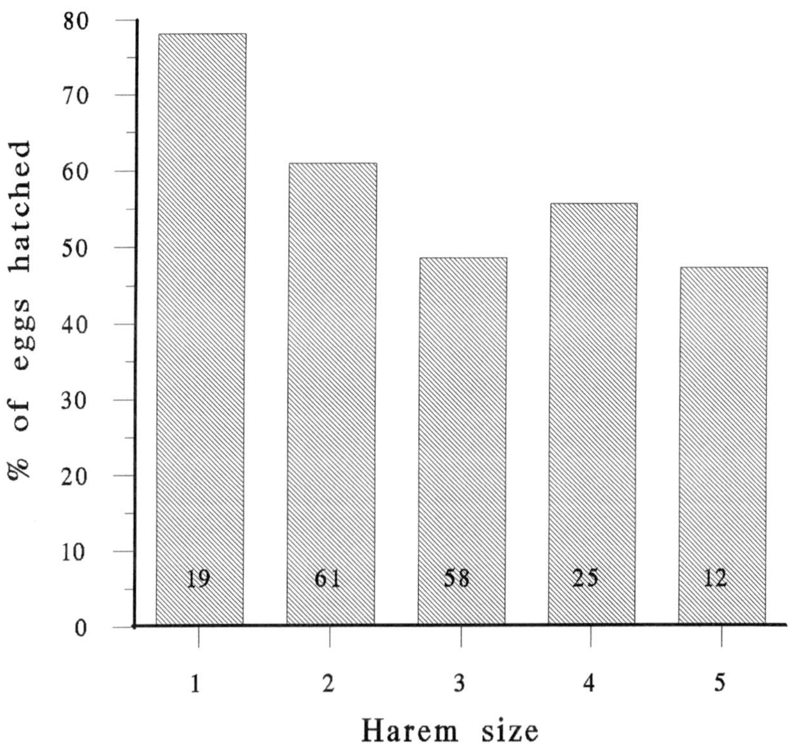

Fig. 5.10 Decline in hatching success of eggs in clutches laid by female Hen Harriers within different sized harems. Numbers are the number of clutches laid. Derived from Balfour and Cadbury (1979).

hatched a greater proportion than β females (72%, n = 150). Since the overall proportion of successfully hatched eggs would decline with an increase in harem size, it is not surprising that Orkney harriers show a parallel decline in egg hatchability (Fig. 5.10).

Despite this support for the assumption that α females suffer little within-harem reduction in success, precondition 2 was violated on all six occasions that harem females were observed settling. Female–female aggression took the form of territorial 'leg-lowering' (above, p. 72) aerial talon-locking, and food piracy. Two particularly aggressive encounters were sufficient to prevent newcomers from settling. Such contests are typically female–female, with little male participation. However, in both Northern Harriers and African Marsh Harriers, lone males may sometimes aggressively chase prospecting females from their territories.

I directly tested the idea of *annual* enhancement of reproduction through cooperation (Altmann *et al.* 1977) by plotting the fledging success of individual females against the size of the harem they were in (Fig. 5.11). Female success declined, not increased, with increasing harem size ($r = -0.28$, $P < 0.1$), violating this idea. Identical trends were calculated from 20 years' data from Wisconsin Northern Harriers ($r = -0.25$. $P < 0.2$: Burke 1979).

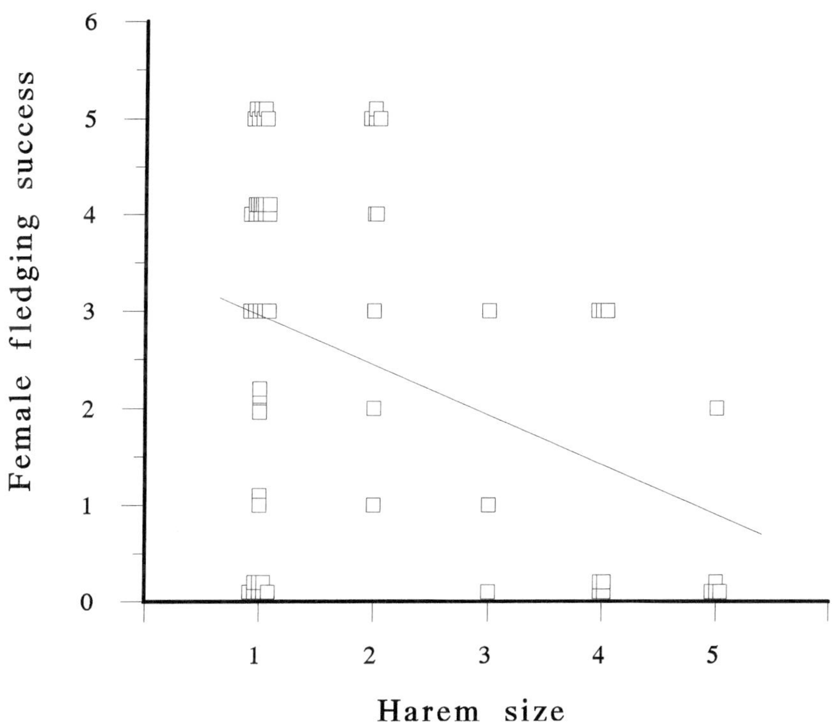

Fig. 5.11 Fledging success of individual female Northern Harriers within harems of different sizes, New Brunswick. The decline was nonsignificant but similar in each year it was measured.

Enhanced lifespan for females in harems?

While these results rule out direct and indirect reproductive advantages, Elliot (1975) originally proposed that harem membership may instead enhance survival for polygynous females. While this may be likely, even with increased longevity, the lifetime success of polygynous females is not likely to equal the lifetime success of monogamous females, because they couldn't survive long enough. For every fledgling reared by polygynous females, monogamous females raised 2.28 young (2.74/1.2: Table 5.2); a consistently polygynous female would, therefore, require at least 12 years ($12 \times 1.22 = 14.6$ young) to equal the lifetime productivity of a monogamous female (~14 young) breeding for a more realistic five years at 2.74 young per year. Because 99% of female harriers are dead by age 12 (Picozzi 1984b), this is clearly unlikely. Hence for Northern Harriers, harem membership is unlikely to enhance reproductive output in an annual or a lifetime sense.

Inheritance of α female position?

We can also discount the possibility that females within harems gradually increase in rank and inherit the α female position. Hamerstrom's work in Wisconsin showed that Northern Harriers do not re-mate on returning to their breeding grounds, and only about 30% of the adults return each year anyway (Hamerstrom 1969, 1986). Thus rising in rank and inheriting the α female position is unlikely for Northern Harriers. However, marked Hen Harriers in Orkney are more sedentary and birds reared there often return to breed. This makes it more likely that certain females could inherit the α position. But do pairs stay together? The answer is provisional on whether the association was successful or not. For females who were successful in one year, 65% remained with the same male ($n = 37$) the next year. If they were unsuccessful, the likelihood that they would be together the next year dropped to 44% (Picozzi 1984a). Since it is generally the α female which is successful and the β female which is not (in both New Brunswick and Orkney), it is more likely that the α females are staying and the β females moving on. This again reduces the likelihood that these poorly reproducing polygynous females are inheriting the α position, even in sedentary populations.

Recouping losses through sexy sons?

The sexy son idea of Weatherhead and Robertson 1979 suggests that while some polygynous females may mate with high quality males and reproduce poorly because they share resources, they can compensate for their poor annual success by fledging particularly attractive sons (because they inherit their father's 'polygyny' genes). These sons may themselves become polygynous at a later stage and make up for their mother's losses with high reproductive success. This is certainly a real possibility with harriers since polygynous males almost always reproduce better than monogamous males (Fig. 5.12) and New Brunswick males attracting two females produce more than twice as many young (6.7) as mono-

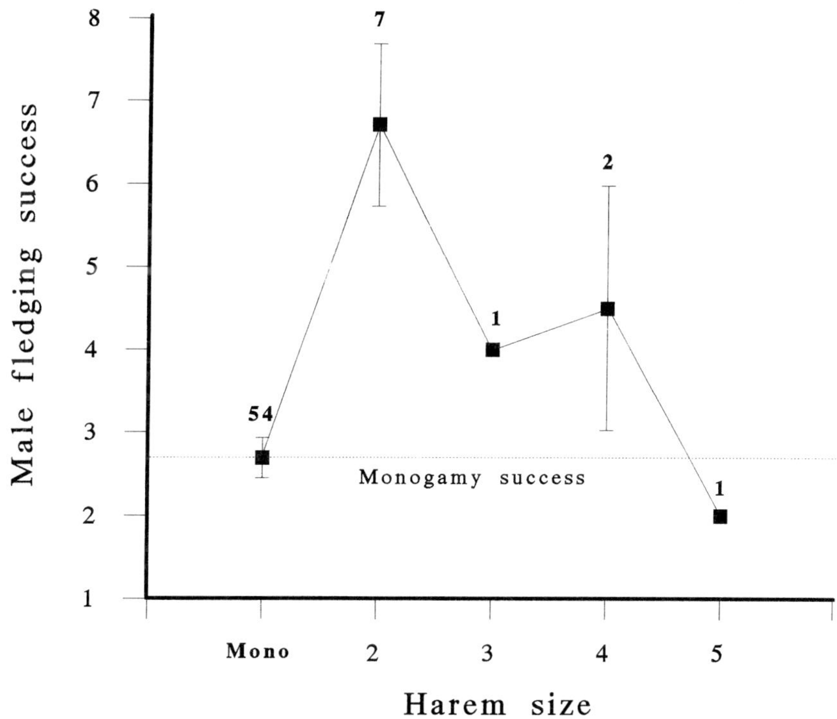

Fig. 5.12 Fledging success of polygynous male Northern Harriers in New Brunswick in relation to their harem size. All except the largest harems fledged more young than monogamous pairings.

gamous males (2.7 young). Indeed these bigamous males produce more young than any other polygynous male (Fig. 5.12). In Orkney, trigamous males had the highest success in later years of the study (Picozzi 1984a: Fig. 5.13) although overall success was lower than in New Brunswick, and proceeded to get increasingly lower as the population became more polygynous (Fig. 5.13). The important point is that polygyny almost always pays for males and it may be in a female's interests to produce 'polygynous' sons.

In order to test this sexy son hypothesis I used Heisler (1981) methods, which takes account of assumptions about the likelihood of monogamous sons becoming polygynous fathers (r), the number of polygynous positions available to males (l), the mean harem size (k), and the breeding sex ratio (s). As such, the model is readily testable because the effects of genetic recombination and random environmental effects (e.g. fluctuating food supplies) make it unlikely that a secondary female could recoup her losses beyond her grandsons' generation (Heisler 1981; Kirkpatrick 1985). This simplifies calculation of the expected ratio of polygynously to monogamously produced sons (x) which we already know from nature (Table 5.2) averaged 0.44. That is, for every one

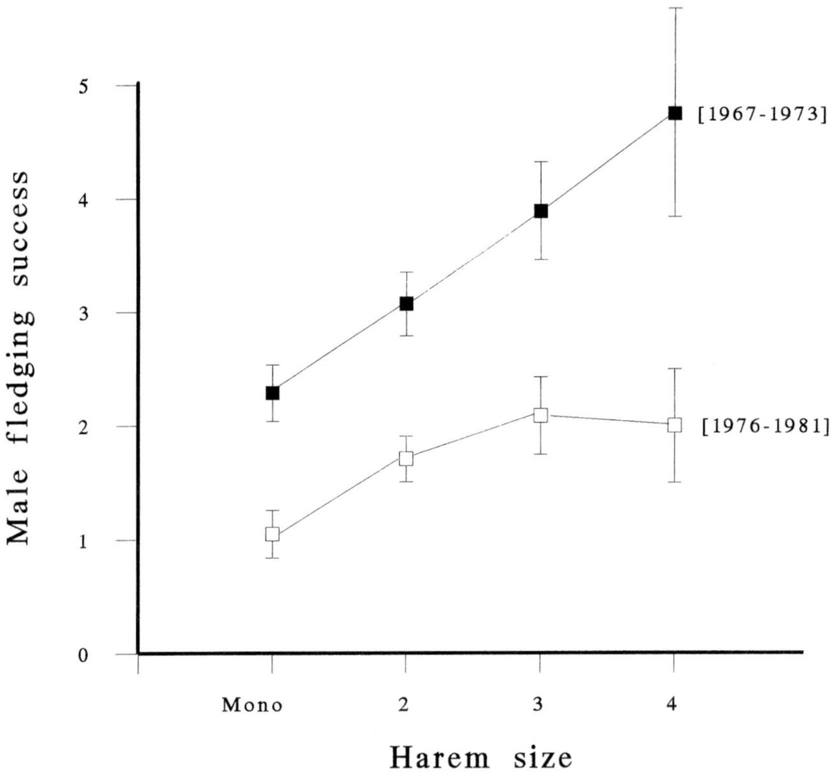

Fig. 5.13 Fledging success of polygynous male Hen Harriers in Orkney, Scotland in relation to their harem size. The top graph gives fledging success for earlier years when polygyny was less prevalent than later years (bottom). Derived from data in Balfour and Cadbury 1979, and Picozzi 1984a.

young fledged by monogamous females, polygynous females produced less than half a fledgling.

While this is a very low production by polygynous females, perhaps it arises because polygynous females began later in the season than monogamists, and their lower production is explicable by the seasonal declines in clutch and brood size of this species. From the seasonal declines (Simmons *et al.* 1986a) it was apparent that polygynous females also averaged 0.44 young for every offspring reared by monogamous females laying simultaneously—seasonal effects could not explain the large differences in reproduction. This value for Northern Harriers, with one of the lowest recorded ratios in avian polygyny (Table 5.1), corroborates suspicions that polygynous female raptors in general fare worse than their passerine counterparts. Some passerine females can compensate so completely that polygynous females rear more young than contemporaneous monogamists. This occurs in no raptors studied to date and is unlikely in a group in which the females rely so completely on their mates for food.

But do polygynous females recoup these losses through their sexy sons? If we assume no monogamous sons become polygynous fathers ($r = 1.0$), and k, s, and f are known from four years' data (Table 5.2), the minimum theoretical ratio allowing a polygynous female to recoup her losses is 0.78. In other words, a polygynous female can afford to rear no less than 78% the number of young reared by monogamists for her sexy sons to compensate for these reductions. Since the observed value of 0.44 is considerably lower than this, the sexy son idea is an improbable explanation for the polygynous choices of secondary females. However, as we shall see later, there are strong hierarchies among the harem females and β females reproduce better than γ or δ females. Perhaps β females can recoup their losses and later-settling females really are constrained in their choices and are mating with older mated males on the chance that they can produce some high quality young? Even this possibility is unlikely because the ratio of young produced by β females to simultaneously breeding monogamists is 0.67, still less than the 'break-even' ratio of 0.78.

Sex ratio manipulation—more sons?

Another possibility is that polygynous females may attempt to alter the number of sons they produce by altering the sex ratio of their brood. This idea is not as unlikely as it sounds because many species do manipulate the sex ratios of their brood with the first two eggs often being of one sex at certain times of the season (females earlier for harriers: Chapter 7). Indeed there is evidence that overall sex ratios are biased towards an excess of sons in several populations of harriers in Scotland (Picozzi 1984b), The Netherlands (Zijlstra *et al.* 1992), and France (Leroux and Bretagnolle 1996), but not in the eastern Canada (MacWhirter 1994). Adaptive manipulation of sex ratios towards an excess of sexy sons is especially possible in the Orkney population because sons started to predominate (58:42 in last five years of study) as polygyny reached levels of 90% in the population (Picozzi 1984b, Chapter 7). Despite the attractiveness of this idea, too few sons are produced in each of these cases because the skew towards sons is relatively slight for the populations as a whole (*C. cyaneus* ~58% males; *C. aeruginosus* 55% males; *C. pygargus* 56% males) in all three areas. Moreover, in a small sample from the New Brunswick population, harem females produced equal numbers of sons and daughters (16:16, $n = 9$) while monogamous females produced slightly more sons than daughters (51:45; $n = 26$), opposite to that expected if polygynous females were favouring sons. These data were taken when the nestlings were first sexable (about 10 to 14 days), and is not necessarily that which the female produced when laying—because there may be selective sex-related death of nestlings before fledging. The same trend was found in a sample of European Marsh Harrier nests by Altenburg *et al.* (1982): harem females produced fewer sons than daughters (86:91) while monogamous pairs produced significantly more sons than daughters (296:209). These data cannot directly support the idea that females manipulated their clutches to produce more sons since not enough is known about post-hatch mortality in relation to sex. We will return to this topic in Chapter 7.

Viability genes?

A subsidiary idea that may have been overlooked in the vast amount of literature generated by the problem of reduced productivity in polygynous females is that females may be choosing certain males because they are older and have thereby proven their viability (Alatalo *et al.* 1991; Møller 1991a). Thus females may be mating with them to pass on viability benefits not only to their sons but also their daughters. This is a more tenable idea than the sexy son hypothesis because unlike the sexy son idea it doesn't rely on just one sex (sons) to recoup losses, and in several studies lifespan of an individual is the most important component contributing to high lifetime success (Newton 1989). This may be true of the harriers because as male Hen Harriers in Orkney become older they attract larger and larger harems (Picozzi 1984a). The relationship is highly significant (Fig. 5.14) and indicates that males of 8 to 9 years old attract on average four females and occasionally up to six. These females may thus be choosing such males because they are long-lived. How females recognize such males is unknown, but as with grouse it may be related to vigour during display (Alatalo *et al.* 1991), or plumage characteristics as in swallows (Møller 1991a). Both possibilities exist with harriers because males differ in their display vigour (Chapter 3), and male Hen Harriers are thought to become paler grey and less brown as they grow older (Watson 1977). Further evidence for this idea comes from Tengmalm's Owls. Few males ever became polygynous but those that did attract two mates survived and bred for 50% longer (2.1 vs. 1.4 years) than monogamous males. This was surprising because like polygynous harriers they foraged for two or three females simultaneously, yet this appeared to have no reproductive cost (Korpimaki 1992). These males thus appeared to be particularly high quality and females may have chosen them because of their higher viability.

If both sons and daughters enjoy viability benefits and live longer than the offspring from monogamous matings, could they compensate their mother's losses? On the down side of this idea is the finding that comparisons of annual fledging success do not take account of immediate (post-nestling) survival of young. One measure of this—the weight of young at fledging—is a relatively good measure of future survival for some species (reviews in Perrins 1979; Simmons 1988a). However, among European Marsh Harrier broods, sons of β females weighed significantly less ($P < 0.001$) than male nestlings of monogamous sons of the same age (Altenburg *et al.* 1982). This suggests, but doesn't prove, that these sons would not survive as well. For daughters, there was no weight difference, suggesting that daughters at least may have a chance to recoup their mother's losses.

In conclusion, it would seem that polygyny in harriers can reach the extremes seen in other species with up to five or six females in a harem; the polygynous females choosing mated males show very low reproductive success, and their choices cannot be explained by enhanced success or survival within the harem, the inheritance of future α status, or by genetic superiority of the sons of polygynous fathers. Skewed sex ratios that accompany polygyny in Hen Harriers in Europe, but less so in North American populations, cannot explain polygyny

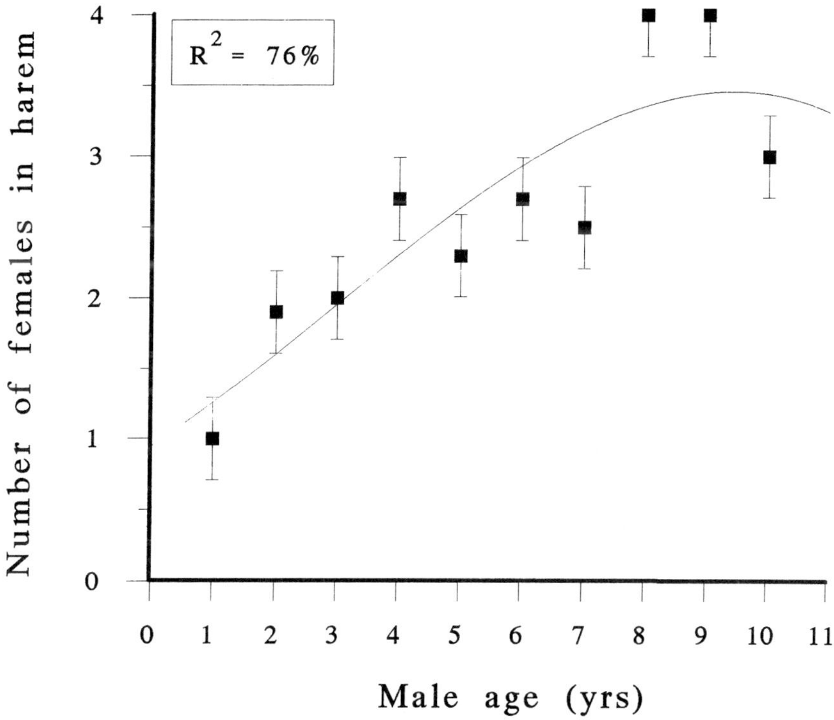

Fig. 5.14 Increasing harem size among individual male Hen Harriers in Orkney as they aged. Males were marked with patagial tags and followed throughout their lives (drawn from Picozzi 1984a).

because females are choosing polygyny over monogamy throughout the season and not just towards the end as would be expected if choices were constrained by mate availability. Moreover, sex ratios are not always skewed in Wisconsin but polygyny is common there. Females are also aware of male mated status because α females actively attempt to dissuade other females from settling and males do not hold widely separated territories and cannot deceive females by polyterritorial behaviour *sensu* Lundberg and Alatalo (1992). Thus all major models of polygyny fail to explain this mating pattern in Hen or Northern Harriers. The next step is to determine what females are using as choice cues in their mate assessment and thus why some females are choosing mated males.

Quality and order of choice: what are females choosing?

The logical approach to this difficult question is to manipulate aspects of both males and territories and see which become more attractive and which less attractive to females. This is difficult in the field, especially with far ranging species like the harriers. We therefore take the best non-manipulative approach—to look for any particular traits of the male or his territory that are:

(i) variable between breeding situations, (ii) assessable prior to settling, and (iii) have a significant impact on breeding success (Lenington 1980).

There are thousands of traits that may affect breeding success, but by asking which significantly influence success we can whittle them down to a few important ones. If these are not assessable by the female (and not correlated with any trait that a female could use), then they too can be sieved out. If any of the remainder do not vary appreciably in quality, then females have nothing to choose between and cannot use them. This sieving process should leave us with a few traits that are important, assessable, and vary in quality. We then assume that a female would *choose the best remaining breeding situation* first. If so, then we would expect to find a correlation between the traits that females are actually using and the order in which females settled.

Potential choice cues

Female harriers potentially choose breeding situations on one or all of the following characters:

(a) *Nest sites*: harrier ground nests are highly susceptible to terrestrial predators and vary in quality (Simmons and Smith 1985).

(b) *Feeding territory quality*: females forage within male-defended territories that vary in food availability (Simmons 1983).

(c) *Paternal investment* in defence and food provisioning: males not only defend nests vigorously, but provide the bulk of food to their incubating mates and young (Hamerstrom 1969; Watson 1977; Simmons et al. 1987).

Mate choice criteria

To determine how females assessed mates or breeding situations, I used the elegant methods first proposed by Lenington (1980) for resource-defending polygynists. Lenington (1980) suggested the following three-part approach:

1. Identify factors that limit female fecundity, either through their effects on nest failure or on brood survival. To be useful and useable in nature such factors must be assessable by females prior to laying(Wittenberger 1976), they must be reliable indicators (or covary with indicators) of future success, and they must vary between breeding situations (Searcy 1979). If they satisfy these simple criteria then one can assume they are potential cues for females.
2. Using multiple regression and discriminant function techniques, combine these determinants of success to predict each female's Expected Fecundity (EF).
3. Presupposing that females pick the best remaining breeding situations first, correlate the order in which females chose males with their Expected Fecundity (EF) to identify which cues, in reality, females used. For example, high rank order correlations between settling order and Expected Fecundity based on territory quality would suggest that females use territory quality cues to pick breeding situations.

Expected Fecundity comprises two parts: survival of the nest, EF1, and the number of young fledging if the nest survives, EF2. Both are evaluated below. While females may not make perfect choices (due to search costs and errors of assessment), I assume that they are sufficiently non-random to detect choice trends.

Defining quality

(a) Nest sites

Nest site quality is an estimate of the probability of nest survival based on its placement in one of three substrates: dry, wet (marshy ground), or very wet (standing or running water). These categories, and the calculations used to assess survival probabilities, are explained in Simmons and Smith (1985).

(b) Feeding territories

Within male-defended territories female Northern Harriers generally forage in a nest-centred area of about 200 m radius. To determine available food resources, and hence quality within these areas, all seven major habitat types on the Tantramar Marsh, New Brunswick were trapped for small mammals (Simmons *et al.* 1986b).The proportion of each habitat within a female's territory was then calculated from a vegetation map of the study area. Availability of small mammals to raptors is more important than their absolute abundance (Baker and Brooks 1981; Bechard 1982; Chapter 4), so I introduced a linear rank reflecting vole vulnerability (i.e. vegetation density). Scores ranged from 2667 to 8376; high scores represent the best quality feeding territories for females.

(c) Paternal investment

Male provisioning rates were determined for 18 males in 1980 and 1981 by recording all items delivered during 257 h of pre- and post-hatch observations at nests. No obvious differences in prey sizes were apparent between males within years, hence provisioning performance is expressed in items, not biomass per day. Provisioning by polygynous males in 1982 was impossible to assess due to the birds' wariness, and courtship provisioning rates remained unstudied because of lack of foresight. Nest defence scores were assigned on the basis of the diving and alarm calling responses of males as a proportion of my visits to nests (Simmons *et al.* 1987).

Assessing male mated status and female settling order

Because birds were unmarked (capture caused nest desertion), male mated status was established from unbroken observations of successive feedings or defence at more than one nest: males were assumed monogamous until proven polygynous. Female position within the harem ('HAREM' in Tables 5.5 and 5.6) was based on the order in which eggs were laid (α = first, δ = last), since other cues were unreliable indicators of pair bonding (Simmons *et al.* 1987). Timing of egg laying is strongly positively correlated with the order in which

Table 5.5 Pearson correlation coefficients of factors associated with Northern Harrier productivity, 1980–81. ( ) gives significance levels in 2-tailed *t*-test; those in bold are significant.

Variable	1980	1981	1980 + 1981
Female territory quality (RANK)	+0.119 (0.618) n = 20	−0.027 (0.933) n = 12	+0.125 (0.495) n = 32
Mate provisioning rate (FOOD)	+0.819 **(0.007)** n = 9	+0.850 **(0.008)** n = 8	+0.722 **(0.001)** n = 17
Clutch size (EGGS)	+0.154 (0.697) n = 14	+0.600 (0.088) n = 9	+0.396 (0.061) n = 23
Female settling date (DAY)	−0.256 (0.277) n = 20	−0.505 (0.090) n = 12	−0.308 (0.086) n = 32
Harem position[a] (HAREM)	+0.361 (0.118) n = 20	—	+0.250 (0.170) n = 32
Total precipitation in nestling stage (RAIN)	−0.432 **(0.057)** n = 20	−0.360 (0.250) n = 12	−0.118 (0.521) n = 32

[a] HAREM for successful nests was constant in 1981.

Table 5.6 Variables used in Multiple Regression equations or 1980 and 1981 and their power in explaining (R^2) the fledging success at harrier nests. FOOD (rate of male provisioning) was consistently the best predictor of success. HAREM = female's position in the harem; RANK = female foraging territory score; EGGS = clutch size.

Year	Variables used	Cases (nests)	Total R^2 explained	Changes in R^2 explained on inclusion of each variable
1980	FOOD HAREM RANK	9 (20)	70%	67.0% (FOOD) 2.3% (HAREM) 0.7% (RANK)
1981	FOOD EGGS RANK	7 (12)	99%	92.8% (FOOD) 5.9% (EGGS) 0.3% (RANK)
1980 + 1981	FOOD EGGS RANK HAREM	15 (32)	70%	54.6% (FOOD) 11.8% (EGGS) 3.2% (RANK) 0.3% (HAREM)

EGGS was not included in 1980 since it was poorly correlated with success. HAREM was constant for successful females in 1981 and could not therefore be included in the multiple regression equation. A limited number of nests was used each year, since not all nests could be studied for food provisioning rates.

females choose mates in passerines (Lenington 1980; Alatalo *et al.* 1984) and raptors (Village 1985) and was thus used as an index of the order of choice by female harriers.

Nest survival and EF1: are females choosing good quality nests?

Two variables potentially contribute to EFI: nest site quality and male nest defence. Male nest defence made no significant difference to nest success, either before or after hatch (Simmons *et al.* 1987), and is thus considered an unimportant female cue. However, among the nest sites, wet and very wet sites were significantly more successful than dry ones (Simmons and Smith 1985). Since moisture levels were assessable prior to settling and did not vary throughout the season, they were potentially reliable cues to EF1. Despite the availability of wet areas in most territories, no significant rank order correlations between order of settling and site quality occurred in any of three years ($r_{1980} = -0.25$, $P < 0.3$; $r_{1981} = -0.2$, $P < 0.1$; $r_{1982} = -0.2$, $P < 0.1$: Simmons and Smith 1985). Hence, female harriers did not appear to choose breeding situations on the basis of nest site quality.

Brood survival and EF2: are females maximizing brood survival?

The second component of fecundity, brood survival, is affected by many factors, some of which females can neither predict nor control (rainfall and laying date), and others which they may use as cues to EF2. Of the latter, male provisioning (food items delivered/14 h), was consistently the strongest predictor of young reared (Table 5.5). Territory quality, reflecting food availability especially to the female, was not significantly correlated with fecundity in any year. Using multiple regression techniques we can combine these factors to determine which ones in combination best explain the number of young fledged. We see once again that the male's food supply was consistently the best predictor (Table 5.6).

Combining variables and years, up to 80.7% of the variance in fecundity could be explained by male provisioning rates (FOOD) and laying date (DAY). We can now use the result that male food provisioning rate was an excellent predictor of success of females to test if females actually used such a cue.

Settling order and mate choice cues

Unlike nest site quality, strong rank correlations existed between settling order and male provisioning ($r_{s\ 1980} = -0.65$, $P = 0.02$; $r_{s\ 1981} = -0.84$, $P = 0.002$). That this occurred in both years it was studied suggests that the correlations were not spurious, and females were indeed basing their choice of mate on provisioning performance. I assessed whether females also took account of harem position and clutch size by adding HAREM and EGGS to the equation. The following multiple regression equations were generated as predictors of the number of young reared (i.e. EF2) in 1980 and 1981:

$$\text{Young}_{1980} = 0.39\ (\text{FOOD}) + 0.56\ (\text{HAREM}) - 0.60 \qquad (R^2 = 70\%)$$

$$\text{Young}_{1981} = 0.24\ (\text{FOOD}) + 0.49\ (\text{EGGS}) - 0.87 \qquad (R^2 = 99\%)$$

By calculating the number of young expected to be reared for 13 nests in 1980 and 9 nests in 1981, and correlating the order of choice with these young, even stronger rank correlations with settling order arose in 1980 ($r_{1980} = -0.80$, $P = 0.001$) and were equally strong in 1981 ($r_{1981} = -0.83$, $P = 0.003$). This suggests that females were able to take variables like their harem position into account. Another way of checking this was to exclude HAREM from the 1980 EF2 equation and correlate the results with settling order. Rank correlations between Expected Fecundity and settling order weakened ($r_s = -0.762$), implying that females not only could, but did assess harem position when settling. Similar results were apparent by adding HAREM (from 1980) to EF2 for 1981: correlations between settling order and Expected Fecundity strengthened ($r_s = -0.85$, $P = 0.002$). Females thus appeared to take into account a certain expected loss of fitness due to harem position in both years. The most important result was confirmed: females appear to choose males mainly on their provisioning performance. But was it a reliable cue for all females?

Courtship provisioning and future male investment

Courtship provisioning per se was not assessed because at the time it was not realized that it might be so important. Thus I indirectly calculated the courtship provisioning of polygynous females from her clutch size and laying date. Clutch size and initial provisioning are causally related in at least two temperate raptors (Newton and Marquiss 1981; Dijkstra *et al.* 1982), and are also closely correlated in African Marsh Harriers (Chapter 9). Furthermore, clutch sizes of monogamous females were significantly related to seasonal (= expected) provisioning rates of male Northern Harriers in New Brunswick (Simmons *et al.*1986b). Since harrier clutch sizes also vary with laying date, one can interpolate expected provisioning rates using clutch size and laying date. Complete data from 13 α and monogamous nests generated the following equation used to predict provisioning rates for polygynous females:

Table 5.7 Provisioning rates expected by polygynous females (predicted from knowledge of their clutch size and laying date) relative to that observed. Observed and expected rates differed significantly.

Female	Harem position	Clutch size	Initiation date	Provisioning rate	
				Expected	Observed
Loop 4	β	5	7 May	10.1	7.6
Alders	γ	5	18 May	9.6	2.5
Culvert 3	β	5	20 May	9.5	4.8
Culvert 4	γ	4	9 June	7.3	1.5
Pb	β	4	8 June	7.3	4.2

Observed vs. Expected, $t = 4.14$, $P < 0.0005$.

$$\text{FOOD} = 3.30 + 1.42\ (\text{EGGS}) - 0.043\ (\text{DAY}) \qquad (r = 0.79, P < 0.01)$$

The observed seasonal provisioning rates for polygynous females (Table 5.7) were significantly lower than those they expected ($t = 4.14$, $P < 0.005$). A major discrepancy is thus apparent between monogamous and α females whose seasonal provisioning rate was reflected in their clutch size, and polygynous females, whose seasonal rate was lower than predicted from their clutch size.

Deceptive acquisition of mates

This is a curious and an important result: the best choice cue that a female harrier can use is the provisioning performance of her mate because it accurately predicts the number of young she is likely to rear. Yet some females—those that join harems and know they are joining an already mated male—appear to receive good supplies of food prior to settling—enabling them to lay a clutch which is no smaller than that predicted by the laying date. (If these females relied on their body condition to lay a large clutch then this might explain their clutch size. This is unlikely because as we shall see in Chapter 9, the provisioning by mates is a better predictor of clutch size than is body condition of females prior to laying.) Clutch size also varies very little among harems of different sizes in Orkney, which is unexpected if polygynous females do not receive good male assistance when egg laying (Balfour and Cadbury 1979). Unlike other monogamous and α females, the initial male-supplied food of polygynous females does not appear to predict the future 'expected' or seasonal supply of food from the male. This essentially means that they are being deceived by the polygynous males. They are fed good initial supplies and later the supply dwindles. I call this the food-deception hypothesis (Simmons 1988b).

Since other harriers are polygynous is there support from any of these species that females may be deceived about food provisioning? Detailed provisioning data from European Marsh Harriers indicate that males provisioned polygynous females at a higher rate than α females during the early incubation period, but these rates dropped to about half that provided to α and monogamous birds in the nestling period (Table 5.8). This data, collected independently of the food-deception hypothesis by Altenburg *et al.* (1982), therefore supports the deception scenario.

One prediction of the food-deception hypothesis is the expectation that polygyny should increase in vole-abundant years, as more males exceed the threshold feeding rate. As we have seen earlier this was so: vole abundance and the number of polygynous males and females were significantly correlated over five years in New Brunswick (Fig. 5.5), a trend corroborated by data from Wisconsin harriers (Hamerstrom *et al.* 1985). But can we be sure that among Northern Harriers the food supply of polygynous females is lower throughout the season, and does it completely explain their lower fledging success?

Table 5.8 Male provisioning effort (items/day) to α and polygynous females of the European Marsh Harriers studied by Altenburg *et al.* (1982). Figures in **bold** indicate higher provisioning rates for the polygynous females.

Phase of breeding cycle	Bigamous male		Trigamous male		
	α_1	β_1	α_2	β_2	γ_2
Incubation	2.8	**3.2**	1.5	**2.6**	**4.4**
Nestling (0–28 d)	8.2	5.6	5.7	4.8	2.0
Nestling (28–45 d)	13.2	4.7	7.2	1.3	4.9
Post-nestling (46–73 d)	0.0	5.9	5.5	0.0	5.7

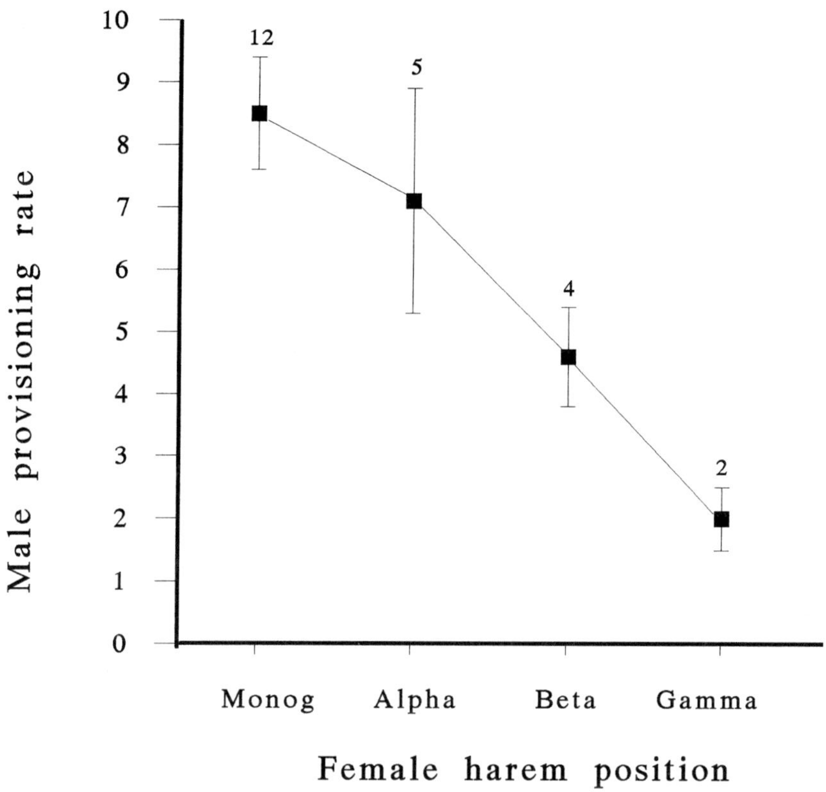

Fig. 5.15 Differences in male provisioning rates per day to females in different positions within the harem. Mean values (± 1 s.e.) for Northern Harriers in New Brunswick from 1980 and 1981 (Simmons *et al.* 1986a). Rates derived from 1982 are not included because, despite reflecting similar hierarchies, they were collected only prior to hatch.

Harem hierarchies

Despite a large number of studies of polygynous harriers, and suspicions of hierarchies within harems (Hecht 1951), only in the early 1980s were hierarchies explored, and then only as far as β females. Work on European Marsh Harriers and Hen Harriers in Orkney found that α females also raised more young per year than β females—as found for New Brunswick birds (Fig. 5.1). Altenburg *et al.* (1982) determined that α females laid slightly smaller clutches and β females slightly larger clutches than expected on the basis of seasonal differences in clutch size (in New Brunswick harriers both α and β females laid slightly smaller clutches than predicted). The fewer number of young raised by β female marsh harriers was not a season-related trend, since fewer young were raised relative to concurrently settling monogamous females (Table 5.1); results for Northern Harriers are identical in this regard.

For Northern Harriers harem females were ordered in a distinct hierarchy based on their mate's provisioning rate (Fig. 5.15). Alpha and monogamous females were provisioned at a significantly higher rate than were polygynous females in 1980 and 1981 (ANOVA $F_{2,20} = 4.72$, $P < 0.05$). On average, α females ($\bar{x} = 7.1$ items/day) received over 1.5 times as many prey as β females (4.6 items/day), and over three times as much as γ and δ females (2.0 items/day). Monogamous females, however, received marginally more (8.5 items/day) than did α females. Perhaps these hierarchies arose because later-settling females generally failed earlier (below), resulting in a lower overall seasonal rate than the typically successful α females? This was not so: the same trend was apparent in 1982, when provisioning rates to harem females were only assessed prior to hatch; α females ($\bar{x} = 4.9$ items/day) once again received twice as many pre-hatch prey items as β females (2.5 items/day), but still fractionally (by 0.3 items/day) less than monogamists. The differences were not significant in 1982, due to the smaller sample of harem females ($n = 4$). Therefore, in all three years in which provisioning rates were studied, hierarchies in food supply (Fig. 5.15) were evident among harems, and paralleled declines in female productivity (Fig. 5.1).

Male behaviour towards different females

Polygynous males not only provisioned their α females more often than other harem females, but also roosted, preened, and loafed more often in the vicinity of these nests. Moreover, α females were the first fed of the harem females at the start of each day. Despite these inequalities, α females did not receive their greater share of attention through aggression, but through male choice. This was evident in the behaviour of males flying directly to the nests that they eventually provisioned—nests that were occasionally furthest from their point of prey capture. In 1980 and 1982, some males were observed soaring and flapping to heights approaching 100 m, directly after prey capture, and flying over one of their nests to provision another. Only once was direct piracy between intra-harem females observed in approximately 2500 h of field observation, and only twice was a male forced to give up his prey (once each by an α and β female) when it was obviously destined for delivery at another nest. Surprisingly few intra-harem interactions took place once early season boundary disputes were settled, even though polygynous females tended to forage a great deal in the harem vicinity from an early stage. Male behaviour and not female aggression was thus responsible for the feeding hierarchies observed.

The parallel between male provisioning rate (Fig. 5.15) and female success within the harem (Fig. 5.1), strongly suggests that male investment plays a central role in who is successful and who is not. But this is only likely to be the case if those females whose male-supplied food decreases cannot compensate for the loss by foraging for themselves in a particularly rich territory. We can check this by making use of a rather unexpected male behaviour seen several times in the course of our work: complete desertion. In a typical situation females begin to forage for their young when they are 2–3 weeks old, supplementing what the male provides. However, 11 of 32 well-studied males either sharply reduced their provisioning or deserted altogether before the chicks were three weeks old. This happened in both monogamous and polygynous situations. By comparing the brood starvation levels in these nests with non-deserted nests we can gauge how important males (and their food supplies) are. The starvation levels in 9 of the 11 nests whose original brood size was known reached 55% of the 33 hatched young. This was significantly higher ($X^2 = 19.3$, $P < 0.005$) than the number starving in non-deserted nests. Thus it seems that females cannot compensate for male desertion, even though some deserted females are able to rear reduced numbers of young single-handed. This ability depends partly on prey size—if a female can capture large prey she doesn't have to spend long periods away, reducing the chances of predation—and partly on rainfall. During heavy rain females do not forage, in order to keep their young dry. However, when they are forced to forage, the young become soaked and mortality rises sharply in wet summers (MacWhirter 1985). The important points here are that females seem so dependent on their male's food support that it explains most of the variation in success—polygynous females do poorly because their seasonal supplies slowly dwindle—and other females who are deserted also exhibit high levels of brood starvation. That polygynous females join harems at all is, therefore, highly surprising, but their behaviour can be explained by the finding that male food supplies early on are high—promising good supplies later on. Since these never materialize they appear to be deceived.

Effects of harem membership on α females

A result common to studies in Orkney (Picozzi 1984a), The Netherlands (Altenburg *et al.* 1982), and New Brunswick was the greater reproductive success of α females than any other females. This was somewhat unexpected in New Brunswick, because α females received slightly less food from their mates than did monogamous females. Three explanations are possible: (i) α females received more biomass per item than monogamists (as implied by the results for European marsh harriers); (ii) α females foraged in higher quality food territories and supplied more biomass to their young than monogamists; (iii) polygynous males attract high quality first mates who invest more in the breeding attempt than they would with a lower quality male. The first idea is the least plausible in this study, because no discernible differences in the mean sizes of the chief prey (*Microtus pennsylvanicus*) provisioned to monogamous and α females were noted between 1980 and 1982. However, the availability of voles in 1980 and 1981, as measured by snap-trap sampling and an assessment of vegetative structure in a 400-m diameter area about the nest (the females' usual foraging area), was significantly different among successful α, β, and monogamous nests (Kruskal-Wallis $H = 6.29$, 2 df, $P < 0.05$). As expected, α females exhibited much higher prey availability scores ($\bar{x} = 7460$) than did monogamous ($\bar{x} = 5068$) or β females ($\bar{x} = 5045$: Simmons 1983). These differences may alone explain the differing successes of harem females in conjunction with male provisioning rates. Among the European Marsh Harriers studied by Altenburg *et al.* (1982) α females were also noted bringing larger amounts of prey to their young than other harem females, supporting this idea. Furthermore, it was notable that over the five years during which harrier polygyny was studied in New Brunswick, polygynous relationships occurred in the same areas in each year and even in vole-poor years the few polygynous nests that did occur arose in those areas where polygyny first occurred in years with abundant voles and polygyny. Since only about 30% of males return to their breeding ground each year (Hamerstrom 1986), it is unlikely that this is due to certain males returning to favoured haunts. It suggests instead that particular areas attracted the best quality males, which in turn attracted the best quality females.

The α females therefore might be expected to nest in the heart of such areas and enjoy maximum benefits. This may also explain the finding that lower hierarchy polygynous females virtually always nested closer to the α female of their own harem than any other female, either within the harem or any other female outside it (Simmons *et al.* 1986a). This may allow such females the possibility of exploiting good food resources close to their nest, especially since polygynous females generally start foraging earlier in their breeding cycle than α females (Altenburg *et al.* 1982) to compensate for lost male assistance. This nest dispersion may also occur so that polygynous males can guard their dependent mates more efficiently, both from extra-pair copulations with other males and from predators at the nest.

The third possibility, that α females invest more in the breeding attempt because they are mated to a high quality mate, arises from results from other studies of mate choice in birds. High quality males are known to attract high quality females to breed with them. These females may be in better condition or they may be genetically superior (Møller 1996). Whatever the case, there is good evidence that some females invest a greater amount of reproductive effort in rearing broods if they perceive their mate as a high quality bird (Burley 1986; Møller 1996). Among swallows this takes the form of greater foraging effort by females, and larger and more frequent subsequent broods for females suddenly given a longer-tailed mate than those given a short-tailed one (Møller and de Lope 1993). It may be that α female Northern Harriers are laying larger clutches and hatching a greater proportion of them, because their mate is particularly attractive for similar reasons. Without experimental manipulation this fascinating idea remains speculative for harriers. While α females reared more young and benefited from greater food resources around their nests, might their higher success also be attributed to better quality nest sites given that they enjoy first choice among available sites?

Nest predation and nest site quality

Nests of polygynous females suffered somewhat higher levels of predation per attempt than either α or monogamous nests (Fig. 5.16). Only 10% of 10 α nests failed through predation, while 18% of 49 monogamous nests, and the majority of polygynous females' nests, were depredated. Differences between these proportions were marginally short of significance (mono vs. α vs. polygynous: $X^2 = 5.5$, $P < 0.07$) but the same trend was also apparent among Hen Harriers in Orkney (Picozzi 1984a). Other forms of failure unrelated to research activity (excluded from this analysis) were also higher among later-settling harem females. Three of four natural desertions occurred among six females; two had laid full clutches, of which one hatched young prior to desertion. Diseased chicks also accounted for the failure of one γ female nest. Through this combination of factors, γ and δ females failed in seven of their eight attempts. Were the predation trends a result of polygynous females choosing inferior quality nest sites because of their later-settling dates?

Quality scores were assigned to each nest in the period 1980–82 and summed, averaged, and compared between all female groups (Table 5.9). Surprisingly, no

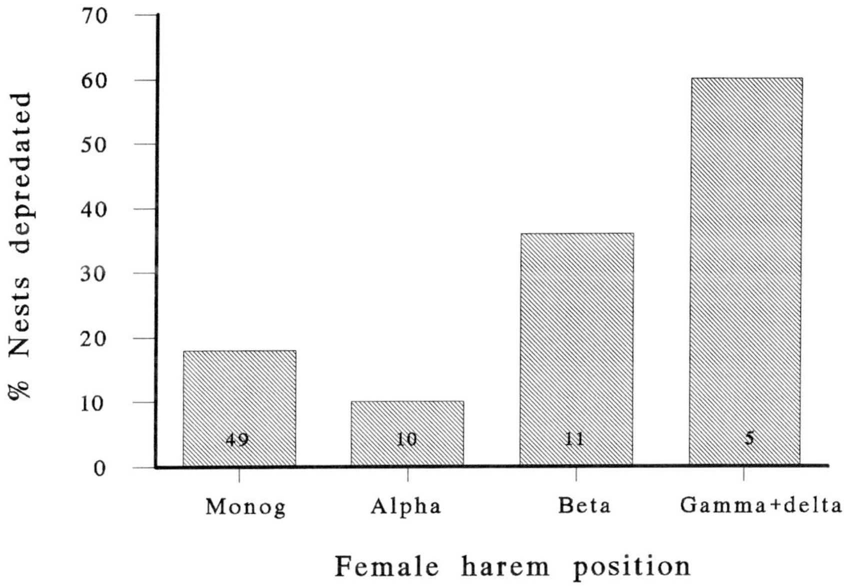

Fig. 5.16 Differences in predation levels of nests of female Northern Harriers in different harem positions, 1980–83, New Brunswick.

Table 5.9 Mean nest site quality scores for all nests of monogamous, α, β, and $\gamma + \delta$ Northern Harrier females over three years (1980–82).* Scores are based on moisture levels (from Simmons and Smith 1985).

	α	β	$(\gamma + \delta)$	Monogamous
Number of females	7	8	4	44
Quality scores	192.6	190.4	179.2	181.9
Range	163–206	161–206	172–198	152–206

[a] Kruskal–Wallis test for difference in means (α vs. β vs. $(\gamma + \delta)$): $H = 0.35$, 2 df, NS.

significant differences existed between the means for each group, suggesting that greater levels of predation were not linked to site quality. Nevertheless, as expected for females exercising first choice among resources, α and β females tended to have better quality sites than γ *or* δ females. Monogamists chose, on average, higher quality sites than later-settling harem females, but lower quality sites than first-settling ones. Only monogamous females used the entire range of nest sites during the three years, and only γ and δ females did not use the highest quality sites. Again, the latter result might be expected for females exercising the last choice among resources. In sum, while nest site quality scores were lower for polygynous females, they are by themselves unlikely to explain the very much higher predation levels on these nests. Are these females generally younger and so does their inexperience explain their low success?

Female age and mated status

Northern Harrier females are relatively easy to age in the field by eye colour. Close-diving females defending their nests during nest visits allowed repeated assessments of age as young (1–2 years) or old (3+ years), depending on whether they have brown or predominantly yellow irises, respectively (Hamerstrom 1968; Picozzi 1981). Of 89 females that bred, 71% could be aged either using this method or from captures (3), hide observations (11), or previous bandings (1). A special effort was made to age each harem female; this indicated that the majority were old (Table 5.10). The lowest proportion of 'experienced' females occurred among monogamists. Furthermore, the younger females breeding on the Tantramar Marsh also raised slightly more young than old females (2.5 vs. 2.3 young per female respectively: Simmons *et al.* 1986b). This finding, in combination with that of fewer young females within harems, means that the lower productivity of late-settling harem females cannot be explained by female inexperience. Senescence, however, cannot be ruled out.

This is an interesting result, with implications for the deception hypothesis also: one might assume that if females learned 'by their mistakes' and avoided polygynous situations later on, predominantly younger females would be found in harems. Since this is not the case, either all females always use the same decision rules ('choose a male with the highest food provisioning') and can never detect a deceiving male, or many of these females die and therefore do not learn, or there are sufficient numbers of old females who have never been in a harem for some to always fill these positions. We cannot distinguish these possibilities and they remain unreported from studies in which female age and harem position were known (Wisconsin and Orkney).

To conclude, the lower success of polygynously mated females was not related to inferior nest sites nor was younger age likely to explain the decreased success. Since male food provisioning performance was the best predictor of success for all other females and the provisioning rate of the polygynous females was significantly lower than that of α females (and decreased with position in the harem: Fig. 5.15), it is most likely that males' behaviour was the chief reason why their late-arriving females reproduced so poorly.

What should a female who finds herself with an ever-decreasing rate of provisioning do? Clearly, she is forced to forage which has several consequences: she leaves the nest for longer and opens up the brood to increased predation and starvation. As ground nesters, harrier nests are potentially open to predation from many sources; a nest left unattended by a female forced to forage for herself should be more vulnerable to both terrestrial and aerial predators. Direct

Table 5.10 Age in relation to mated status of female Northern Harriers, 1980–83.

Female status	α	β	$(\gamma + \delta)$	Monogamous
Number of breeding females	11	11	8	59
Number of known age	11	9	7	36
Number of females > 2 years (% of known age)	11 (100)	7 (78)	6 (86)	21 (58)

evidence for this was noted twice when two females that had left their nests unattended both lost their clutches to common ravens *Corvus corax*, even though their nests were in sites designated as 'concealed' (Simmons and Smith 1985). Daeman and Lorij (in Nieboer 1973) also noted that higher predation levels were characteristic of harrier nests left unattended by less vigilant females. Since neither inferior quality nest sites nor female inexperience could account for the greater predation levels, female absence seems the most plausible explanation. In its most extreme form, a poor male food supply may have also led to the desertion of viable nests by δ females. This is likely because δ females were the least provisioned of any harem females (Fig. 5.15), and probably needed to forage for themselves to avoid starvation.

Cost and benefits to polygynous males

So far we have dealt exclusively with female-related costs and benefits because of the importance of female choice in polygyny studies. However, male strategies, once males have attracted and started a nest, are of interest since males, too, are assumed to be maximizing their yearly fitness even if the two sexes' goals sometimes conflict. Here we have determined that males not only fed more prey items to their first-settled females, but also that these females were the first fed in the day and the male spent more time near the α nests. This greater investment in the α nest seems a logical tactic. The α nest probably not only has a greater chance of avoiding predators because of the slightly higher quality of such sites (Table 5.9), but nests started early in the season have inherently greater chances of high reproductive successes (for undetermined reasons). Pooled, these advantages to the male would probably outweigh advantages in preferentially feeding later-arriving females. However, this raises two questions: (a) why can't polygynous males feed their other dependent females to the same degree as they do their α females?; (b) why don't polygynous females desert their mates on realizing there is a reduction, or at least a failure to increase the rate as found in monogamous birds? The probable answer to the first question is that males are simply unable to; for European Marsh Harriers where time budgets were accurately recorded, all males, regardless of mated status, spent equal proportions of the day hunting (~29%). These proportions were similar to that found for African Marsh Harriers at similar stages of breeding. Both suggest that it is possible to allocate only a certain amount of time to foraging and demands above that cannot be met.

Second, polygynous females cannot desert poorly provisioning males and re-nest because they generally receive adequate supplies throughout incubation (~31 days) and only then suffer reductions (Table 5.9). At this stage it is too late for a female to lay a replacement clutch since the egg laying season is likely to be over. For Northern Harriers egg laying covers about 50 days and the median egg starting date for polygynous females (24 days from the end of laying) makes it impossible for a female to desert and relay. Once caught in this situation a female's best strategy is to try to rear young with minimal assistance, or desert to enhance her chances of surviving to the following season. The first strategy can work if food is superabundant and rainfall is minimal. Entirely deserted females have been known to rear some young in such situations (Van der Kraan and van Strien 1969) but food was then superabundant.

Additional costs: unhatched eggs

Another fitness-related cost noted among Orkney harriers (Balfour 1957) was the high incidence of unhatched eggs among harems. Balfour noted that unhatched eggs were more common as polygyny and population density increased. That polygyny rather than population increases led to greater nonhatching was suggested further by the finding that the proportion of unhatched eggs of harem males increased as harem size increased (Fig. 5.12). In the last seven years of the Orkney study, the hatching success of fully incubated eggs varied from 73 to 83%—a low proportion relative to figures for other species (Koenig 1982). In the New Brunswick population, nonhatched eggs accounted for only 7.3% of 302 eggs incubated to full term. Despite the lower incidence, once again this seemed to be related to polygyny rather than food or population density because polygynous males gave rise to 1.6 times as many unhatched eggs as did monogamous males. Furthermore, even though α females raised a greater proportion of their young than monogamists, they suffered high nonhatching rates, along with other harem members. Because polygynous males are copulating simultaneously with several females, it is possible that sperm depletion effects account for these trends. We will explore the interplay between food and sperm depletion as the source of unhatched eggs further in Chapter 6. Whatever its cause, in polygynous populations unhatched eggs are costs that polygynous males endure, in among the reproductive advantages that polygyny usually brings them.

To conclude, the four models proposed to explain polygyny in other species are insufficient to explain the poor reproductive success of polygynous female Northern Harriers. By determining mate choice cues of female harriers it

appeared that all females were choosing males with high provisioning rates. However, some females were fooled by males who first provisioned them well and then neglected them in favour of their α females. This deceptive acquisition of mates by polygynous male harriers provides a better explanation of polygyny in harriers than the skewed sex ratio idea, because settling patterns of polygynous females and non-skewed populations in several areas of North America are inconsistent with predictions of the sex ratio model. Polygyny is more prevalent in northern latitudes where vole population are generally higher, and within populations it is more prevalent in years with high vole populations. Ultimately, polygyny in harriers is associated with their ability to economically monopolize resources in the open habitat they breed in.

6 *Copulation patterns and sperm depletion*

Copulation patterns of birds may seem a somewhat perverse subject to study (and is not half as exciting as it might sound). Brief cloacal contact between mates, usually preceded by few preliminaries, is all the reward an observer gets for patiently observing birds for many hours. Yet this particular aspect of behavioural ecology has attracted more attention in recent years than virtually any other single subject. Why the interest?

The reason is not hard to find: it is a central assumption in most studies of reproduction, particularly those measuring annual and lifetime fitness, that the parents attending the brood are the biological parents of youngsters in their care. More and more this premise is being shown to be false. Several studies using sophisticated DNA finger-printing techniques have shown, for example, that anywhere from 15% to almost 100% of broods in single nests may not be the offspring of the attendant parents (Gowaty and Karlin 1984; Westneat 1987; Møller and Birkhead 1993a). This finding changes downwards our estimates of annual and lifetime fitness for individuals who have been cuckolded and increases fitness estimates for the cuckolders. Furthermore, it alters our concept of monogamy which is assumed, probably incorrectly, to be the predominant mating system in the majority of birds. The laying of extra eggs in other nests, known as 'egg-dumping', may add to this parasitism of parental care and the phenomenon (which is difficult to detect) has been detected in female ducks, starlings, coots and several other species (e.g. Andersson and Eriksson 1982; Karlsson 1983; Jamieson *et al.* 1999). More often extra-pair copulations (those that occur outside the pair bond) contribute to this cuckoldry. The present record is held by two species of cooperatively breeding Australian Fairy Wrens (*Malurus* spp.), in which as many as 100% of the brood may be fathered by males outside the group! (Brooker *et al.* 1990). These super males, are affectionately dubbed 'Big Blue Bonkers' by the researchers, and the high rate of extra-pair paternity, averaging 78%, has since been confirmed in this genus (R. Mulder in Møller and Birkhead 1993a). It has been suggested in this extraordinary group that females seek copulations elsewhere to avoid inbreeding in a species in which the sons and daughters stay behind to help subsequent breeding events. This is not the only or even the usual reason invoked for extra-pair copulations (EPCs).

Reasons for extra-pair copulations

Females may accept extra-pair copulations for several reasons: they may gain increased paternal investment from the cuckolder or they may receive sperm from a higher quality male than the one they are paired to. In the first case they can gain immediate benefits because the cuckolding male may help feed at the nest where his offspring are reared. Although a male is unlikely to recognize his own offspring directly, there is strong evidence that his rate of parental investment in the form of food provisioning is directly related to the chances that he is the father of one or several young (Davies 1992; Sheldon *et al.* 1997). In the other case females may gain genetic benefits by mating with males who have indicated their viability by beginning their own brood earlier, or through plumage characteristics (Kempenaers *et al.* 1992). More simply, females may be insuring that they gain some sperm to fertilize all their eggs (Wetton and Parkin 1991). This may occur if their mates prove to be infertile as found in north American blackbirds that were vasectomized (Bray *et al.* 1975), or possibly if sperm depletion occurs in some males which copulate in rapid succession, as confirmed in captive finches (Birkhead 1991). Because data for sperm depletion in wild birds are non-existent, we will return to this idea for male harriers towards the end of this chapter.

It is not surprising to learn that intense selection has favoured behaviours which apparently reduce the chance of a male being cuckolded. There are three main methods identified thus far by which male birds may do this. (Birds have not been quite as inventive as insects in this regard since some insects remove rival sperm from their mates prior to insemination with specially modified penises; two African buffalo weavers (*Bubalornis* spp.) show external structures which were thought to function in a similar way (Birkhead *et al.* 1993), but which are now known to function as a stimulatory organ only (Winterbottom *et al.* 1999). The closest that certain birds have come to this bizarre method of sperm competition is by cloacal pecking. Prior to copulation a male pecks vigorously at the female's extended vent or cloaca through which sperm are passed, a behaviour that is curiously invited by the female. This eventually stimulates her to eject some of the last ejaculate received, prior to a copulation by the attendant male. The behaviour is thought to increase the chances that the pecking male fertilizes the remaining eggs, and it has been confirmed in two species: the Dunnock *Prunella modularis* (Davies 1992) and the House Sparrow *Passer domesticus* (Møller 1987a). However, this is not the main method of reducing cuckoldry.

The most frequently observed behaviour is mate guarding (Beecher and Beecher 1979; Birkhead 1979), whereby males follow their mates closely wherever they go, especially during the fertile period before egg laying. With some important exceptions, virtually all species that have been studied in this regard show mate guarding behaviour during the critical fertilization window period (Birkhead and Møller 1992). This is a period—shortly after a female has laid an egg, when the next follicle is released from the ovum into the infundibulum and is available for fertilization—of about 30 minutes. Mate guarding certainly doesn't prevent a male being cuckolded (Kempenaers *et al.* 1992), but it may reduce the chances of extra-pair fertilization.

Reasons for high copulation rates in raptors

The third method occurs most frequently in those species which are less able to guard their mates continuously because they are away foraging. These species include colonial sea birds, which forage a few kilometres out to sea leaving their mates surrounded by many potential interlopers (Birkhead *et al.* 1987; Birkhead and Møller 1987). As we have seen, raptorial birds do precisely the same; the males hunt for long periods to bring essential egg-forming food to their mates. Moreover some of these raptors are colonial. Tim Birkhead and Anders Møller were struck by the fact that these species showed extraordinarily high copulation rates: over 400 copulations for one clutch of eggs for European Goshawks *Accipiter gentilis*—a fact that also puzzled the original observer of goshawk copulation behaviour, V. Holstein—and 160 copulations for Ospreys (Birkhead and Lessells 1988). This is in contrast to the single copulations needed to fertilize a similar number of eggs among species like grouse, sparrows, and larks. Birkhead and Møller reasoned that the very high copulation frequency evident among colonial sea birds, vultures, and raptors is one means by which males can at least devalue the likelihood of extra-pair paternity arising from extra-pair copulations.

A prediction of the Birkhead–Møller paternity assurance hypothesis is that the highest copulation frequencies should be found among colonially nesting raptors, and lower frequencies among solitary species. Although their reviews have provided evidence for this, they often had to rely on qualitative assessments of copulation frequency. For raptors, the importance of the colonial lifestyle versus solitary nesting in the evolution of copulation rates can be separated by looking within populations at pairs that are clumped versus those that are solitary. Higher copulation frequency is clearly expected for the clumped nesters. Harriers fortuitously provide an excellent opportunity to test these ideas since they nest both singly and in loose assemblages, and for species such as Montagu's Harriers in very dense assemblages. To analyse these different ideas for harriers I observed four pairs of African Marsh Harriers per year, for a total 1670 h. Data for other pairs were incomplete due to poorer visibility into their nesting territories. Because all pairs were observed from raised vantage points between 60 and 100 m from their nests, it allowed good views of copulation behaviour, both in and out (up to 1 km) from the territory. The identity of each bird could be confirmed by the patagial wing marks given to them early in the year, or in previous years. I compare my findings with detailed studies recently published from two Montagu's Harrier populations.

Copulation behaviour

Like vultures and hawks, female marsh harriers have a far-carrying whistling 'psssew' call, repeated more frequently as the male approaches to copulate. This made the task of identifying copulating pairs easy and also informed other pairs in the area that the birds were copulating. Only the females call. Copulations were usually initiated by this incessant calling from the females as their mates flew over, with or without food. It sometimes appeared that females demanded copulations, judging by the vociferous calling, head lowering, and tail fanning that accompanied it. This is a curious behaviour on the face of it because females probably require just one insemination to fertilize their eggs especially as they probably have specialized tubules, as found in other species, where sperm can be stored. Possibly it is the food that they are more interested in because the call is indistinguishable from that used when copulations had ceased.

To copulate, male harriers hover over their mates, landing gently on the female and laying their long tarsi along her back and clenching the potentially damaging feet and talons into a fist. Receptive female hold their tails sideways or slightly elevated, and males effected cloacal contact by lowering their tails while flapping to maintain balance. Copulations were of relatively short duration averaging 5–6 s (5.6 ± 1.8 s, n = 129), the longest being 11 s. After pairs dismounted, both male and female preened vigorously, ruffling and shaking their feathers and preening around the cloacal area. This happened without fail and is commonly recorded in many other species. One interesting suggestion for its occurrence is that birds are preening away parasites that may have been

Fig. 6.1 Copulating Pied Harriers. Most attempts among harriers occur on the ground and often take place close to the eventual nest site.

transferred during copulation—this being the only time when adults are ever in direct contact.

Copulations followed both nest building and food presentation (below). Male harriers often build their nests stick by stick rather than bringing in several sticks or grass stems at a time. Because copulation sometimes follows these visits one function may be to allow a male more opportunities to copulate. An alternative, that frequent copulations stimulate egg laying, could be ruled out because no relationship was apparent between early egg laying and copulation rate ($r = 0.20$, $n = 11$) for the African harriers.

Most mating occurred on the ground (92% of 191 copulations) and more rarely on low fences or single posts (7%), or bushes (1%). All copulations between known pairs took place less than 150 m from the nest site, 80% of them less than 50 m from the nest. Of the remainder, 3% occurred on the nest itself, and 2% occurred outside the territory. The latter were all extra-pair copulations. Females were not indiscriminate but clearly recognized their mates; the food–copulation call was accompanied by low alarm calling when other males approached. The ability of females to recognize their mates in this way illustrated the legendary vision that raptors have. Grounded females gave these calls to males that flew over head well outside my range of vision (about 700 m against a bright sky); on descending into the territory it became clear that females were not only seeing but *identifying* their own mates at these distances.

Within-pair copulations: diurnal and seasonal fluctuations

Copulations were much more likely to occur in the early morning than at other times. The peak occurred between 07h00 and 08h00, and at 0.23 copulations/h was fourfold higher than at other times of day (Fig. 6.2). This is typical of many raptors and may coincide with the fertilization window which occurs early morning. However, Montagu's Harrier copulations in a large Spanish population did not follow this pattern. Arroyo (1999) found that copulations peaked from mid morning to midday, and almost no copulations took place early in the morning when little flight activity was evident. This is confirmed in an Italian population of Montagu's Harriers in which copulations were also most frequent between 10h00 and 14h00. Unlike African harriers, Montagu's Harrier copulations are associated with food deliveries (below) and since the rate was low in the early morning, this may explain the differences (Arroyo 1999). Seasonally,

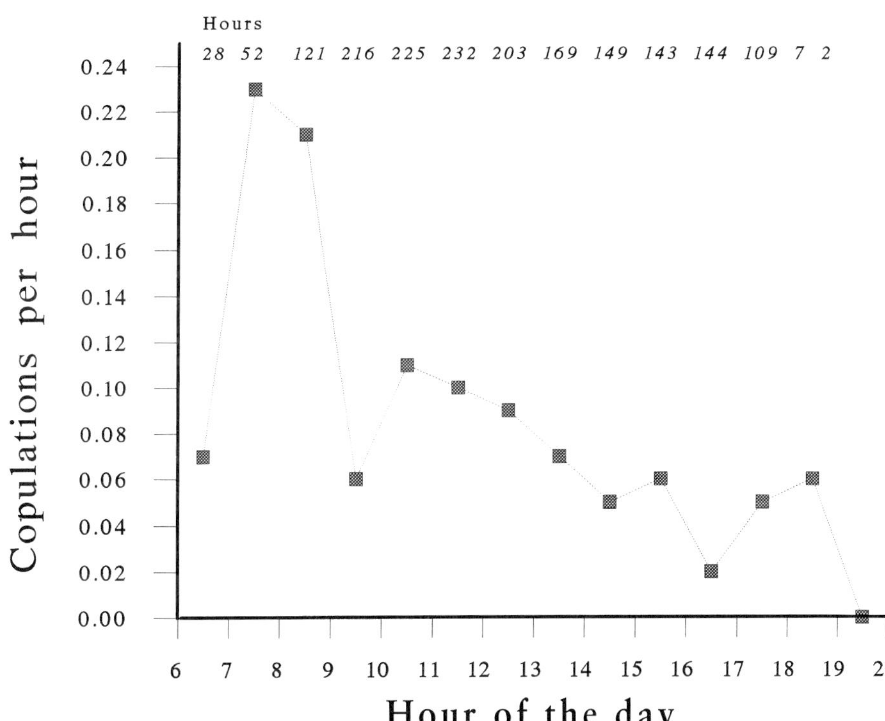

Fig. 6.2 Diurnal fluctuations in copulation frequency of African Marsh Harriers, based on 196 copulations recorded between 1984 and 1986. Hours of observation for each hourly block are shown at top.

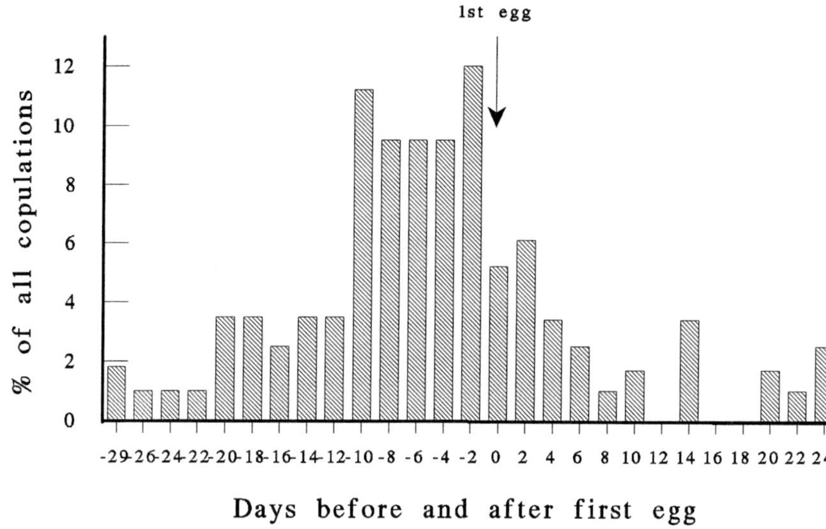

Fig. 6.3 Copulation frequency in relation to egg laying, the fertile window is assumed to start 7–8 days prior to egg laying when these Montagu's Harriers copulate most often. Based on 117 copulations (B. Arroyo).

copulations for both African Marsh and Montagu's Harriers were clumped around egg-forming. This is best illustrated by data on Montagu's Harrier from Italy and Spain. In Italy, Pandolfi and co-workers (1998) showed two peaks for successful copulation: the first, 4 weeks prior to egg laying and the second, as expected in the week when eggs were being formed. Montagu's Harriers in Spain were also studied 4 weeks prior to egg laying (Arroyo 1999) but showed only one peak in the proportion of total copulations, from 10 to 2 days prior to the first egg (Fig. 6.3). The latter pattern is expected if fertilization occurs shortly before eggs are laid (Birkhead and Møller 1992).

Most African Marsh Harrier pairs ceased copulating about 120 d after they started. For two other species, the curious bimodal peak of copulations found by Pandolfi's team was also found. In goshawks, peaks occurred about 35 days and 10 days before egg laying and in Lesser Kestrels 65 days and 5 days before first eggs appeared. These patterns are not observed in marsh harriers, or Ospreys (Birkhead and Lessells 1988). Their function in other species may be to stimulate breeding or enhance the pair bond. This may explain why it doesn't occur in Ospreys or African harriers: rainfall is the most obvious proximate stimulus for breeding in the latter species and both species enjoy longer term pair bonds than kestrels and goshawks. Pandolfi's team suggested that male quality may be tested by the female's soliciting copulations well before eggs are fertilizable and in good time to perhaps change mates, if males prove incompetent. This suggestion was made because females were observed refusing copulations, and did so most often (18% of 72 copulation attempts) during the courtship period. This may be a test of male persistence, or it may simply mean

Table 6.1 Copulation rates per 12-h day for four intensively monitored pairs of marsh harriers over three years. Rates are separated for the pre-laying and the post-laying periods.

Year	Pair	Copulation rates/day	
		Pre-laying (12-h day)	Post-laying (13-h day)
1984	EV1[a]	0.9	0.5
	EV2	4.3	1.2
	EV3	4.0	1.3
	RONDV 2.5	—	
1985	EV1	1.7	0.9
	EV2	1.8	0.4
	EV3	—	1.3
	RONDV	2.2	—
1986	EV1	1.5	0.5
	EV2	2.7	—
	EV3	1.0	1.8
	RONDV	2.4	—
Means	± 1 SD	2.3 ± 1.1	0.9 ± 0.5

[a] EV1 female changed mates in each year.

the females are themselves not ready for copulation. Refusals declined in the fertile period amounting to 3% of 29 copulations. The fertile period is taken to be a period 7–8 days before the first egg is laid through until the last egg.

The rate of attempted copulations in African harriers in the period before and during egg formation varied considerably between pairs, but averaged 2.3 copulations/pair/day (Table 6.1). For Spanish Montagu's Harriers rates were virtually identical during this period at 2.2 copulations/day (Arroyo 1999), but they were much higher for Italian birds at about 6.0/day. The highest levels occurred during the presumed egg forming/laying period, when some male African Marsh Harriers copulated up to 7 times daily ($\bar{x} = 3.53 \pm 1.6$, $n = 12$). Rates following egg laying decreased 2.6-fold to just below 1.0 copulation/pair/day. At these rates a pair of harriers copulates about 50 to 220 times to fertilize one clutch of eggs. Over the entire season a pair copulate up to 300 times. These rates are about 70% of the reanalysed rates for goshawks, but similar to those reported for Ospreys. Since 73% of all attempts were estimated to be successful, 37 to 160 ejaculates go towards fertilizing a clutch of 3 to 5 harrier eggs. Ospreys exhibit similar trends with 20 to 97 successful copulations per clutch. Several other raptors have now been studied since these data were collected and they show similar patterns: high copulation rates with relatively low extra-pair copulations (Table 6.2). Pandolfi *et al.* (1998) documented a minimum of 31 and a maximum of 252 successful copulations per pair within small colonies of Montagu's Harriers in Italy. Despite these high rates they found relatively high non-hatching rates (20% of 25 eggs). Spanish populations studied by Arroyo had relatively low copulation rates averaging 39 copulations per clutch of four eggs.

Table 6.2 Copulation patterns of 12 species of raptor in relation to breeding dispersion and risk of cuckoldry. EPC = Extra-pair copulations.

	Average rate per day before laying	(Max)	Total copulations per clutch	EPC (%)	Mate guarding?	Source
Solitary						
Goshawk	5.1	(12)	421*	?	Some	1
Osprey	9.8	(17)	160	~1%	Some	2
Black Vulture	?	?	?	0%	?	8
Egyptian Vulture	?	(9)	55	4.6%	?	12
Mean	**7.5**		**212**	**1.9%**		
Semi-colonial						
Af. Marsh Harrier	3.5	(7)	220	2.1%	Some	3
Merlin	7.5	(~11)	60	Yes?	No	4
Bearded Vulture	?	(8)	?	?	No	5
European Kestrel	11.5	(?)	230	1%	Yes	9
Mean	**7.5**		**170**	**1.6%**		
Colonial						
Cape Vulture	2–3	(8)	131	~1%	?	6
Lesser Kestrel	4.8	(11)	326	6.7%	No	7
Montagu's Harrier	6.0	(9)	252	3.4%	Yes	10
Montagu's Harrier	2.2	(5)	39	3.5%	Yes	11
Black Kite	?	(11)	196	2.7%	?	13
Mean	**3.9**		**188**	**3.5%**		

1. Moller (1987). * Recalculated from this reference based on a 13-h day; 2. Birkhead and Lessells (1988); 3. Simmons (1990); 4. Sodhi (1991); 5. Brown (1988); 6. Robertson (1986); 7. Negro *et al.* (1992, 1996), J. J. Negro *in litt.*; 8. Decker *et al.* 1993; 9. Korpimaki *et al.* 1996; 10. Pandolfi *et al.* 1998; 11. Arroyo 1999; 12. Donazer *et al.* (1994); 13. Koga and Shiraishi (1994).

Extra-pair copulations

Apparent extra-pair copulations have been reported in the polygynous Northern Harriers in Wisconsin (Hamerstrom 1969) and polygynous Hen Harriers in Orkney (~3 in 59 (5%): Picozzi 1984a,b), so cuckoldry is a potential cost of clumped breeding in this group. Among the 196 attempted copulations observed over 12 pair years for African Marsh Harriers, only four were extra-pair copulations (2.1%). Two occurred during the fertile period of one renesting female and two occurred during the nestling dependence period of one other female—outside any fertile period. Both the males involved were unpaired young adults who had set up neighbouring territories. For both females, only one copulation each was probably successful judging by length of time the male stayed on the female's back and the tail movements during the attempt. The female actively solicited, and cooperated in, all these copulations. This is important because it seems that females are the sex that both invites copulation and, more often than not, dictates whether it is successful or not.

The reaction of the mated males in two instances was unexpected. At no time was the younger courting male attacked despite being a mere 200 m away.

Instead, one male undertook intense aerial evolutions high over the pair, and plummeted into his own nest area, skyspiralling as he descended. This display was not sufficient to prevent the two extra-pair copulations during the fertile window, but did induce his mate back within the same morning. As expected his own copulation rate thereafter was very high. This is compatible with the idea that males copulate often to dilute any other sperm present—sperm competition—and that the male copulating last tends to father a greater proportion of the eggs (Birkhead *et al.* 1988). That male African harriers did not attempt to physically prevent extra-pair copulations is rather puzzling, but a similar phenomenon has been observed in Merlins *Falco columbarius* (Sodhi 1991), and Lesser Kestrels *Falco naumanni*. In the latter species extra-pair copulations are more frequent than in harriers, constituting 6.7% of all attempts, and on 14 of 18 occasions when paired males could have directly prevented cuckoldry they made no effort to do so (Negro *et al.* 1992). This is both unexpected and inexplicable. It seems unlikely that larger more powerful cuckolding males prevented such interference as suggested for kestrels, since the mated males merely need to disturb the extra-pair copulation, as commonly seen in smaller species (Trail 1985; Barnard and Markus 1989). Only one study reports male harriers physically interfering with possible EPCs: in her four-year study of colonial nesting Montagu's Harriers, Arroyo (1999) observed 11 possible extra-pair attempts of which three were interrupted by a second male flying towards and then pursuing the copulating male out of the area. No such harassment was observed in similar study of the same species in northern Italy (Pandolfi *et al.* 1998). The differences may be explained by the clumping of nesting birds, which in Arroyo's study was higher than for any of the species. It may be that because of the close proximity of nests in her study, either other males incidentally break up the copulation attempt, because they are more likely to detect it, or more plausibly, other close nesting males are more at risk and thus more vigilant in preventing stolen copulations.

It is of interest that Poole (1985), studying colonial Ospreys in North America, noticed that females paired with males who were not good provisioners were more likely to seek extra-pair copulations. Although extra-pair copulations in raptors are relatively rare compared with other groups (Møller and Birkhead 1993b), and constitute less than 7% of all copulations observed (Table 6.2), male inexperience or low quality clearly may cause a female to seek sperm elsewhere. Of course cause and effect may be reversed here and a female with a poor food-provisioning mate may simply be more open to EPC because she is left alone for longer periods by a mate who spends longer periods pursuing food.

Recent studies of North American blackbirds (Icterids) and British tits show that those females which seek extra-pair copulations specifically seek them from higher quality males. Such males also do more 'sneaking' than other males (Gibb *et al.* 1990; Kempenaers *et al.* 1992, 1997). This was not so for African harriers since the females who sought EPCs left to copulate with same age or younger males. The important point here, to which we will return later, is that extra-pair copulations are surprisingly *uncommon* among raptors.

Does copulation rate increase with nest clumping and threat of EPCs?

Because three of the marked pairs of African Marsh Harriers were present in all years, and nest spacing changed, it was possible to assess the effect that variable nest spacing had on copulation behaviour while keeping pair effects constant. Copulation rates for the same pairs between years varied up to fourfold (Table 6.1), and as predicted co-varied with different nest spacing. In 1984, three pairs nested a mere 40–50 m apart, but separated in 1985 and 1986 when the central pair, EV2, moved 2 km away to a solitary site. For two of the three clumped pairs in 1984, copulation rates were 1.9 and 4.0 times higher than their own rates in the following two years when they were solitary. Combined, these males also exhibited copulation rates 2.2 times higher than the average for all other males. Moreover, just before egg laying, the EV2 male copulated 6–7 times daily with his mate for both his first and his replacement clutch. These rates were equalled only by the same male in response to EPCs by his mate (below).

While these increased rates lend support to the idea that nest spacing (and particularly the presence and proximity of neighbouring males) effects mating frequency, an alternative is that 1984 rates were intrinsically higher than other years. I could rule out such a possibility because two pairs whose spacing didn't change between years showed rates no higher in 1984 than other years (EV1 and RONDV in Table 6.1). An additional reason to suspect that the high copulation rates of the two males in 1984 were related to paternity assurance was that they were equalled by the cuckolded male in 1986. Following these copulations, the female rejoined her mate and he copulated 4–5 times a day immediately thereafter. These results therefore suggest that male harriers do indeed copulate more frequently when they are more susceptible to cuckoldry from close nesting neighbours.

Male age and inexperience added some variation to the differences in copulation rates but body size did not. For three of the four pairs monitored, the same individuals remained together. The fourth female, however, changed her mate in every year. In contrast to the three experienced pairs, the average copulation frequency of the fourth female before egg-laying (1.37 ± 0.4 copulations/day), was consistently lower than that of her neighbours (2.6 ± 1.1 copulations/day). Despite these lower copulation rates, this female suffered no greater incidence of unhatched eggs (10%) than the other pairs ($\bar{x} = 11\%$). These patterns differ from Ospreys in which inexperienced pairs copulate most frequently in order to achieve the same number of successful copulations as older pairs (Birkhead and Lessells 1988; Poole 1989). Neither male body size ($r = -0.18$, $n = 6$) as determined by wing chord, or female body size ($r = 0.34$, $n = 4$) influenced copulation rate.

Recent confirmation of the propensity for harriers to copulate more when the risk of cuckoldry and sperm competition is greatest comes from studies of Montagu's Harriers. Not only did clumped groups copulate for a greater number of days, but copulation rate during the critical egg laying period were significantly higher with an increase in the number of neighbours (Fig. 6. 4).

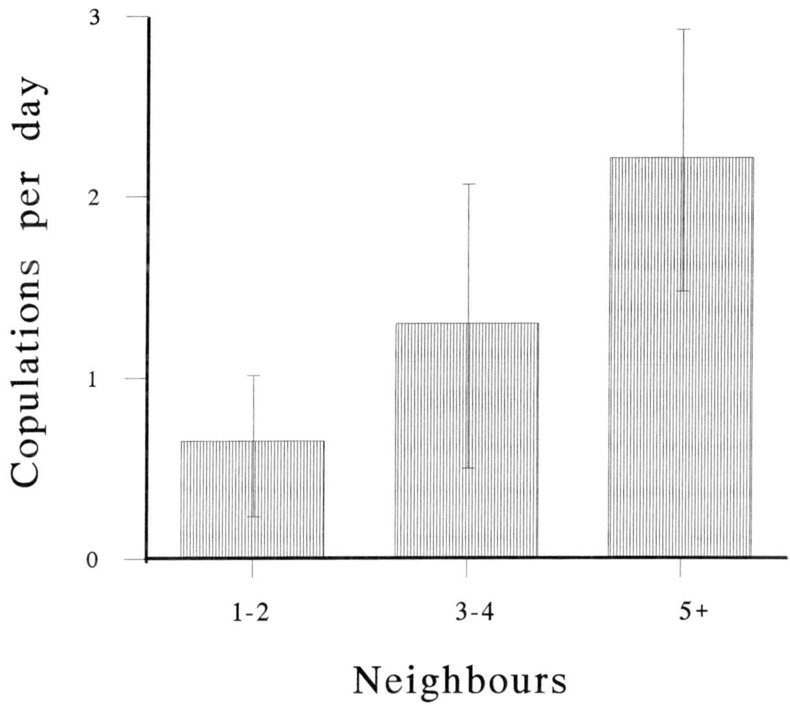

Fig. 6.4 Copulation rates in relation to the number of close neighbours in Montagu's Harrier breeding in central Spain (B. Arroyo). An increase in the number of neighbours increases the probability of extra-pair fertilizations and the trend for more frequent copulations may be a response to dilute the probability of being cuckolded.

Do copulation frequencies among raptors support the paternity assurance hypothesis?

We can now use these findings for harriers and those reported for other raptors with colonial breeding patterns to determine if birds at greatest risk from rival sperm copulate more often as predicted. At least seven species do not fit the pattern of a greater frequency of copulations with higher nest clumping (Table 6.2). Two colonial species, the Cape Vulture *Gyps coprotheres* and the Lesser Kestrel which should be most vulnerable to stolen copulations, copulate no more frequently than solitary or semi-colonial species. Ospreys nesting solitarily in Scotland exhibited the highest rates of all. Bearded Vultures *Gypaetus barbatus*, ranked here as semi-colonial because their pothole nests are within sight of each other, do not feed their mates and the females self-forage, allowing the possibility of EPCs. Yet males do not apparently guard their mates because Brown (1988) reports 45–55% of all foraging birds were single. However, these vultures copulate at maximum only 8 times/day/pair, less than two solitary species for which good data are available (Table 6.2). These results are difficult to explain in light of expected sperm competition trends arising from spacing alone.

One trend to emerge which is predicted from their colonial lifestyle is that extra-pair copulations are about twice as frequent among colonial raptors (mean 3.1% from 5 studies) as they are in semi-colonial (1.6% from 2 studies) and solitary species (1.9% from 3 studies: Table 6.2). From the same table it is apparent that the average number of copulations immediately before the first egg in the same groups is lower not higher in colonial species (3.9 copulations/day) than it is in semi-colonial or solitary species (7.5 copulations/day). More species need to be studied and phylogenetic relationships accounted for these trends to be verified. We would predict that colonial species, as a group open to greater threat of extra-pair copulations to their partners, would copulate more often than non-colonial birds. The results show a non-significant inverse trend ($r = -0.38$, $n = 8$) with body size. However, they suggest that a constraint is operating on colonial species: more EPCs are experienced, and fewer overall copulations are performed to counter the EPCs, possibly because birds are away more often from the colony and it is simply one of the hazards of colonial living. But are EPCs actually resulting in unrelated young being introduced into the nests?

Extra-pair paternity resulting from observed EPCs are also low in this group. No extra-pair paternity was found with DNA fingerprinting techniques in either vultures (Decker *et al.* 1993) or Merlins (Sodhi *et al.* 1993), and only 5% of all sparrowhawk broods were found to contain extra-pair young (Møller and Birkhead 1993b). It is useful to know how observations of EPCs actually translate into extra-pair parentage. This has been assessed by Korpimaki *et al.* (1996) who observed a 1% proportion of successful extra-pair copulations in kestrels, and later looked at genetic evidence for 319 young over three years. Just six young, or 1.9% of all those sampled, were not fathered by the attendant male; in at least one case the bird was a know poor food provider and had a low copulation rate. For colonial Lesser Kestrels, where the opportunity for fathering other chicks is probably greater, observed EPCs (6.7%) were greater than the actual parentage achieved outside the male's own nest (3.4% of 87 nestlings: Negro *et al.* 1996). So it seems that intense observations around the time of egg laying are a good guide to the actual degree of extra-pair paternity for open country species such as kestrels and should therefore be relatively good for harriers too.

A second, more recent hypothesis is that the frequency of extra-pair paternity is related to the degree to which males feed their offspring (Møller and Birkhead 1993a). In other words, for those species like raptors in which males frequently feed their offspring, extra-pair copulation and paternity should be low. This is certainly the general case judging by the low proportion of extra-pair copulations in this group. It may arise because males will only expend considerable effort feeding their young if they are certain that they fathered the brood. If they cannot be certain they will show less interest in feeding. From a female's perspective it may be costly for her to indulge in extra-pair copulations since she may lose the services of her mate who is vital to the breeding effort. This may be one reason that females rarely seek or receive extra-pair copulations in raptors,

yet may do so in species where fathering is not as important. What Møller and Birkhead are saying is that the certainty with which males can assess their fathering ability may have affected their whole lifestyle. But how can a male raptor be certain that he has fathered the brood when he is away so frequently?

Food for sex?

Food is often delivered shortly before copulations occur, suggesting that a 'food for sex' trade-off occurs. That is, the female may demand food in return for copulation, while the male may provide food to obtain a copulation—a two-way compromise. If so, is this one way in which a female may test a male's commitment to the breeding schedule? Alternatively, is it a subtle form of paternity guarding whereby a neighbouring male *must* provide some food before he can expect to get a copulation? Is food parcel size perhaps a measure of male quality and motivation? Does the presence of food improve the success of each copulation? And are females likely to be more faithful if their mates are better providers, as suggested for Ospreys?

In common with many birds, particularly raptors, harrier copulations often accompany the provision of food parcels. However, of 125 copulations for which duration was estimated, food was present in 62 instances and absent in 63 instances. Thus food presentation was not a pre-requisite for copulation activity in harriers, contrary to findings for Scottish Hen Harriers (Picozzi 1984a) in which food was present at 73% of 59 copulations. By way of confirmation of this result for African birds, there was no relationship between pre-egg copulation frequency and provisioning rate (r = 0.15, n = 10 pairs, n.s.). Does food improve success as measured by the length of the copulation? In total, the success of copulations was estimated at 73% of all attempts, similar to that for kestrels (89%: Negro *et al.* 1992) but higher than the 39% estimated for Ospreys (Birkhead and Lessells 1988). Copulation duration in African harriers was unaffected by the presence or absence of food. The average duration of copulations with food present (5.3 ± 1.7 s) did not differ significantly (t = 1.63, 123 df, $P > 0.1$) from that with food absent (5.9 ± 2.0 s). This is different to that found for Montagu's Harriers in Italy in which copulation duration with food present was significantly longer (4 s) than with food absent (0 s—median values: Pandolfi *et al.* 1998). Moreover, Spanish Montagu's Harriers showed strong correlations between copulation rates and provisioning rate in the early pre-laying period, although this relationship disappeared during the late pre-laying and laying period (Arroyo 1999).

Perhaps food size assists males in gaining successful copulations or females may assess male parental motivation on the size of the prey package supplied? I estimated the size and biomass of provisioned items by ranking each item relative the purveyor's foot and comparing them with known weight items (Simmons 1986a). However, this measure could also be ruled out as a factor influencing copulation success. First, there was no correlation (r = –0.03, n = 53) between prey size and copulation duration, and second, the prey present during successful copulations $\bar{x}$ = 38.2 ± 29 g, n = 53) was no larger than that present during unsuccessful copulations $\bar{x}$ = 38.6 ± 30.7 g, n = 17). Food size per se, therefore, played no role in the success of copulations in African birds but did in Montagu's Harriers.

There was one important exception, among the African birds, when food did appear to affect copulation behaviour. For a subset of copulations for which I could be certain that the attempt was unsuccessful (because there was no attempt at cloacal contact), there was a difference with respect to food presence. A total of 31 unsuccessful copulation attempts occurred when food was absent, in contrast to 16 unsuccessful copulation attempts when food was present. The difference was significant (X^2 = 4.18, Yates corrected, $P < 0.05$). The difference was unassociated with the behaviour of the female; only nine within-pair copulation attempts were ever rebuffed by the female—seven with food and two without food. Thus, food is not a pre-requisite for copulation activity, nor does it affect duration, but copulations are more likely to be unsuccessful when food is absent. Similar results have recently come to light in an Italian population of Montagu's Harriers in which significantly more copulations were attempted when food was present than when it was absent (Pandolfi *et al.* 1998). In that population, females also rebuffed a high 12% of 114 attempts.

How can the differing results between African Marsh Harriers (no trend for food-for-sex behaviour) and Montagu's Harriers, in which food and copulations are related, be reconciled? Since we have assumed that females are testing a male's ability to provide food by allowing him to copulate when he does so, are the differences between African Marsh and Montagu's Harriers related to mate assessment? One important difference pointed out by Arroyo (1999) is that there is less need for mate assessment in African birds because they typically enjoy pair bonds of more than three years while Montagu's Harriers re-pair each year. This difference would also explain the pre-requisite of food before copulations in Hen Harriers which also tend to re-pair each year. It further explains the trend for a strong relationship in Montagu's Harriers between copulations and food presentation in the early pre-laying period (when mate assessment is occurring) and the lack of such a correlation after mate choice has occurred (late pre-laying and egg laying). Might this also have some bearing on extra-pair attempts? It was worthy of note that both of the EPC attempts that occurred during a fertile window were accompanied by food, and in the other cases the male had provisioned several food items prior to attempting any copulation. Thus food might not be necessary for copulation in African Marsh Harriers but it certainly helps!

To summarize these findings briefly: it is clear that a food-for-sex trade-off is not found among African harriers but does occur among harriers that re-pair each year. The most significant finding was that the closer the nest spacing, and thus the greater the likelihood of increased cuckoldry, the more often the pairs copulated in both African and Montagu's Harriers. These trends therefore support the idea that males attempt to devalue the chances of being cuckolded.

Can males also guard their mates?

One may regard it as somewhat odd that raptors in general are expected to tolerate high rates of cuckoldry because of their lifestyle, and yet actually suffer a low rate of extra-pair copulation (Table 6.2). How can this be reconciled? We will now look at the possibility that male harriers do indeed insure paternity by *also* guarding their mates. It has been assumed that mate guarding does not occur in species like raptors because males are constrained by their lifestyle to forage away from the nest at critical fertile periods. Møller and Birkhead (1993a) argued that high frequency of copulations may be a 'best of a bad job' compromise for species unable to mate guard. It is a poor option because the male is not *assured* of paternity by frequent copulations—last-male precedence or sperm dilution is not a 100% cure, merely the best alternative. So is it strictly true that male raptors do not mate guard? Or are there more subtle ways of mate guarding than simply staying close to ones mate at all times?

There are two potential methods open to male raptors. It is a well-established fact that eggs are normally fertilized in the early morning hours because this is when the follicle descends from the ovary shortly after egg laying (Cheng *et al.* 1983). Reviews have shown that many species do indeed show their highest copulation activity at this time (Birkhead and Møller 1992), and African Marsh Harriers are no exception (Fig. 6.2). Thus the first possible means of reducing cuckoldry is for harriers to spend longer periods of time with their mates during the morning period. By recording activity of focal birds every two minutes I could categorize their activities as those inside (within 150 m) and outside the territory, for all daylight hours of the day, during their fertile periods. One male whose mate was re-laying spent 42% of his total time perched in the territory and only 9% perched outside it. Two other males observed simultaneously, who were close to hatching their clutches, spent an average of only 16% perched in the territory and 34% perched outside it. Thus the male who had a greater chance of being cuckolded spent 2.5 times as long perched in his territory near his mate as the two males who no longer needed to mate guard.

This occurred *despite* the greater foraging effort of males during the egg-forming period discussed in Chapter 4. This suggests male harriers do guard their mates during the most critical portion of the day and this may reduce extra-pair fertilizations. This has recently been confirmed by French researchers undertaking experimental work on Montagu's Harrier. Mougeot *et al.* (1998) presented stuffed decoy males at nests during and after the fertile periods and found that males stayed longer around their nests during the fertile period than at any other time. Even more striking was the increased copulation rate of males presented with decoys of male intruders, and the longer copulation periods (Mougeot *et al.* MS) only during the laying period.

Similar patterns were uncovered by Møller (1987b) who reanalysed the extensive observation on goshawks made by V. Holstein. He showed two trends: male goshawks stayed closer to their mates during the period 20 to 5 days prior to egg laying (approximating the fertile period)—and they specifically stayed close to their mates in the early morning period (04h00–07h00) than during incubation. Hence there is good evidence that at least two other raptors show mate guarding behaviour, contrary to that normally assumed for these birds.

These trends have recently been confirmed in the European Kestrel by Erkki Korpimaki and his team (1996). They found that, like harriers, kestrels are very tuned into the cyclic population fluctuations of their vole prey and this affects the degree of mate guarding males undertake. In high and increasing vole years, when nest density was over twofold higher than in the low vole year, mate guarding and copulation rates were also significantly higher. Thus these small falcons attempted to insure paternity both by more guarding and more copulating, paralleling findings with harriers.

A second, more obvious, method of mate guarding in harriers was evident in an increase in vigilance when other males were in the area. This was evident for two close-nesting males which frequently flew over each other's territory and engaged in territorial displays (as did their mates). Females often food-solicited these males as they flew near them. However, subtle differences occurred between these calls and those given to their mates; calls to intruders ended with a short cackling alarm. These calls were sufficient to alert their mates, who returned from afar, even when they were clearly departing to forage. This was a female-initiated warning to which mates responded instantly. Other female behaviours such as direct aggression, or a lack of cooperation during attempted copulations have been reported for female vultures, and for two species of kestrel preventing copulation attempts (Robertson 1986; Negro *et al.* 1992; Korpimaki *et al.* 1996).

A third and unexpected method of mate guarding was the increased aggression of males towards their mates when neighbouring males were in the vicinity. On several occasions males were observed, when another male was in the area, striking their mates in the back, sending them squealing into the vegetation below. This unusual behaviour declined to zero as the egg laying period came to an end. Harriers are not the only raptors to exhibit such behaviour. Close following, and males swooping low over and touching their mates back 'as if urging a return to the nest' are also reported in Ospreys (Poole 1989). Thus the statement that raptors cannot mate guard needs some qualification. Clearly they do so within the constraints imposed by their lifestyle of foraging for their egg-bound mates. This may be the reason why surprisingly few extra-pair copulation are observed in a group in which males are apparently absent for long periods.

Another obvious reason is that if all males are busy guarding their own mates and thereafter have their work cut out foraging for their dependent mates, there's little opportunity to go hunting for copulations anyway! Only unpaired males are likely to feature at all, and this is precisely the trend found for African harriers and Montagu's Harriers (Arroyo 1999).

Reasons for extended copulation periods

Several species of raptor show large numbers of copulations which often continue long past the date of laying. Some of these are clearly in response to human activity or interspecific intrusions (Ellis and Powers 1982; Gargett 1990) but many are not. Nor can post-laying copulations be attributed to reproductive synchronization, as suggested for pre-laying copulations in other species (Erickson and Zenone 1976). The only adaptive function that seems likely is one of insurance. Many of the raptors showing this trend are able to lay replacement clutches on the failure of the first. In conflict with this are declining sperm counts (found in American Kestrels) during the refractory period of spermatogenesis, particularly when replacement clutches are laid (Bird and Lague 1977). Males may thus insure parentage, if they are forced to relay, by leaving sperm in the female's sperm storage organs. They have little to lose and much to gain by doing so. One can assess this idea by looking at the seasonal occurrence of copulations past the last egg in relation to dates when birds typically re-lay. Among marsh harriers, copulations continued for an average 40 days after the last egg in a clutch. Shorter periods of non-fertilization copulations characterized newly formed pairs (11 days) and longer periods, up to 68 days, occurred among well-established pairs. This was similar to the interval between first and replacement clutches which averaged 55 days for five pairs, with the longest being 74 days. Thus it is plausible that apparent non-fertilization copulations are a way that males can insure some sperm for their re-laying mates when their sperm count is declining. Montagu's Harriers are unusual in this regard in that copulations decline to zero just two weeks after the clutch is complete (Pandolfi *et al.* 1998). Perhaps in this species, particularly the Italian population studied, replacement nests are so rare or non-existent that copulations past the first complete clutch never have a fertilization role and this explains the sudden cessation of copulation in this migratory species. Arroyo (1995) also reported seeing few copulations after egg laying in Montagu's Harriers, and only one copulation after hatching in a migratory central Spanish population. These trends, therefore, support the notion that species in which replacement nests and pair fidelity are fairly common partake in copulations well past initial egg-laying.

A non-adaptive alternative explanation is that hormonally both sexes remain open to copulations until brood-caring duties start to dominate. At this time the pair is both less likely to meet and less likely to copulate because of a lack of internal motivation. That the cessation of copulations occurs about 40 days after the last egg means that this period does indeed coincide with increasing foraging activity for nestlings that are 10–18 days old. This does not account for the almost complete cessation reported in Montagu's Harriers (Pandolfi *et al.* 1998; Arroyo 1999).

Infertility avoidance, sperm depletion, and unhatched eggs

African Marsh Harriers, vultures, goshawks, and kestrels each copulate up to 200–400 times to fertilize one clutch of eggs. Despite these high copulation frequencies and the high probability of sperm storage tubules in raptors (Bakst and Bird 1987), most harrier pairs produced some unhatched eggs, as do some vultures (Robertson 1986). I have argued thus far that the high copulation frequencies are mainly a form of paternity guard in line with many other species. However, the fact that non-hatching still occurs prompts the question of whether high copulation rates might also be an attempt to avoid unfertilized eggs, but paradoxically may result in infertility. Sperm depletion and its possible effects on egg fertility have proven difficult to study in wild birds because other factors such as mating competence, infertility in the female, and food resources during egg formation may all increase the chances that eggs don't hatch. For field studies there is the added difficulty of deciding whether a well-incubated but unhatched egg was infertile or simply died during early development.

On reading Donald Dewsbury's (1982) seminal paper on the costs of sperm production and the widespread sperm depletion effects in other organisms, I made a special study of this effect in polygynous Northern Harriers. Contrary to the view that sperm are inexpensive and cheap (or in Richard Dawkins, words 'to a male excess has no meaning'), Dewsbury pointed out that sperm come in specific packages (ejaculates) and frequent copulation can not only lead to depletion in the number of sperm per ejaculate, but can lead to lowered fertilization rates. This has recently been confirmed for captive finches by Tim Birkhead (1991a). (For reasons as yet unknown more and more human males suffer from low sperm counts; this is one of the main causes for the inability of couples to reproduce). Male raptors and harriers are surely perfect subjects for assessing these trends in wild birds because they copulate so frequently. Moreover, polygynous male harriers mate simultaneously with several females at a time, *and* they must also presumably squeeze their copulations into the early morning fertilization window. Therefore, they offer unprecedented opportunities for examining possible sperm depletion effects in a species outside the laboratory.

We know from some very excellent work carried out by Eddie Balfour on the island of Orkney that the Hen Harriers there showed a slow but steady increase in the degree of polygyny. That is, the number of females per male increased to the extent that some males were paired with five or six females at a time (Balfour 1957). Coincident with this increase in polygyny was a parallel increase in the proportion of unhatched eggs—to levels above 30%. More importantly, nonhatching increased as harem size increased (Balfour and Cadbury 1979). Later studies of the Orkney population by Nick Picozzi revealed that the unhatched eggs were indeed 'infertile or rotten with no detectable development'. These are tantalizing correlations which argue against the idea that possible food stress affects the size and thus hatchability of the eggs. They do not, however, discount the likelihood that these later-settling females were poorly provisioned during incubation and thus neglected their eggs to forage for themselves. We must start our investigation, however, by looking at species-wide general trends for nonhatching eggs in birds.

General trends for nonhatching

The most significant findings for this investigation are those by Walter Koenig (1982), who collated data from 113 species (including falconiforms) and determined that nonhatching was most prevalent in species with 'sophisticated' mating systems and independently common among species breeding at low latitudes; Ian Newton's 1979 review confirmed that unhatched eggs were widespread among the falconiforms, and these authors proposed several hypotheses explaining their findings. While their approach has been criticized because they did not control for phylogeny, possibly biasing their findings, they favoured inadequate food supply and extra-pair interference during copulation as the most plausible explanations for nonhatching. Koenig explicitly dismissed the possibility of sperm depletion because ' the relatively low cost of sperm production makes it likely that males will be fecund at all times of female receptivity'—a view refuted by Dewsbury's (1982) review. However, for raptors we can reject the copulation interference idea because for both Lesser Kestrels and harriers even cuckolded males do not attempt to disrupt copulation attempts. From the remaining food stress and sperm depletion ideas we can now make some direct predictions and use them to explain the patterns of nonhatching that my coworkers (Bruce MacWhirter and Gay Hansen) and I uncovered in our polygynous Northern Harrier population in Eastern Canada. These data were never published because they were disbelieved by referees of three prominent journals they were sent to in the early 1980s!

We were aided in our attempts to separate food effects from sperm effects by the fact that harems of polygynous males are hierarchical (Chapter 5). The hierarchies dictate that first-settling or α females receive more male-provisioned food throughout the season than secondary females; they are thus expected to suffer less embryo death through egg neglect than other harem members. Data were collected over a four-year period, during which polygyny and the bird's main food source—voles—fluctuated synchronously among years (Simmons *et al.* 1986b). Unhatched eggs were classified as those that were incubated to full term (31 days) but failed to hatch; no more than two such eggs ever occurred in a clutch. Normally nonhatching data were collected within 10 days of hatch unless all eggs and chicks could be accounted for after this interval. Of 101 clutches, 72 were used for analysis; the remaining 29 were either destroyed before hatch or did not meet the criteria described above. Full clutches ranged from 1 to 8 eggs, including 27 from harem females. This sample was representative of the population because hatching data were collected from both early and late nests, from harem and monogamous nests, and with equal frequency in all years. Let us look first at the predictions arising from food and sperm depletion ideas.

Food-limitation hypothesis and predictions

Newton argued that Eurasian Sparrowhawks and Buzzards tend to produce more unhatched eggs in poor rather than good food areas, and in late rather than early clutches, thus 'addling' is linked to food. A proximate mechanism through which food may affect hatchability was not proposed but two pathways are possible: (i) albumen limitation—small eggs produced in food-poor years may suffer embryo death because of inadequate albumen for development (Romanoff and Romanoff 1949) and (ii) egg neglect—incubating adults poorly provisioned by their mates may spend greater periods self-foraging leaving their eggs open to possible chilling or over-heating. Unfortunately these findings may also arise from poor condition or younger birds suffering reduced sperm production, as found in domesticated species (Birkhead and Møller 1992), and giving rise to infertile eggs. We will consider this possibility in greater detail below. Nevertheless, the following predictions arise from these ideas:

1 Effects of season

The food-limitation hypothesis predicts a higher frequency of unhatched eggs as the season progresses. For harriers this is possible because seasonal reductions in clutch size parallel male provisioning rates (Simmons 1988b). Hence, late-laying females may be prone to egg neglect or lay smaller eggs.

In total, nonhatching accounted for 7.3% of 302 harrier eggs incubated to full term over four years—a relatively low frequency for this species (Balfour 1957). Despite fluctuations in food abundance and polygyny between years (Simmons *et al.* 1986b), no relationship existed between the annual frequency of nonhatching and either prey abundance or the number of breeding harem females (Table 6.3). There was, however, a strong ($r = -0.76$) but non-significant ($P > 0.05$) inverse correlation between unhatched eggs and small clutch size. The trend appeared to arise from high nonhatching (12.5% of all eggs) in the food-poor year of 1982. Because only four years were assessed the relevance of this near-significant result cannot be explored without further data. However, we can explore the food link further by assessing the seasonal prevalence of nonhatching. Hatching success did not parallel the significant seasonal declines

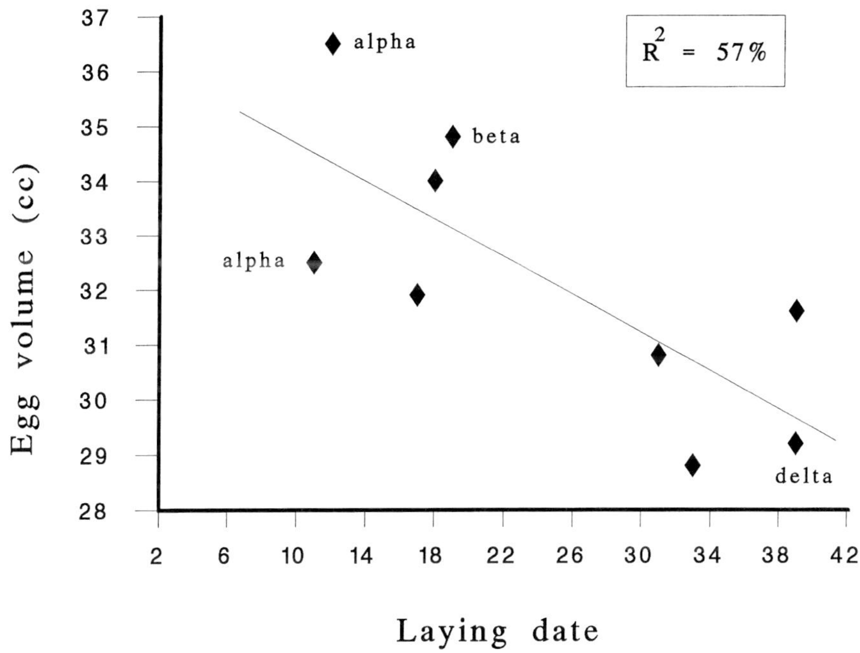

Fig. 6.5 Significant declines in egg volume with laying date of monogamous and polygynous Northern Harriers in New Brunswick. Each point represents the average of one clutch of eggs and the Greek letters denote the position of the female in the harem. If harem females laid smaller eggs than predicted because of reduced food supplies this could explain their higher incidence of unhatched eggs. This does not seem to be the case from this small sample.

Table 6.3 The relationship between nonhatching of Northern Harrier eggs over four years and vole abundance, the proportion of females in harems, and mean clutch size in New Brunswick.

	1980	1981	1982	1983	Correlation with nonhatching
Nonhatching (%)[a]	4.9	3.9	12.5	9.3	
Vole abundance[b]	6.3	1.9	5.2	6.3	0.45 (n.s.)
Harem females (%)[c]	40.0	11.0	33.0	42.0	0.46 (n.s.)
Mean clutch size	4.9	4.3	3.6	4.4	–0.76 (n.s.)

The power of the significance tests is low because of the sample size of four years; the coefficients are given as guide only.

[a] Unhatched eggs/all eggs given chance to hatch

[b] Voles captured/100 trap days

[c] Total number of females in the harem

in egg volume (Fig. 6.5). Contrary to the food-limitation hypothesis, clutches containing unhatched eggs were not laid significantly later than clutches with 100% hatch (Table 6.4).

Table 6.4 Mean initiation dates of completely and partially hatched clutches of Northern Harriers in New Brunswick.

	Year			
	1980	1981	1982	1983
Mean initiation date of hatched clutches (n)	19 May (18)	17 May (10)	5 June (6)	21 May (19)
Mean initiation dates of partially hatched clutches (n)	21 May (5)	18 May (2)	2 June (5)	28 May (7)
Difference in means[a] (days)	2 d	1 d	−3 d	6 d

Years are kept separate because mean initiation dates differed by up to 19 days precluding their amalgamation.

[a] Yearly difference in means tested with Kruskal–Wallis test and none were significant: $H = 2.1$; 0.05; 0.03; 1.56, respectively.

2 Effects of female mated status

The food hypothesis predicts that secondary females are more likely to produce unhatched eggs than α females. This arises as a consequence of the unequal distribution of food among harem females; secondary females gain less food throughout the season and thus may lay smaller eggs or be prone to egg neglect.

Neither result was apparent. Just as many α females (45% of 11) gave rise to unhatched eggs as secondary females (44% of 16). Secondary females also laid eggs that were not consistently smaller than those predicted from the seasonal decline (Fig. 6.5). Thus food stress for secondary females seems unlikely to account for the greater incidence of unhatched eggs among polygynous harriers. In hindsight this was not surprising since other analyses show that these females, while poorly provisioned after laying, probably receive just as much food during the critical laying period as do α females (Chapter 5; Simmons 1988c). So can sperm depletion effects explain nonhatched harrier eggs?

Sperm-depletion hypothesis

Dewsbury's review of sperm depletion included mammals, amphibians, fish, and insects and showed that as copulation frequency increased sperm numbers per ejaculate decreased to a level at which egg fertility was affected. Although missed by Dewsbury's review, Bird and Lague (1977) showed that the sperm count of American Kestrels *Falco sparverius* was also significantly correlated with egg fertility ($r_s = 0.65$, $P < 0.05$). The predictions which apply to unhatched harrier eggs are:

3 Effects of copulatory overlap

A comparison of unhatched eggs laid by harem females experiencing overlapping and non-overlapping copulation periods should indicate greater nonhatching in overlapping females. Functional copulation periods (i.e. when fertilizations occur) were taken as the interval from seven days prior to clutch

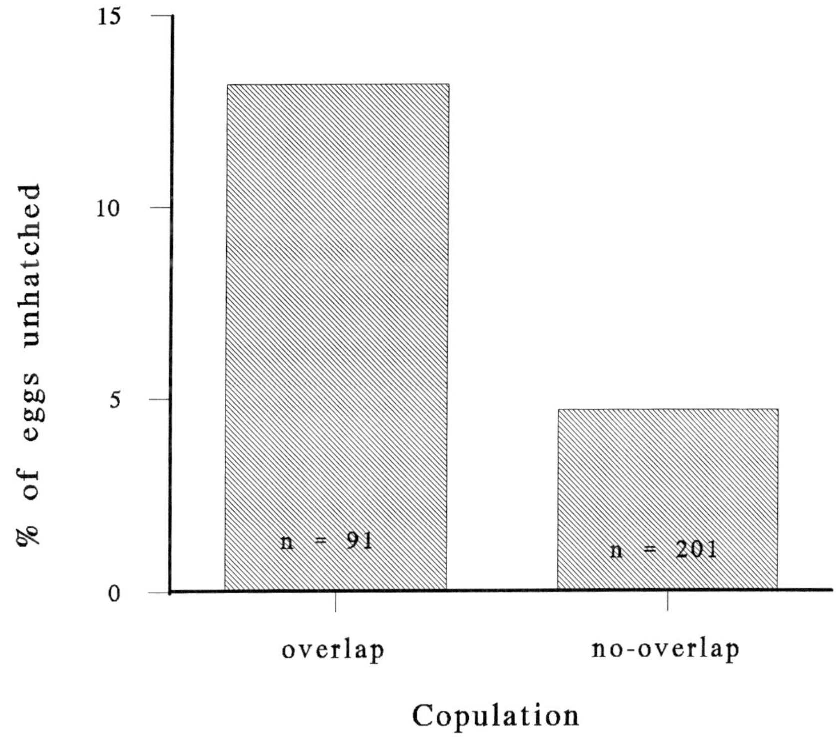

Fig. 6.6 Possible effects of sperm depletion are apparent in eggs of females within harems whose mates experience copulation overlaps with other harem females. These females give rise to 2.8 as many unhatched eggs as non-overlap females. n = total number of eggs assessed in each category ($X^2 = 20.8$, Yates corrected, $P << 0.01$)

initiation until clutch completion. Other females copulating within this interval may suffer possible sperm depletion effects arising from multiple copulations by their mates. This may be especially so if most copulations are performed in the early morning period and not spread equally throughout the day.

Among the 27 harem-produced clutches, 13 unhatched eggs occurred. Of these, 12 eggs arose during 12 (of a possible 16) copulatory overlaps. If we add monogamous birds who also suffered no competition for sperm with other females, the difference in frequency of unhatched eggs among the two groups is highly significant (Fig. 6.6). Because eggs from the same female are not strictly independent data points, the statistic reported is used as guide only to the biological significance of this result. Poor food was unlikely to have been responsible for unhatched eggs between 1980 and 1982 since 4 of 5 unhatched eggs were the product of males with provisioning rates ranked medium-high to high (> 8 items/14 h day). Thus, the data not only support the sperm-depletion hypothesis but provide support for the specific mechanism that repeated copulations can lead to unhatched eggs because of insufficient sperm.

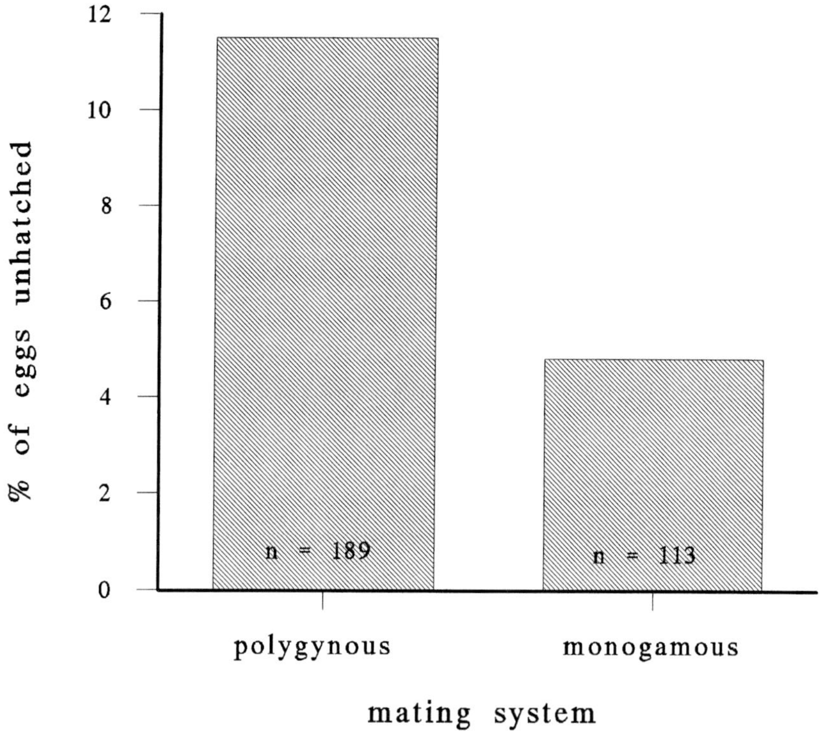

Fig. 6.7 Polygynous males who copulate with up to five females in a season give rise to 2.4 times as many unhatched eggs as do monogamous males. The difference is significant ($X^2 = 4.76$, $P < 0.05$), although eggs in a clutch are not strictly independent and differences cannot easily be tested statistically.

4 Effects of male mated status

Polygynous male harriers in New Brunswick paired with up to five females and often later into the season than monogamous males. Both factors suggest that polygynous males should suffer greater infertility than monogamous males. This was verified: mates of polygynous males produced 2.4 times as many unhatched eggs as mates of monogamous males. That is, 11.5% of 113 eggs vs. 4.8% of 189 eggs (Fig. 6.7). The food-limitation hypothesis predicts no relationship because egg size differs only with season, not with mated status (see Fig. 6.5), and egg neglect by harem females was never apparent (Simmons *et al.* 1987).

The crux of the investigation was to verify that infertility was the main source of nonhatching by examining unhatched eggs in 1982 and 1983. We searched for embryonic development, apparently visible 24 h after laying, and also measured the internal volumes of five eggs as an index of the albumen fraction (Boersma 1982). Of 18 eggs examined, 89% showed no development, while two others contained a viscous gel indistinguishable from true development. Infertility was strongly inferred, although not proven in this sample. Further support for the infertility conclusion comes from the small sample of measured eggs. Internal volumes were not significantly smaller ($\bar{x}$ = 31.18 cc, t = 0.75) than the average volume of 40 eggs measured at random over two years ($\bar{x}$ = 32.23 cc); hence inadequate albumen fraction is unlikely to have caused their nonhatching.

So of the four tests distinguishing the food-limitation from the sperm-depletion hypothesis, two failed to support the food hypothesis, while two others supported the sperm hypothesis. The additional lack of embryonic development among 89% of examined eggs, in conjunction with the 'normal' egg volume of several unhatched eggs, leads to the conclusion, albeit circumstantial, that sperm depletion reduced egg fertility and led to the nonhatching of eggs among New Brunswick harriers. This is in accord with Dewsbury's review, but at odds with those of Newton and Koenig.

Only among harriers could such an analysis probably be undertaken because they are the only raptor in which polygyny is regular and copulation overlap common.

Despite these results apparently pointing strongly towards sperm depletion, there are two alternative explanations which could explain the trends found. First, individual infertility among some males or females may produce unhatched eggs, and second, poor food resources leading to poor condition in males can result in reduced testis size and sperm production, as found in domesticated species (Birkhead and Møller 1992). Circumstantial evidence suggests that individual infertility would have led to many more eggs in single clutches being infertile. Since the maximum was two unhatched eggs in any one clutch this seems unlikely unless females were pursuing extra-pair fertilizations as vasectomized Redwinged Blackbirds apparently did in the study of Bray *et al.* (1975). This, too, seems unlikely since extra-pair copulations in other harriers are very rare, especially in the fertile window (above). Thus, while infertility among individual birds may exist, it is likely to have produced a different pattern of unhatched eggs in the clutches examined. The body condition—reduced sperm production idea is more difficult to evaluate without specific data on the condition of male harriers. Nevertheless, since the main evidence for a role for sperm depletion comes from within one group of males—polygynous birds—who are known to be particularly efficient food providers (Chapter 5), poor body condition in this group leading to reduced sperm production seems remote. Furthermore, since there was no seasonal increase in the incidence of unhatched eggs which one might expect if poor condition males were producing most clutches at the end of the season, this idea is most unlikely among monogamous birds, too.

Sperm depletion remains the most parsimonious explanation for unhatched Northern Harrier eggs and I doubt that it is a phenomenon peculiar to the New Brunswick harriers. Orkney populations suffered correspondingly higher nonhatching rates (up to 35% of all eggs) when the incidence of polygyny was such that over 90% of all females were involved in a harem (Balfour 1957). In New Brunswick the comparable figures were 7% nonhatching and 34% of females involved in polygyny. Similar findings are apparent in polygynous trios of Herring Gulls (*Larus argentatus*) studied by Fitch and Shugart (1983): trios suffered significantly higher infertility rates than monogamous pairings. This may be the reason that Koenig's review revealed that birds with 'sophisticated mating systems' suffered high rates of nonhatching—sperm-depletion effects are more likely in these species. Koenig's other major finding was that birds at higher latitudes show high rates of nonhatching. I will end this section by speculating that the latitudinal nonhatching trends might also have a sperm-related origin.

Sperm production and latitudinal trends of nonhatching

Koenig found that hatchability was enhanced at higher latitudes. The obvious difference with latitude is that day length and photoperiod clearly vary; interestingly, so to does spermatogenesis, the sperm production process. Specifically, photoperiod affects the onset and duration of sperm production and also the size of the developed testes (Murton and Westwood 1977). Birds enjoying longer light periods have larger testes, and produce sperm for longer periods. High latitude, unlike lower latitude breeders, are also known to produce sperm while on migration to their breeding grounds (Quay 1985). Thus, despite their apparent shorter periods of spermatogenesis, higher latitude breeders may minimize sperm depletion effects through (i) the relatively rapid onset of spermatogenesis as they migrate to higher latitudes; (ii) the constricted breeding seasons (with the decreased chances of copulation during the refractory period of sperm production); and (iii) the larger testis size and the increased sperm production that large size confers (Birkhead and Møller 1992). Each factor may lead to a reduced potential for sperm-related nonhatching and conceivably account for the latitudinal trends found by Koenig (1982); these suggestions are speculative but testable.

To conclude, we began the section by asking whether females might be demanding frequent copulations to avoid the deleterious effects of infertile eggs. Harriers suffer high nonhatching rates and the incidence increases with increasing levels of polygyny. Thus, far from decreasing their chances of infertility, female harriers, particularly those sharing a mate, increased their chances of having infertile eggs. That this arose from sharing sperm and not from sharing male-provisioned food was revealed by the subset of females involved in copulatory overlaps with other harem females. They showed higher incidences of nonhatching than non-overlapping birds, despite the fact that both sets may suffer from reduced food resources from an overworked mate. That polygynous males in general suffer higher rates of nonhatching than monogamous males and that most eggs examined were apparently infertile suggests that sperm depletion is the most plausible explanation for unhatched eggs in the Northern Harrier. Since neither food nor copulation interference can explain the prevalence of unhatched harrier eggs nor large-scale latitudinal trends in nonhatching of other birds, I conclude that sperm depletion may offer an alternative explanation of the latitudinal trends of nonhatching via the effects of day length on spermatogenesis. Ideally since harriers can be induced to copulate more frequently and for longer by the introduction of decoy males (Mougeot *et al.* MS) the trends reported here could be experimentally checked very simply.

7 *Sex ratio and egg size manipulation*

In this chapter we will look at the ability of harriers and other raptors to manipulate the number of sons and daughters they produce (sex ratio manipulation) and determine whether harriers can adaptively change the size of the eggs they lay. Both are relatively controversial topics because there is presently no known mechanism for females to alter the sex ratio of their eggs—but growing evidence that they can do so—and little experimental evidence that birds can adaptively change the size of the eggs they produce in relation to changing environmental conditions.

For over 100 years it has been known that most vertebrates produce equal numbers of sons and daughters. However, unequal sex ratios are also apparent in many taxa and, as first tabulated by Darwin (1871), they are common among insects, fish, birds, and mammals, including man. Darwin attributed differences mainly to the larger size of male babies (itself a result of sexual selection on males) and their demise during birth itself. He also noted that among polygamous human societies more females than males were thought to be born; however, scrutiny of harems in Siam by one of his many correspondents convinced Darwin that this was not always the case, and monogamous unions gave rise to the same sex ratios as polygamous ones. In a survey that then ranged from the polyandrous hill tribes of India to the island societies in the Pacific, Darwin concluded that infanticide was a second major influence on infant ratios distorted to an excess of males—the reverse of that usually seen in human societies. On pondering the likelihood that sex ratios might be naturally skewed at birth (secondary sex ratios) in one direction or the other, Darwin concluded that 'a tendency of this kind [to skew sex ratios to an individual's advantage] could not be gained through natural selection' and that 'the whole problem is so intricate that it is safer to leave its solution for the future.'

Most authorities credit statistician Ronald Fisher (1930) with providing the first good answer to the observation that sex ratios in most broods or litters of animals were one male for every female. He argued that, in general, parents should invest equally in the two sexes, and the sex ratio at the end of that period of parental care should be the inverse of the costs of producing individuals of the each sex. Since the costs in rearing sons and daughters are generally the same, sex ratios should be roughly equal. However, let's assume that daughters are easier to rear than sons (my own conclusion from limited experience); daughters, then, should predominate slightly in the population. This is because while the future reproductive success of sons and daughters should be equal, for parents the gain in fitness from rearing (expensive) sons is achieved at a greater cost than that through rearing daughters. Natural selection should thus favour the rearing of daughters because they are cheaper (Clutton-Brock and Godfray 1991). This may, in turn, be countered by the fact that as one sex becomes rarer in the environment it would be advantageous for any set of parents that were able to invest more resources in rearing that (rare) sex for the simple reason that their offspring would have a much better chance of successfully mating and producing their own young. For example, imagine that 60 men and 100 woman remained after a nuclear winter. Each man could attract a mate but at least 40 of the women would either have to share a mate or remain celibate. It would be adaptive if the couples producing children in the next generation could produce (or invest more in rearing) the rare sex—sons. All their sons would then be assured of mates in the next generation, increasing their success. Those producing daughters would not. Of course as more sons are born, so the sex ratio of sons and daughters drifts back towards 1:1, and there would be no advantage in favouring sons anymore. One can thus expect an oscillation around a sex ratio of 1:1, dependent on the cost and future benefit of rearing each sex. Indeed one circumstance in which one sex becomes rarer in human societies is during and following world wars: one would predict that more sons should be produced following wars because population sex ratios should be biased towards females by the death of so many males, for which there is some evidence.

It is a quirk of history that Fisher is in fact wrongly credited with the first explanation of equal sex ratios—Darwin (1871) had provided the first explanation in his first edition of *The descent of man and selection in relation to sex* (Edwards 1997) but he later removed his explanation because he did not believe it! It was the second edition that Fisher apparently owned and it is thus he who is credited with this theory.

Harriers are unusual and possibly unique in that selection for more sons should arise for two reasons: they are both polygynous (which should influence sex ratios in one direction through the future benefits of polygynous sons) and males are *smaller* (and presumed less expensive to raise) than females, which should influence the sex ratio in the same direction. They therefore provide the possibility for seeing some strong selection towards the rearing of more sons than daughters. Since four populations of harriers have now been assessed we can determine if this is so—particularly since some are polygynous and some are not.

In a review of sex ratios in birds of prey testing the Darwin–Fisher theory, Olsen and Cockburn (1991) found that sex ratio deviations in birds were common and biased towards more females. This was contrary to prediction: more males were expected because they are presumed cheaper. Their analysis was questioned and criticized by Krakow (1993) who concluded from his reanalysis of a larger data set that the majority of raptors studied (with some exceptions) show no deviation from parity in their sex ratio. This is sex ratio parity at the population level. However, individual parents of differing quality or under different environmental conditions may find it adaptive to vary the sex ratio of their young, as predicted by Trivers and Willard (1973).

Facultative manipulation of sex ratios

Sex ratios may be manipulated by parents selectively investing more in daughters or sons according to the reproductive value of each sex (Gowaty and Droge 1992). Thus if sons are more valuable because they are (i) rarer in the population, (ii) have the chance to attract many females in polygynous mating systems, (iii) survive better under the prevailing circumstances, or (iv) are more valuable to the parents in helping rear subsequent broods (in cooperatively breeding species), then more sons may be produced. This may occur in mammals in the following manner. Among deer, sons provided with increased maternal care can achieve large body size and become powerful competitors as adults; they are likely to acquire large harems and thus repay their mother's greater investment. For Red Deer, therefore, the rules might be: invest in sons when conditions are good (or your dominance rank is high such that you have access to high resources) so that your high quality sons sire many offspring. If conditions are not so good, natural selection will favour hinds that invest in daughters which will always breed as adults. It will act against those investing in sons because poor quality sons are unlikely to sire many grandchildren (Clutton Brock 1991).

Without the bias that polygyny brings, however, parents may simply rear more of the sex that costs them less to rear. So for deer, since sons demand more energy in terms of length of time suckling, the amount of milk taken, and the total weaning period (Clutton Brock 1991), they are more expensive than daughters. Thus, in the population females should end up rearing more daughters because they are less expensive. In this case the costs are less in terms of future survival of the parents.

There are several stages at which sex ratio manipulation can proceed: (i) at conception due to sex chromosome segregation (primary sex ratio); (ii) at hatching or birth where sex biased mortality of embryos or eggs may occur (secondary sex ratio); (iii) the ratio at the end of the independence period in which parents or prevailing environmental conditions may favour one sex over another (tertiary sex ratio). Biologists assume that among birds it is females that manipulate the primary sex ratio because they have one male and female sex chromosome (heterogametic). In mammals it is likely to be the (XY) male since they are the heterogametic sex.

It seems, therefore, that we have two outcomes: in general, females should be rearing the sex that is the least expensive to rear, but for those able to do so, some should be rearing the sex that is going to provide them with the best returns (i.e. sons who will become polygynous). Both may vary with the prevailing conditions during brood rearing. Because female nestling raptors are generally larger and stay longer in the nest, it is assumed that they cost more and may be more prone to starvation; males then predominate. Alternatively, if food is scarce, then the smaller males may be less able to compete with their larger sisters and females should then predominate (Newton 1979; Bortolotti 1986). What has to be determined is if sons really are less expensive.

Which sex is more expensive?

One of the first studies of sex ratios in raptors assessed the relative feeding rates of sons and daughters from nest observations (Newton 1978) to determine if females were more expensive as predicted. Newton and co-workers measured the proportion of bites taken per meal by females and males, and also looked for any preferential treatment of young sparrowhawks. Unexpectedly, they found no greater intake by young female sparrowhawks and in a subsequent paper (Newton and Marquiss 1979) found a sex ratio that did not deviate from parity. Thus in one of the world's most sexually dimorphic birds, no evidence was found that daughters cost more and no sex ratio bias in over 1000 nestlings was evident.

There are now two studies of harriers in which the relative cost of rearing sons and daughters has been assessed and both suggest a similar result. In the first, in New Brunswick, MacWhirter (1994) assessed relative provisioning of Northern Harriers not only in the nestling period but also during the pre-independence period after young had left the nest. Females tended to spend longer in the nest but the brood broke up within three days at independence,

Table 7.1 Sex ratio variation (proportion of males to females) in four species of harrier. The results represent the sex ratio of broods when they were first sexable, approximately halfway through the nestling period.

Species	Sex ratio (% males)	Sons:daughters	Region	Source
Montagu's	55.1 [a]	203:165	W. France	Leroux and Bretagnolle 1996
European Marsh	54.8[a]	1244:1026	N. Holland	Zijlstra *et al.* 1992
European Marsh	54.2	137:116	Poland	Witkowski 1989
Hen [1953–78]	46.9[a]	579:655	N. Scotland	Picozzi (1980)
Hen [1979–84]	~ 60.9	n.d.	N. Scotland	Picozzi (1984b)
Northern	50.7	114:111	S. New Brunswick	MacWhirter 1994

[a] Significantly different from 50:50 ($P < 0.05$).

so neither sex enjoyed a longer period of parental care. During the nestling period, females ingested an estimated 4.77 kg of food each while males ingested 3.90 kg of food per bird. Thus daughters consumed 28% more food. Had the analysis ended here, one might conclude that females were fed more and the reproductive value of daughters was higher. However, in the pre-independence period thereafter, males consumed about 3% more than females (2.84 kg vs. 2.24 kg each), which, when combined with the earlier figures, resulted in females ingesting about 5% more food than their brothers. This was not a significant difference among the 6 broods intensively studied, and it did not reflect the average difference in body mass of nestlings at fledging. Since females are on average 34% heavier at fledging than males in this New Brunswick population (MacWhirter 1994) one would expect females to consume appreciably more than they did. MacWhirter concluded that the provisioning of sons and daughters in this Northern Harrier population was little different to other size-dimorphic species, and the results imply that the reproductive value of the sons and daughters was equal in the eyes of the parents. A sex ratio of 1:1 was thus expected, and found in this population (Table 7.1). Thus, no facultative sex ratio manipulation was occurring and these data support Fisher's predictions for equal allocation of parental care between sons and daughters. It is useful to note that the same conclusion was reached by Witkowski (1989) in his field study of relative biomass ingested by female and male nestlings of European Marsh Harriers. The following study, however, suggests that field estimates underestimate the energy intake of female harriers.

The second study is a just completed one by Serge Daan's group in The Netherlands on European Marsh Harriers. They used a more accurate assessment of energy intake—doubly labelled water. This is a radioactively labelled isotope of water injected into the body cavity of the bird; depending on the energetic output of the bird, the amount then measured in the blood is a measure of the energy expenditure. Riedstra *et al.* (1998) calculated the metabolized energy in both males and females matched for age and nest environment and found that males used 20% less energy as nestlings than females. This was also the difference in body mass between the siblings, hence energy turnover was proportional to body mass, as assumed by other authors. The researchers then determined the population sex ratio, using the assumption that if males are cheaper to rear then more males should occur than females. This was indeed the case in their population—55% of fledged birds were male and 45% were female in 735 broods comprising 2260 nestlings (Ziljstra *et al.* 1992: Table 7.1). Thus both studies on harriers support Fisher's contention that the population sex ratio should match the relative costs of rearing males and females. (It should be noted that because this species is polygynous the preferential rearing of males could have arisen from the benefits of rearing sons who later attract several females; this was not considered, and we will assess the possibility later.)

Using Riedstra *et al.*'s important result that energy turnover was proportional to body mass in their European Marsh Harriers, we can predict that in sexually dimorphic species in which males are smaller, the sex ratio should always be biased towards males—they should always be cheaper. This prediction has been tested by Leroux and Bretagnolle (1996) who plotted sex ratio (male:female) against sexual dimorphism expressed in terms of body weight (male weight: female weight) for 17 species of raptor. They expected to find a trend for males to predominate more and more in those species that were the most dimorphic. However, they found no trend supporting this prediction, and they concluded that sex ratio variation in raptors does not fit Fisherian predictions and other mechanisms must be giving rise to the biases seen.

Environmental influences on sex ratios

In certain birds of prey environmental conditions have been shown to have a pronounced effect on sex ratios. The egg's position within the clutch and when in the season the clutch is laid influence the male:female ratio.

During a seven-year study of European Kestrels in northern Holland, ringers were asked to collect one tail covert from all nestlings they ringed. This allowed identification of the sex of each of 3765 nestlings. Simultaneously the ringers were asked to record the laying date, clutch size, and brood size, so that variations of sex ratios with laying date could be determined. Researchers Cor Dijkstra, Serge Daan, and J. Buker of Groningen University then showed that the number of male kestrels (sons) decreased the later in the season the clutch was laid. That is, early season clutches were dominated by sons (58:42) and late season clutches were dominated by daughters (68:32). Dijkstra *et al.* then looked closely at the laying order within the clutch to assess any differences in male vs. female eggs. Taking just *early* season clutches, they noted that more sons hatched from eggs 1 and 2 than expected and fewer sons hatched from eggs 5, 6, and 7 than expected. The reverse applied for *late* season clutches: more daughters hatched from eggs 1 and 2 than expected and few daughters hatched from eggs 5, 6, and 7.

Because later laid eggs were more prone to mortality than eggs 1 and 2, this could explain why sons predominated in early clutches and daughters in late clutches. To understand why this complex manipulation of sex ratio took place they looked at the subsequent breeding of sons born at different times of the season. They found that sons born early were more likely to breed the following year than those born late. This was not because they were more likely to survive, but because they were reared in good food conditions and thus more likely to secure a territory and breed the very next year. Sons born late, on the other hand, were generally reared in poor conditions and tended to skip breeding for one year. Daughters were likely to breed the next year whatever their hatching date. Thus it was adaptive for breeding female Kestrels to increase the number of early born sons because those sons were more likely to breed early and enhance their lifetime success. Precisely how breeding females manage such a manipulation is unresolved but both Dijkstra *et al.*'s data and that of Wiebe and Bortolotti (1992) on American Kestrels suggest that differential mortality of chicks in the nest is *not* the reason. Complete broods with no mortality were biased in similar ways suggesting that mothers were altering the primary sex ratio. The mechanism remains unknown.

When the same seasonal analysis was undertaken with Lesser Kestrels breeding in Spain, a 1:1 sex ratio was found (Negro and Hiraldo 1991). However, when Tella *et al.* (1996) reanalysed these results with a larger sample they also found in most years that the sex ratio was 1:1, but discovered that there was also a seasonal decline in the sex ratio when all years of data were pooled: more males were born early in the season and more females were born late in the season.

Now consider that, unlike these falcons, some male harriers are polygynous and it would be doubly advantageous to rear sons *in good conditions* because like Red Deer they have a good chance of becoming polygynous, and like kestrels it is probable that those reared early in the season would benefit from the good conditions and be able to breed earlier than those not reared in good conditions. One should thus expect strong selection for sons to be produced early on.

Are sons preferred among polygynous species of harriers?

Fortunately we have tests of this hypothesis from five studies of four species of polygynous/monogamous harriers: two from Europe by the Groningen group (Zijlstra *et al.* 1992) and Witkowski (1989) on European Marsh Harriers; one from North America on the New Brunswick study area, by Bruce MacWhirter, one from Orkney and the long-term Hen Harrier data set of Eddie Balfour and Nick Picozzi, and the last from a monogamous population of Montagu's Harriers in western France studied by Vincent Bretagnolle and co-workers. European Marsh Harriers are mildly polygynous (13% of nests: Chapter 5) with up to three females to one male (Altenburg *et al.* 1982) while Northern and Hen Harriers can be strongly polygynous with five to six females per male (Balfour and Cadbury 1979; Simmons *et al.* 1986a). Thus for those parents that get it right, the returns in the form of particularly high quality sons who become polygynous and may produce up to ten offspring per year (Simmons 1988c) are great indeed.

In the first example, European Marsh Harriers offer no support for the idea: the proportion of males produced in The Netherlands and Poland increased as the season increased (Zijlstra *et al.* 1992; Witkowski 1989), contrary to that expected if the more rewarding sex is reared early in the season when food and conditions are good. We also saw previously (Chapter 5) that there was no evidence that polygynous female European birds reared more sexy sons than other females in the Dutch study area (Altenburg *et al.* 1982).

Among Montagu's Harrier populations in western France, pairs are monogamous and little selection is expected for more sons. However, this population showed a significant bias towards males (Table 7.1) also contrary to the 'advantage of polygynous sons' idea. There was also no significant correlation of sex ratio with increased food resources; indeed the trend ($r = -0.30$, $n = 6$) was for more daughters to be produced in good food years (Leroux and Bretagnolle 1996).

Among the two most polygynous populations of harrier, one, the Northern Harrier, showed no significant deviation from a 1:1 sex ratio (Table 7.1) but in partial support for the 'early-sons' idea, sons predominated early in the season (58%) and daughters predominated later in the season (63%). This is opposite to the seasonal effects found in the Dutch European Marsh Harriers. In the Northern Harrier population the trend was also apparent in broods in which all hatched offspring survived, and thus the trend for more females towards the end of the season could not be explained by increased mortality of sons. These results give some support to the idea that early sons are favoured, although it does not directly support the idea that these sons are being produced because of the premiums on future polygyny.

The polygynous Hen Harrier in Orkney is the most interesting and the population with the longest-term sex ratio data. Starting in 1953, brood sex ratios at fledging were recorded by Eddie Balfour, and continued by Cadbury and Picozzi through to 1982. The 30-year data set (Fig. 7.1) shows a strong trend away from a predominance of daughters in the fledged brood (~ 65% females), to a predominance of males (~ 58%) in the last 5 years of the study. The latter levels are similar to that found in other harrier populations in Europe. Coincident with this trend were three others: polygyny rose to very high levels of over 90% of all females: breeding males steadily dropped in numbers from 44 in 1977 to 21 in 1981; and density increased as measured by the inter-nest distances of females in bigamous relationships (Picozzi 1984a). Thus again we cannot attribute the production of sons to an increase in polygyny because it is equally plausible that the decrease in breeding males (rare male effect), or increased population density (and presumed greater competition for food) leading to the rearing of the cheaper (male) sex, may both explain the swing away from daughters to sons in this population. What is unknown from this population is whether the results arise from differential mortality of nestlings of one sex, or

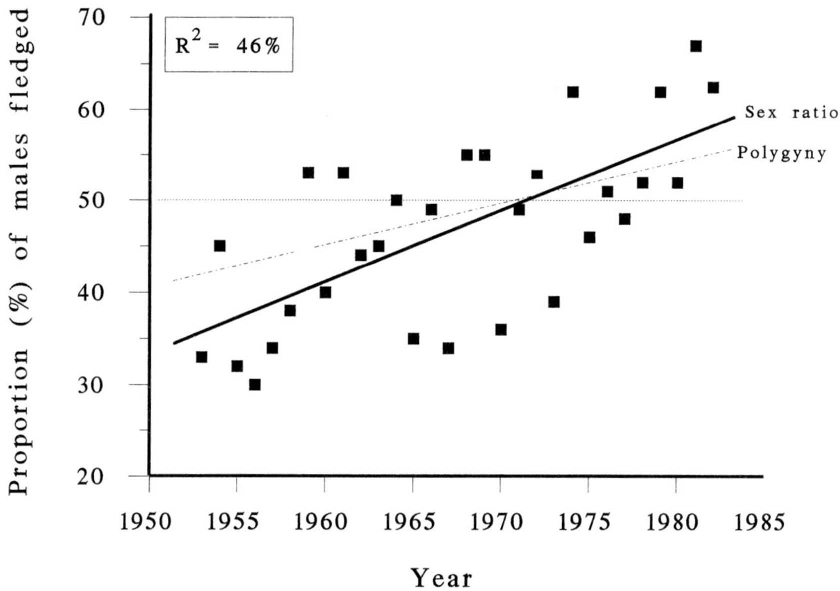

Fig. 7.1 The significant long-term increase in sons fledging from broods of Hen Harriers hatched in Orkney, Scotland (Picozzi 1984b). In the last seven years of the 30-year data set, in which daughters predominated, sons became more common. This paralleled the increasing proportion of polygyny in the population as indicated

whether they arise at hatching. Present day monitoring would be useful because on present trends (Fig. 7.1) one would expect 74% of all fledglings to be sons by the year 2000!

In sum, little evidence exists for the idea that adaptive variation towards an excess of sons occurs in any of the harriers. Only the increasingly polygynous Hen Harrier in Orkney provides evidence for this, but the data are equally explicable by the notion of constraint: the cheap sex (males) are being reared as competition for food increases in the denser and denser population. It would be worthwhile watching this population because polygyny among males has decreased to mainly bigamous relationships (S. Thirgood pers. comm.).

Other trends in sex ratio variation

In an attempt to understand if any sex ratio trends are consistent between populations, data for four species of harrier and two well studied kestrels are tabulated in Table 7.2. The independent variables are: food source (mainly increase or decrease in small mammals), season, clutch size, and differences with laying order. Other effects are sometimes apparent for different species (e.g. differences with colony size in Montagu's Harriers) and we will explore some of these, too. Rarely are all trends reported for each species but some

Table 7.2 Sex ratio variation in six species of raptors in relation to ecological and morphological variables.

Species	Food resource	Season	Clutch	Laying order	Other	Source
American Kestrel [259 broods]	Sons increase as food declines	None	None	None	Small mothers produce more sons	Wiebe and Bortolotti 1992
European Kestrel [684 broods]	—	More sons early (57%); more daughters late (57%)	More sons in smallest and largest broods	More sons in first egg (early); more females in last egg	—	Dijkstral *et al.* 1990; Daan *et al.* 1996
Montagu's Harrier [145 broods]	More sons as food declines (weak trend)	None	More sons in smallest brood (weak trend)	More females in first egg; more sons thereafter	More sons in smaller colonies	Leroux and Bretagnolle 1996
European Marsh Harrier [735 broods]	—	More daughters early (57%); more sons later (58%)	More sons in smallest and largest broods	—	—	Zijlstra *et al.* 1992
European Marsh Harrier [82 broods]	—	—	—	More females in first egg; more sons 2nd, 3rd egg	Large eggs males; small eggs females	Witkowski 1989
Northern Harrier [225 broods]	None	More sons early (58%); more daughters late (63%)	Fewer sons in smallest broods	More females first egg; 2nd, 3rd hatched, males	Young mothers produce more daughters	MacWhirter 1994
Hen Harrier [> 500 broods]	—	—	—	—	Long-term increase in sons as polygyny and population increase	Picozzi 1984b

consistent differences are apparent. For both the Montagu's Harrier and the American Kestrel, females produced more sons as annual food resources in the environment declined. The simplest explanation for this is that males are cheaper and given the poor-food circumstances, males will cost less for females (or pairs) which are somewhat food stressed. For the American Kestrel this was a strong trend, but for the Montagu's Harrier it was weak and non-significant.

Hypotheses explaining seasonal effects

Seasonal effects were pronounced in three of the six species, not present in one (Montagu's Harrier) and unreported in two other species. However, there was no consistent pattern: for two species, European Kestrels and Northern Harriers, sons were significantly more likely in early broods (sons comprised from 57% to 58% of first clutches in both cases). However, for European Marsh Harriers, daughters predominated early on (57%). Since Daan's Dutch group found the same trend for two species, they have explored the reasons why in detail and offer the following adaptive 'early-maturation' explanation (Daan *et al.* 1996). The sex that predominates early in the season is that which benefits most from a short maturation time such that it can breed in the next breeding season. Thus, for example, male kestrels which are born early in the season are over-represented among the yearling breeders in the very next season (Dijkstra *et al.* 1990). This is not so for females; an early or late birth makes no difference to the start of their reproductive lives. Since there is evidence that those birds beginning earliest have higher lifetime success (Newton 1989), kestrels should benefit by starting sons earlier. This is a subtly different argument to that provided by Trivers and Willard (1973) because it does not rely on parental body condition and sex ratio effects. What Daan *et al.* had to explain is why some raptors produce sons early, as does the kestrel, while others, such as the marsh harriers, produced daughters early on. They argue that the early sons idea will only apply to those species in which maturation is fast (i.e. there is only one year before breeding commences). For those larger species in which maturation is slow (two to three years before breeding occurs) it is generally the females that breed first, not the males, and males rarely breed in sub-adult plumage (Newton 1979). Thus for these species it would be advantageous for females to be hatched earlier. Their 'early-maturation' hypothesis can best be generalized in this way: *the gender whose maturation time is more strongly reduced by an early birth date should be produced first* (Daan *et al.* 1996). Since larger species are different from smaller ones in the maturation time of females (females tend to breed earlier in these species) it is they which should predominate in early broods. This was indeed the case for European Marsh Harriers (Table 7.2) and Peregrines (Olsen and Cockburn 1991) and data for Goshawks presented in Daan's paper. For smaller species, such as the Lesser Kestrel (Tella *et al.* 1996), males predominated slightly and there was a steep decline towards females thereafter. They too fit the trend.

Despite the model's ability to predict these seasonal differences between species, data from the Northern Harriers do not fit this general early-maturation

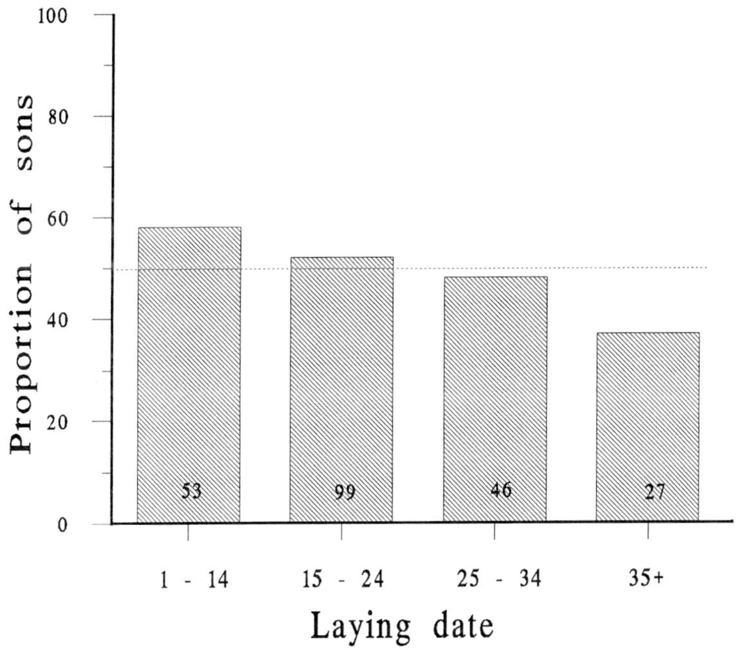

Fig. 7.2 The decline in the proportion of sons occurring in late-season broods of Northern Harriers in New Brunswick (MacWhirter 1994). Laying date, 1 = May 1, and broods checked are shown at the base of each bar.

model. In Bruce MacWhirter's study of 225 Northern Harrier broods collected over 10 years, he found that more sons (58%) occurred in the earliest broods, and more daughters occurred in late broods (Fig. 7.2). According to Daan's model, this should mean that young males breed earlier than young females. However, being a 'larger' species this is unlikely. It is also unlikely based on a small sample of tagged males and females returning to breed on Langholm Moor in the UK (Redpath and Thirgood 1997). In that study, females breeding 1 or 2 years after they were born outnumbered males doing so by 9 to 1 (n = 10 tagged returnees). In the Orkney population of Hen Harriers yearling breeders occurred in equal numbers: 31% of 19 tagged males and 26% of 35 tagged females bred in the following year. While no comparable data are available for Northern Harriers, these data from their closest relatives suggest that daughters should be favoured in early clutches or no bias should occur. It would appear that Daan *et al.*'s model is not able to predict the early predominance of sons in Northern Harrier populations in New Brunswick, but specific data on early breeding in relation to sex from Northern Harriers is lacking.

The (unpublished) data set of Diermen *et al.* on sex ratio variation with season for the small Eurasian Sparrowhawk appearing in Daan's paper also does not sit easily with the early-maturation hypothesis. The data indicate an extreme preponderance of females (64% of 25 broods) in the earliest broods, changing quickly to approximately the same bias of sons at the end of the season. As a small species one would expect males to begin breeding earlier, counter to that predicted by the bias towards females in early broods. In fact in this species, young males and young females breed with about equal frequency in United Kingdom populations (Newton 1986). Daan *et al.* explained the preponderance of females in early broods as being a result of a strong positive relationship between recruitment in females and early birth date, which was less strong for males. The strength of their model will depend on its ability to predict the seasonal sex ratio bias of other species for which early breeding in one sex is known.

Clutch size and laying order effects

Most studies in Table 7.2 have reported some bias in relation to both clutch size and the order in which eggs are laid. For three species—European Kestrels, Montagu's Harriers, and European Marsh Harriers—more sons were apparent in the smallest broods. For two of these species more sons were also apparent in the largest broods. Thus median-sized clutches gave rise to more females. The Northern Harrier was again different in this respect with the smallest broods producing more daughters (61% of 33 young of broods of 1 to 2 at fledging). Since male nestlings were more likely to die than females in this population (MacWhirter 1994), this may have arisen from differential mortality. However, the bias was present at all levels (fully hatched broods that were successful) although sample sizes were small. It may have been that the seasonal effect (Fig. 7.2) of an increasing preponderance of daughters in later clutches in conjunction with smaller clutches towards the end of the season (Chapter 9) gave rise to this result. It is an unusual result given that males should be the cheaper sex to rear and food resources at the season's end are presumed lower. It suggests that an adaptive shift is in operation, but since it is at odds with the early-maturation hypothesis of Daan *et al.* (above), other reasons will have to be sought.

The only sex ratio trend that is common to all the harriers for which data exist is that of female bias in the first egg laid. For European Marsh Harriers 57% of 30 hatchlings were females among first eggs (Witkowski 1989), while only 43% of 28 second eggs were females—a trend that continued for the later eggs, too. Among Montagu's Harriers in France, 60% of 86 first hatched eggs were females, which again switched to a preponderance of males (> 60% of 81 eggs) in second and later eggs. Trends for Northern Harriers were identical: 57% of 23 first eggs were females, switching to 35% of 23 second eggs. Clearly these trends cannot be due to differential mortality (they are from complete broods) and it thus arises from a manipulation by the parent birds. Its adaptive significance is unknown but it clearly allows the parents to change investment to their advantage by possible preferential feeding of certain chicks. One might argue that either sex could predominate, so why should it be the female? One answer is that males develop faster than females, and the staggered hatching common in most birds of prey means that sons born second have the chance to catch up on their larger sisters, equalizing the chances of survival for the first two chicks. A second reason for suggesting this comes from the finding by Witkowski (1989) that males were hatching from heavier eggs than females. Since heavier or larger eggs provide a survival advantage for the chicks (Chapter 9), this may also provide a chance for the second (larger male egg) to overcome any initial disadvantage.

In comparison with harriers, kestrels differ completely. In these falcons, sons were more likely in first eggs, at least in early clutches, and dominated in last eggs too (Table 7.2). In American Kestrels however, no trends were apparent. To determine whether this trend of females first is genus-specific to the harriers we must await further data from other species. In particular, if large eggs are more likely to be males (Witkowski) and size and laying order can be altered by experimental supplements (Simmons 1994), an experimental approach to sex ratio manipulation in relation to food might be a worthwhile avenue for future researchers.

One final trend that is both curious and would repay further investigation among the harriers is the finding that broods are strongly male-biased in the smaller colonies of Montagu's Harriers (Leroux and Bretagnolle 1996). They explained this strong trend in the following way. Males disperse less than female Montagu's Harriers, and thus males are likely to return to increase the size of the smaller colonies. This may increase the success of such colonies if larger colonies are less prone to predation as they grow (Chapter 2). It is a simple exercise to test whether males are indeed returning to the smaller colonies from which they were hatched, and if this in turn increases the success of these colonies. Data are lacking on both points.

What is clear from this rapid survey of sex ratio manipulation in harriers and other raptors is that the trends are strong at many different levels (laying order, clutch size, season, female size) but simple constraints (rear cheap sons when conditions are poor) or sophisticated adaptive explanations (rear more sexy sons as polygyny increases in the population) cannot explain many of the apparent trends. It is a field wide open for experimental manipulation and the work of Daan and his colleagues provide some of the best avenues for further testing. What researchers must beware of, however, is looking for adaptive explanations where more parsimonious physiological ones exist. It is a sobering thought that researchers in this field still have no idea how females can manipulate the sex of their young at the physiological level, even though there are numerous ideas of how and why they should.

Eggs and egg-size variation

We will now turn to the question of egg-size variation and consider whether harriers might also be manipulating the size, and thus energy apportioned to each egg, to their advantage. Eggs come in many shapes and sizes, from the two-gallon monster of the now-extinct Elephant Bird *Aepyornis maxima* to the pea-sized 0.5 g offering of the Jamaican Emerald Hummingbird *Melisuga minima*. Eggs are discrete parcels of energy, closely related to the dimensions of the bird laying them, that must be formed and deposited outside the body of the female before the food resources which are going to support them are known. A female must decide, long before the chicks emerge, how much energy she will apportion to each, and how many of each she will lay. She may lay and incubate them as they appear, in which case they will hatch some days apart—disadvantaging the last chick—or she may wait until the last is laid and then incubate them, resulting in synchronous hatching. Many altricial species adopt the first (asynchronous hatching) strategy and lay eggs with relatively small amounts of energy (5–10% of female body weight) compared with species whose chicks can walk or even fly as soon as they hatch, and that weigh 10–15% of their mother's body mass (Rahn and Ar 1974). The greater the precocity of the chick the more energy—yolk—it receives and kiwi eggs, with 70% yolk, contain more than twice that typically found in raptor eggs (Sotherland and Rahn 1987). Eggs within a clutch may receive different portions of energy, which as we shall see with some raptors is decisive in the outcome of their hatchability and survival (Williams 1993; St Clair *et al.* 1995). We can see from these examples of extreme cases that, within limits imposed by body size, a female can manipulate the size of her eggs, or incubate them differentially to alter the outcome of their hatchability and subsequent survival. How do harriers apportion energy into their eggs?

Since egg size is a reflection of the amount of food available for the chick at hatch, and often governs both the hatchability of the egg and survival of the chick (Parsons 1970; O'Connor 1977; Williams 1993; Simmons 1997b), we might expect a female to change her strategy under different food regimes or population densities to maximize her reproductive fitness. For example, in food-rich years it may be beneficial to lay many small eggs, while in food-poor years a few high quality eggs may be a better strategy. Furthermore, if food resources are unpredictable in some regions, it may be beneficial for a bird to lay small final eggs: these 'cost' little in terms of energetic investment, and when they hatch the chick may be lost or sacrificed early on if food resources prove to be poor (Lack 1954; Clark and Wilson 1981). Alternatively, since later-hatched chicks have to compete against already developed siblings, in regions in which food is predictable, it may be an advantage to provide extra energy resources for late-hatched chicks to maximize their chances of survival in the nest. These adaptive scenarios do not, however, take account of the possibility that egg size may be constrained by clutch size, or by the size of the female laying them.

Egg-size trends: allometry or adaptive variation?

While larger species lay larger eggs, eggs become proportionately smaller as female body size increases (Rahn *et al.* 1975). Raptors are no different in this regard even though their eggs are slightly larger than other birds of a similar body size (Newton 1979). For female harriers, averaging 400–600 g, a fresh egg weighing 30–40 g represents 7.5 to 6.7% of their body weight, very similar to other raptors of this body weight (Newton 1979, fig. 19). Tabulating the average size of the eggs being laid by each species of harrier (Table 7.3) we expect to see a close correlation between the egg size and the size of the female

Table 7.3 Egg volume[a] and body mass statistics of the world's harriers. Where possible female body mass and egg volume are derived from the same study area. No data exist for the Papuan Harrier.

Species	Female body mass (g)	Egg volume (cc)	Source (mass)	Source (egg vol.)
African Marsh	570	32.71 ($n = 121$)	Simmons 1989	Steyn 1982
Swamp	880	38.90 ($n = 373$)	Baker-Gabb 1986	Marchant and Higgins 1993
European Marsh	720	36.33 ($n = 544$)	Witkowski 1989	Witkowski 1989
Eastern Marsh	860	38.70 ($n = 91$)	Fefelov 1996	Fefelov 1996
Hen	520	29.93 ($n = 901$)	Balfour 1957	Balfour 1957
Northern	513	29.78 ($n = 125$)	MacWhirter and Bildstein 1996	MacWhirter and Bildstein 1996
Cinereous	496	29.30 ($n = 40$), 25.9 ($n = 28$)	Jimenez and Jaksic 1988	Jimenez and Jaksic 1988, M.S. Bó *in litt.*
Pallid	440	28.10 ($n = 90$)	Nieboer 1973	Cramp and Simmons 1980
Montagu's	345	23.20 ($n = 121$)	Nieboer 1973	Arroyo *et al.* 1998
Longwinged	612	38.70 ($n = 16$)	Brown and Amadon 1968	M.S. Bó *in litt.*
Spotted	671	39.53 ($n = 10$)	Marchant and Higgins 1993	Marchant and Higgins 1993
Black	574 ($n = 1$)	33.40 ($n = 20$)		P. Chadwick 1997
Pied	455 ($n = 1$)	27.05 ($n = 50$)	Brown and Amadon 1968	Brown and Amadon 1968
Reunion	—	32.06 ($n = 3$)		Clouet 1978
Madagascar	850 ($n = 1$)	36.13 ($n = 12$)	R. Thorstrom *in litt.*	R. Thorstrom *in litt.*

[a] Volume calculated from length × breadth^2 × 0.51.

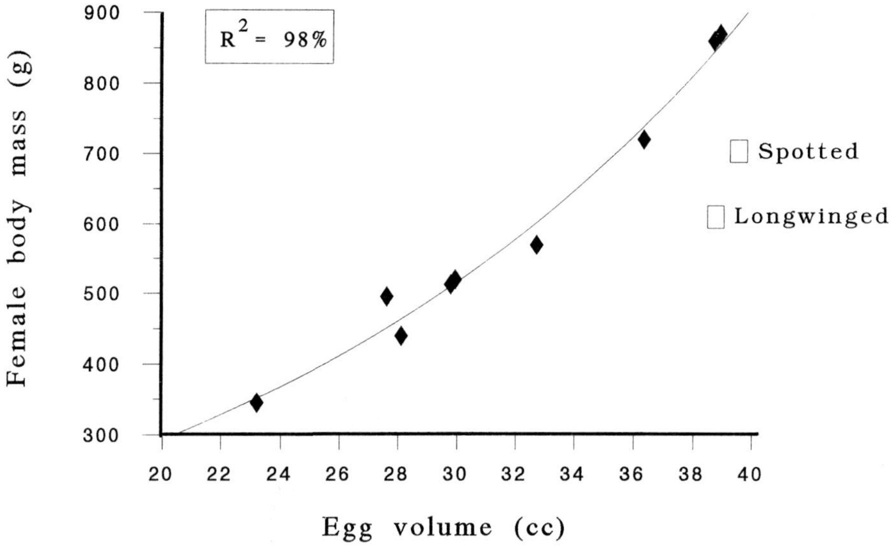

Fig. 7.3 The relationship between mean egg volume of nine species of harrier (Table 7.3) and the mean body size of the female. The two outliers (Spotted and Longwinged Harriers) are not included in the exponential equation (body mass = $89.91 \times 10^{(0.0251 \text{ egg vol})}$ and body mass then explains virtually all of the variation in mean egg volume R^2 = 98%, $P << 0.01$) of more modern harriers (see text). With the outliers included (n = 11 species), the exponential equation becomes: body mass = $122.6 \times 10^{(0.020 \text{ egg vol})}$, and the variation explained drops to R^2 = 85.4%.

laying them if eggs are constrained by body size. As expected, very strong correlations are apparent between egg volume and female body mass (Fig. 7.3). Inspection of the data reveals two outliers that do not conform to the trend: these two species are the tree-nesting Spotted Harrier of Australia, and the Longwinged Harrier of South America. On removing these two species the correlations become so close (R^2 = 98%) that we can use them to predict the body mass of harrier species for which only egg size is known.

But why should these two species lie away from such a close correlation? Put another way, the Spotted Harrier lays eggs (39.53 cc) that are 13.7% larger than that predicted by body mass (34.8 cc), and the Longwinged Harrier lays eggs averaging 38.7 cc, 20.2% larger than predicted (32.2 cc: Fig. 7.3). It is tempting to explain the large eggs of the Spotted Harrier as a trade-off with clutch size which may have been reduced as it adapted to tree-nesting behaviour. In other words, if clutch size became smaller to reduce the risk of eggs falling from its flimsy tree nest, the relative size of the eggs may have increased to increase the chances of chick survival. It is noteworthy in this regard that sibling aggression is unknown in this species (D. Baker-Gabb pers. comm.) and the species may have adopted a 'brood survival' strategy (below). In a similar fashion, while the Longwinged Harrier is not tree-nesting, it, too, lays a relatively small clutch for its size. Latitudinal equations derived from northern species and applied to southern species (Chapter 9) indicate that the Longwinged Harrier's predicted clutch size (3.9 eggs) is higher than its known one (3.4 eggs: Bó unpublished) at 37° S. It also lays smaller clutches than sympatrically nesting Cinereous Harriers at this latitude (3.4 eggs vs. 4.1 eggs: Bó unpublished). The relatively small clutch may thus be adequate in explaining the relatively large eggs of both these species.

One other trait distinguishes these two species from other harriers—they are both somewhat ancient harriers as determined by genetic similarities (Chapter 2). Thus their large-egg 'strategy' may be no more than an atavistic genetic trait that is not detrimental to the species in its present day setting, but one which changed with the Montagu's Harrier and all subsequent harriers. Without other 'ancient' harriers it is not possible to draw a second regression line, but possibly the two lines parallel one another and ancient harriers evolving under different selection pressures all laid smaller clutches and larger eggs.

Nevertheless, these body-size—egg-size correlations alone suggest that simple allometric rules govern egg sizes and we need look little further for any adaptive variation. Yet two findings suggest that, within these constraints, some adaptive variation can occur. The first is that, at the far end of the body-size scale, we

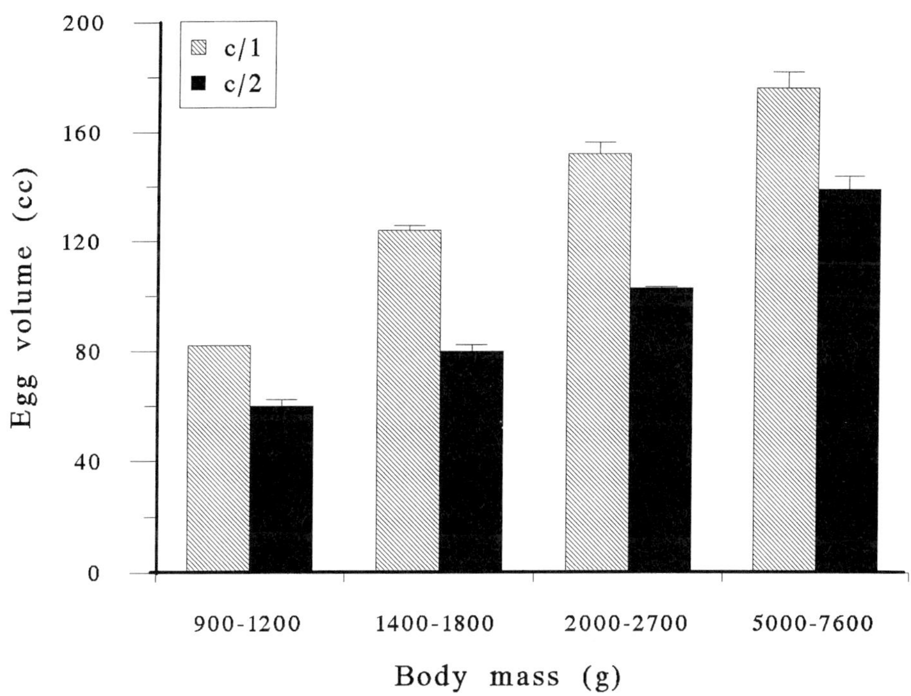

Fig. 7.4 Differences in egg sizes of eagles (and one vulture) laying single eggs relative to those laying two eggs (size of first egg used). For species of the same body mass, single-egg layers produce significantly larger eggs than first eggs of c/2 species (Simmon 1997b).

find some raptors are laying single eggs and some are laying two eggs (Newton 1979; Steyn 1982). This has a significant effect on the size of the first egg laid: for the single egg layers their egg is 8–12% larger than the first egg of species *of the same body size* which lay two eggs (Fig. 7.4). Since it is generally only the larger egg that survives in these species, it seems that some species are opting for a single large (high quality) egg while others are hedging their bets and laying a smaller first egg followed by a second egg which some believe may act as insurance (Meyburg 1974; Mock *et al.* 1990). Thus two different strategies are apparent for larger raptors, which may be related to the birds' longevity (Simmons 1997b).

The second finding suggesting some adaptive manipulation was found by Tore Slagsvold and co-workers in 1984. They determined egg-size variation within clutches for 67 species of birds and found that large species generally have small final eggs, supporting a brood-reduction hypothesis. Most small species, on the other hand, had relatively large final eggs (relative to the mean for the clutch) supporting a brood-survival hypothesis. This means that the females expected to rear this egg and attempted to decrease any disadvantage in its late hatching by providing it with relatively more resources. Slagsvold and co-workers hypothesized that differences between large and small species arose because of the presumed greater likelihood of starvation in large species, hence they lay optimistic clutches and selectively cull later-hatched young when food resources are poor (Slagsvold *et al.* 1984).

Slagsvold's fascinating findings and hypotheses are of course adaptive explanations. Might these within-clutch egg-size hierarchies just as easily be explained as energetic consequences of the temperature around egg laying (Jarvinen and Ylimaunu 1986; Magrath 1992) and/or the food resources available at the time of laying (Jarvinen and Väisanen 1983; Slagsvold and Amundsen 1992; Nilsson and Svensson 1993b)? Females laying eggs during rapidly increasing food conditions, or who receive increasing amounts of 'courtship' provisioning from their mates during egg formation, may lay increasingly larger eggs, and vice versa for females laying in poor conditions. Slagsvold *et al.* (1984) initially rejected this view in an inter-specific comparison because individuals within species tended to adopt the same strategy, regardless of environmental conditions (e.g. Parsons 1976; Graves *et al.* 1984). However, species can vary their strategy depending on food resources, at least as far as asynchronous hatching is concerned. Asynchronous hatching, which has been shown to contribute more to nestling size hierarchies than differences in egg size in some species (Stockland and Amundsen 1988; Magrath 1990; Jover *et al.* 1993), is known to vary with food resources. For example, among European Blackbirds *Turdus merula*, females with experimentally increased hatching spread are more productive when food is scarce, because, as predicted, asynchrony facilitates quicker brood reduction. Synchronous broods, on the other hand, are more productive when food conditions are good (Magrath 1989). Thus, an instructive way of looking at egg-size asymmetries to test their adaptiveness or otherwise, is to look within species under differing conditions of food availability.

Predictions of the adaptive and constraint hypotheses

1 If (large) species can adaptively vary the size of their final eggs, these eggs should be smaller than the mean if the general strategy is brood reduction and food is typically unpredictable. Conversely, eggs may be larger than the mean if food supply is predictable. Predictability of food resources (i.e. accurately basing the level of future supplies on level of resources at the time of egg laying) may be more common in small species, because there is a short time lag between clutch size 'planning' and brood rearing, than in larger species where the time lag is two or three times longer. As an example, in *Parus* tits, this period is approximately 20 days from egg laying to midway through the nestling period, and for a harrier of 500–600 g this period is about 50 days. Thus there is less need to propose the greater likelihood of starvation in large species as a decisive factor in the evolution of egg-size hierarchies as did Slagsvold *et al.* (1984). This is not to say that starvation probability plays no role.

2 If the reduction/survival strategy is not body-size dependent, but results from an interaction between initial egg-laying food resources and their future predictability, pairs may switch from one strategy to the other. So a female may opt for brood survival (large last eggs) in small clutches when food is predictable but poor; alternatively, she may opt for brood reduction (small final eggs) if food is initially plentiful but unpredictable resulting in less certainty for pairs attempting to rear many young. In other words, a female laying eggs during a period of abundant food may lay an optimistic clutch, with the possibility of brood reduction (facilitated by smaller last eggs). Clearly, the seasonality of the environment and the nature of the food resources may both affect a female's ability to predict future resources.

3 If egg size hierarchies are simply dependent upon energetic constraints during laying (as now implied for small species: Slagsvold and Lifjeld 1989; Slagsvold and Amundsen 1992; Nilsson and Svensson 1993b), then egg-size differences within the clutch should mimic the prevailing environmental conditions at the time of laying.

Much of the debate over the adaptiveness of egg-size hierarchies arises because few experimental manipulations of food resources have been undertaken to evaluate this egg-size variation. Exceptions are the work of Bolton *et al.* (1992) on European gulls, and Nilsson and Svensson (1993a) on European tits. By experimentally providing pairs of African Marsh Harriers with extra food before and during laying and determining the resultant variations in egg sizes I could directly determine the influence of food on harrier egg size.

Natural variation between clutches

African Marsh Harriers laid natural clutches of 2–4 eggs with a mean egg volume ($\pm$ 1 s.e.) of 29.51 $\pm$ 0.22 cc. The natural range in egg volume was large, varying by 10 cc from 23.53 to 33.53 cc. Mean egg volume differed significantly between females ($F = 12.12$, df = 11,17, $P = 0.0001$) and was highly correlated

Table 7.4 Mean egg volume variation (cc) of African Marsh Harriers in relation to year, clutch size, nest status, and season. Means are given with 1 s.e.

Variable	Mean	s.e.	*n*	Variable	Mean	s.e.	*n*
Year[a]				Clutch[b]			
1984	28.71	0.77	4	2	25.01	0.00	1
1985	30.63	0.67	8	3	28.87	0.74	11
1986	29.18	0.89	12	4	30.05	0.73	14
1987	29.12	1.56	5	5	31.17	2.36	2
Nest status[c]				Season[d]			
First nest	29.65	0.56	24	Early	29.28	0.62	16
Re-nest	28.81	1.14	5	Mid	30.51	1.14	8
				Late	28.62	1.11	5

[a] F = 0.66, df = 3.28, $P = 0.59$; no significant effect.
[b] F = 1.63, df = 3.27, $P = 0.21$; no significant effect.
[c] F = 1.71, df = 1.28, $P = 0.21$; no significant effect.
[d] F = 0.88, df = 2.28, $P = 0.43$; no significant effect.

with female wing chord: smaller females laid larger eggs but not larger clutches than other females. However, neither year, replacement vs. first nest status, clutch size, or season (early vs. mid- vs. late) explained further variation in the mean volume of eggs. Despite these non-significant trends, eggs in five replacement nests averaged 2.8% smaller than those in first nests, and eggs in larger clutches tended to be increasingly large (Table 7.4). The latter result is important since it indicates that females laying large clutches did not trade small eggs for large clutches—in other words they were not constrained in egg size by the number of eggs they laid. Precisely similar trends were found for Montagu's Harriers: Spanish birds laying larger clutches also laid significantly larger eggs (Arroyo *et al.* 1998).

With an average difference of 5.3% between the smallest and largest eggs in natural clutches (Table 7.5), egg-volume variation within clutches was generally lower than that between clutches. The same trend was found recently for Montagu's Harriers, in which the smallest egg was a mere 38% the volume of

Table 7.5 Average differences (in cc) between the largest and smallest eggs in natural (c3 and c4) clutches of African Marsh Harriers. First eggs were consistently the smallest and last eggs the largest.

Position	Means	s.e.	*n*	Difference (%)
Three-egg clutches				
1	27.75	0.79	9 smallest	
2	28.70	1.03	9	5.3
3	29.22	0.81	9 largest	
Four-egg clutches				
1	29.02	1.57	6 smallest	
2	30.23	1.34	6	
3	30.21	1.52	6	4.8
4	30.43	1.23	6 largest	

the largest egg (Arroyo *et al.* 1998). Interestingly, Arroyo and co-workers found that shape played a major role in the size of the egg, and elongated eggs were more common than round ones. For African birds this was not measured but the difference between the mean of the smallest (first) and largest (last) eggs was apparent in both three- and four-egg clutches of African Marsh Harriers (Table 7.5). Thus different sized clutches were similar in the size hierarchy evident between the first (smallest) and last (largest) eggs, and they were combined for further analyses. The largest individual difference was 17.7% between first and last eggs in a three-egg clutch, similar to European harriers (Witkowski 1989), although this was unusual.

African Marsh Harriers laid increasingly large eggs with their position in the laying sequence (Fig. 7.5)—a relatively unusual pattern for a large species. Only in the Montagu's Harriers in food-rich years when clutch size was high, were first eggs smallest, and the last eggs larger (Arroyo *et al.* 1998). In most other species of this size, natural patterns are for increasingly smaller eggs later in the laying sequence (Slagsvold *et al.* 1984).

Effects of food supplements on egg-size variation

Fed pairs, which laid significantly larger clutches (0.52 more eggs) than unfed controls (Simmons 1989), did not produce significantly larger eggs (ANOVA F = 0.63, df = 1,28, P = 0.44). Mean egg volume (30.10 ± 0.67 cc, n = 9) was only 2.9% larger than that of unfed controls (29.24 ± 0.66 cc, n = 20). Thus females were not adjusting mean egg volume in response to increased food resources, supporting the idea that this trait may be constrained by body size in harriers.

However, marked differences were apparent in the manner in which females apportioned the increased energy into different eggs within the clutch. The natural variation in control pairs was for last eggs to be larger than first eggs. Last eggs were on average 1.8% (3rd) to 1.7% (4th) larger than the mean for the clutch while first eggs were always the smallest, averaging 3.0% below the mean volume (Fig. 7.5). While these eggs were significantly different from one another (ANOVA), eggs within a clutch are not independent and the analysis is confounded because the 3rd, 4th, or 5th egg can be the last. I thus plotted percentage deviations of just the first, second, and last eggs from their own clutch mean, which revealed the same trends; the last eggs ($\bar{x}$ = 0.58 + 0.24 cc, n = 15) were significantly larger (F = 9.38, df = 1,44, P = 0.004) than the first and second eggs combined ($\bar{x}$ = –0.33 ± 0.17 cc, n = 30).

By contrast, females who enjoyed excess food and laid larger clutches adopted the opposite strategy. Egg size diminished following very large second eggs and their fourth or fifth eggs were 2.7% and 4.2% smaller than the mean (Fig. 7.6). I also tested whether last eggs were different from the mean by plotting first, second, and last eggs as before. Last eggs ($\bar{x}$ = –0.79 ± 0.40 cc, n = 8) were significantly smaller than the first two eggs ($\bar{x}$ = 0.31 ± 0.40 cc, n = 16). Because of the smaller samples and greater variation, the comparisons were significantly different only at P = 0.098 (F = 2.98, df = 1,23). Nevertheless, the trade was for a decrease in last eggs for these experimental birds.

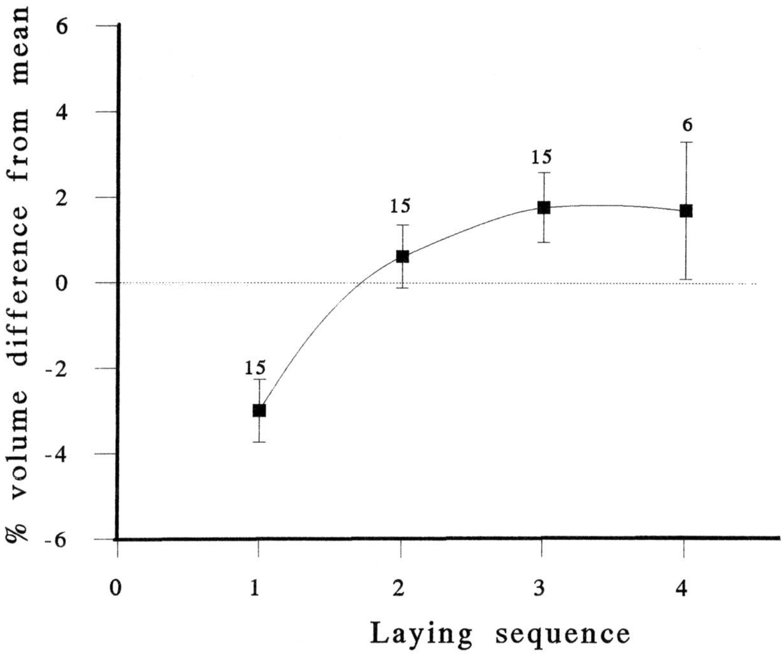

Fig. 7.5 Proportional differences in mean (± 1 s.e.) egg size with laying sequence of African Marsh Harriers indicating a natural increase in egg size with laying order. Most species of this size lay increasingly small eggs after the first egg.

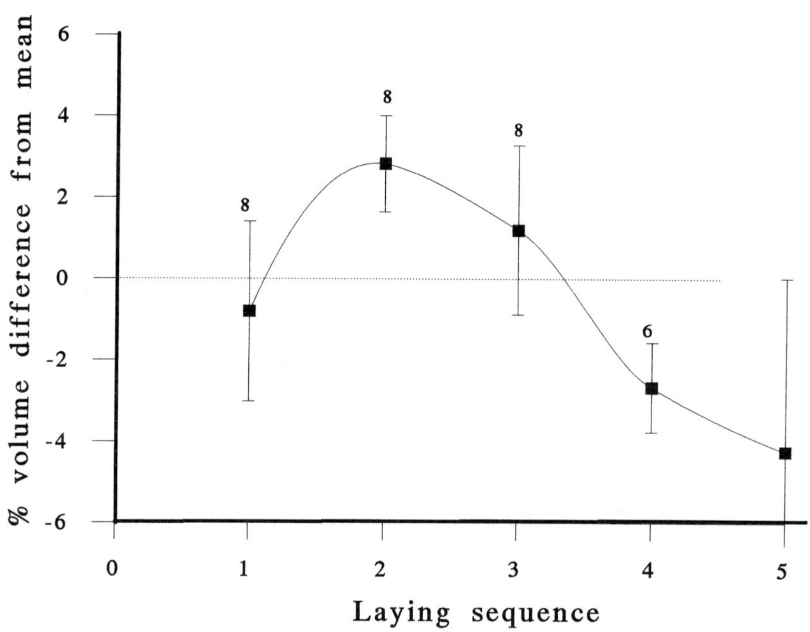

Fig. 7.6 The change in sizes of eggs with laying order for food supplemented African Marsh Harriers. The feeding regime reversed the typical trend (Fig. 7.5) and females laid increasingly smaller eggs from the second laid egg, mimicking egg patterns of temperate harriers.

Table 7.6 Average size differences of the last egg in a clutch relative to their own mean egg volume for fed and unfed African Marsh Harriers. Means are given in both percentage differences and real (cc) differences and differed significantly between fed and control birds.

	Average difference from own mean	s.e.	*n*
Percentage differences			
Fed	–2.55	1.32	8
Control	2.11	0.89	15[a]
Volume differences (cc)			
Fed	–0.79	0.40	8
Control	0.58	0.24	15[b]

[a] Difference highly significant: $F = 9.18$, df = 1,2,3, $P = 0.006$.
[b] Difference highly significant: $F = 9.59$, df = 1,2,3, $P = 0.006$.

To test whether females changed strategies while avoiding violation of statistical independence I compared the deviation of only the last eggs from the average of their own clutch mean, for both fed and unfed pairs. Fed birds consistently laid final eggs that were about 2.5% below the mean while controls laid last eggs that were about 2% above the mean. The differences were highly significant whether they were measured as percentage or actual volumetric differences (Table 7.6).

Egg-size variation *between* harrier clutches was significantly affected by only one factor, the size of the female herself. Small females laid the largest eggs, possibly because they could divert more male-provisioned food energy from self-maintenance into their eggs. This size dependence was reflected in the response of individual females to supplementary food; egg size showed a slight (2.9%) but non-significant increase. This result is common to virtually all experimental supplementation studies that have reported egg-size differences (Horsfall 1984; Poole 1985; Hochachka and Boag 1987; Arcese and Smith 1988; Carlson 1989; Korpimaki 1989; Magrath 1992; Kallander and Karlsson 1993; Nilsson and Svensson 1993a; Simmons 1993). The only species that seems to break this rule is the Magpie *Pica pica* in Europe and North America. In experimental supplements Hogstedt (1981b), Clarkson (1984), and Hochachka (1988) each found increases of 4%, 8%, and 3% in egg volume for fed relative to unfed birds.

Results presented in the next chapter corroborate the experimental data given here: egg size appears to be affected by the body size of the female and an intrinsic quality of the female herself—presumed to be a genetic component. Given that egg-size variation seems to be relatively invariable, is there any latitude for adaptive variation?

A vast amount of literature has appeared in recent years on the adaptiveness of egg size and its variation within and between clutches (e.g. Parsons 1976; Slagsvold *et al.* 1984; Pietiainen *et al.* 1986). It is remarkably to note that few of these studies (an exception is Jarvinen and Väisanen 1983) have documented the size of the egg in relation to the size of the female, despite the fact that correlations are known from poultry studies (e.g. Festing and Nordskog 1967). Several studies have shown that egg size is highly heritable (Ojanen *et al.* 1979; van Noordwijk *et al.* 1981; Moss and Watson 1982; Wiggins 1990). These correlations alone suggest that egg size per se has little adaptive significance, because high heritability is generally opposite to that expected (Falconer 1981) or found (Gustafsson 1987) for a fitness-related trait. This fact has caused researchers intent on finding adaptiveness in egg size to seek other adaptive explanations of the high heritability values (Moss and Watson 1982), or to disbelieve that low heritabilities are characteristic of fitness-related traits (e.g. van Noordwijk 1984; Pietiainen *et al.* 1986). An increasing number of studies suggest that egg-size variation is a non-adaptive consequence of physiology or allometry (e.g. Saether 1985; Wicklund *et al.* 1987; Nilsson and Svensson 1993a; Jover *et al.* 1993).

Among Montagu's Harriers in France very similar findings were evident: Arroyo *et al.* (in press), found a high repeatability of egg size between years which could not be explained by females returning to the same territories, laying eggs under the same food regime, or staying with the same mate. They concluded that the high repeatability reflected a high genetic heritability (as found in many other birds: Lessells and Boag 1987). They also found that egg size was 'slightly related to female wing-length $r = 0.34$, $n = 29$, $P = 0.069$, but not to tarsus length'. This relationship with wing but not tarsus length is identical to that for African Marsh Harriers, and may be more important than Arroyo *et al.* suggest. If wing rather than leg length is the best measure of the structural size of a female harrier it may possibly effect body constraints on the length and breadth of her eggs. This they found when comparing chicks hatched from large eggs—it was the wing length which best reflected this large size.

Thus the general trend is for no, or tiny increases in egg size, suggesting a physiological or anatomical ceiling to egg volume. Thus increasing the mean size of all eggs does not appear to be a strategy open to many females enjoying good food resources during egg formation.

Since the supplemented pairs could not have suffered from reduced food resources during the forming and laying of their last eggs, food limitation cannot explain the small final eggs. Thus the energetic constraint hypothesis (Slagsvold and Lifjeld 1989; Slagsvold and Amundsen 1992; Nilsson and Svensson 1993b) can be excluded in these birds. Protein quality deficiency (Bolton *et al.* 1992) was not investigated but it is also unlikely in view of the high protein supplements provided to experimental pairs. Recent work by the same Glasgow group has shown some striking costs of egg laying that were not apparent from previous experimental work. By inducing some female gulls to lay an extra egg but not incubate it, Monaghan, Nager, and Houston (MS) found that this group reared significantly fewer young merely as a result of having laid one extra egg. This surprising result suggests that many studies have underestimated the costs of laying eggs. An additional experiment by Monaghan *et al.* (1995) found that the additional egg and chick produced by a female whose first egg was taken had 25% the survival rate of last-laid eggs in a control group rearing clutches of a similar size. This was because the size of the hatchling was smaller than in the control group. Thus extra eggs not only have a cost for the parent but for the young emerging from them.

Egg-size patterns in harriers

The increasingly large eggs laid by unmanipulated pairs of harriers is unusual in large birds, particularly predatory ones such as these, because it is the reverse of usual trends. In their review of egg-size variation in mainly temperate species Slagsvold *et al.* (1984) found that large species such as raptorial birds show decreasing egg size with laying order, implying brood reduction strategies. There may be two reasons for the discrepancy between harriers and other birds of prey. First, their review included only eagles, well known for their brood reduction strategies (Simmons 1988a; Gargett 1990). So egg-size differences among eagles may be atypical in the falconiforms. This is the least likely explanation since species such as Ospreys (Poole 1989) and Eleonora's Falcons *Falco eleonorae* (Wink *et al.* 1985) show identical patterns to eagles—last-laid eggs are always the smallest (but see Horvath 1955). Egg-size hierarchies are rarely reported among raptors, but are worth further analysis.

Second, as suggested by the experiments here, food resources may override apparent tendencies for large species to lay small final eggs. Hence small last eggs in other large species may arise from pairs laying optimistically large clutches when food supply is good but unpredictable. This is expected in large predatory species, where the planning of the brood size precedes the brood rearing period by as much as two months (Steyn 1982). This would explain the larger clutches, yet of increasingly small sized eggs, laid by harriers in conditions mimicking (good) feeding conditions in more temperate zones. This view is supported by egg-size hierarchies among European Marsh Harriers. First- and last-laid eggs are frequently the lightest in this species, and second eggs are often the heaviest (Witkowski 1989), exactly the same as for their supplemented African counterparts (Fig. 7.6).

It is also possible that sex related differences affect egg size in harriers because sex ratio manipulation discussed earlier is known in these and other raptors (Table 7.2) and Witkowski (1989) suggests that male European Marsh Harriers hatch from large eggs and females from small eggs. However, this effect was not found in Northern Harriers where egg size, gender, and hatching sequence were studied (MacWhirter 1994). If this is true of African Marsh Harriers, females may be preferentially altering sex ratios rather than adopting brood-survival or brood-reduction hypotheses. Hence, during food-rich years, when first eggs are large, more males may result. The opposite may be true in food-poor years assuming that first hatched eggs have a greater chance of survival than last laid ones. Present data are insufficient to evaluate this possibility.

Nevertheless, I suggest that a brood-reduction strategy is more likely to occur where pairs can lay large clutches because of initially good food resources, but the chances of rearing later nestlings is unknown because of the unpredictability of future resources. This switches the emphasis, from brood reduction as a body

European Marsh Harrier.

size related trait, to brood reduction via small last eggs as dependent on the initial amount and future predictability of food resources. Since harriers switched between the two strategies, this rules out body size (or phylogeny) as a determinant of egg-size hierarchies. A testable prediction of this change in emphasis is that apparently brood-reducing temperate species which lay small clutches under poor-food conditions may reverse the usual order of egg-size variation. That is, last eggs may be larger than the mean when food resources are predictable but poor. Data from temperate zone Ural Owls *Strix uralensis* strongly support the idea of egg-size variation dependent on the predictability and initial level of food resources. In food-poor years, clutch size is reduced and eggs increase from first to second eggs. In good or intermediate food years, females lay increasingly large clutches and egg size decreases significantly with laying order (Pietiainen *et al.* 1986). A change in egg mass with laying order was also evident with coots *Fulica atra*, but in this case the main effect was an increase in the relative size of first eggs in a clutch, not the last (Horsfall 1984).

Clearly egg variation will also depend on the predictability of the environment (i.e. greater in tropical environments), the generation time of the prey base itself (smaller prey would be more prone to sudden increases or decreases), and the 'worth' of the final egg. For some species, the final egg is clearly of insurance value (Graves *et al.* 1984; Mock *et al.* 1990; Forbes 1991) and may not alter under variable food conditions (e.g. Bolton *et al.* 1992).

But how can we interpret the atypical pattern of increasingly large eggs in unmanipulated broods of African Marsh Harriers? The pattern is a natural extension of the idea that the level and predictability of the food resources influences egg-size variation: in the poor food conditions which these harriers typically face, clutches are small because pairs are unable to rear any more young (Simmons 1989). It is relevant in this context that rainfall during all four study years was about 25% below average, hence the small mammal prey base, whose breeding is rain-influenced (Perrin 1986), might be expected to be slightly less variable than normal. Furthermore, because resources are typically less variable in subtropical regions (e.g. Ricklefs 1980), parents are more assured of rearing the small number of young in these clutches. Thus they may increase the chances of later young surviving their asynchronous hatching by providing them with more *in vitro* resources. It is important to note that the last eggs were not merely the largest, but they were about 5% larger than first eggs, which were consistently small. Hence, the entire clutch hierarchy and not just the final eggs appear to be strategically altered in these birds.

Whether hatching asynchrony and subsequent fledging success was also influenced by increased food resources is poorly known because nests could not be visited frequently to mark young as they hatched. However, some evidence suggests that hatching intervals were slightly longer among supplemented birds: hatching intervals for three fed birds were 6.7 ± 2.5 days $\bar{x}$ ± 1 SD) in contrast to 5.6 ± 2.1 days for nine controls. The greatest hatch spread was 9 days in two nests and in both of these the smallest nestlings died. These differences, however, are marginal and asynchrony may add little to brood reduction relative to brood survival strategies in this species. Hatching success, was not significantly different between fed and unfed pairs (Simmons 1989).

To test whether food level and predictability are more important than body size trends, one would expect brood survival strategies (large final eggs) in naturally food-stressed pairs, and brood-reduction strategies (small final eggs) in those species allowed the luxury of laying optimistic clutches under good food conditions, yet unable to predict whether high resource levels will persist. This appears to be the case for African Marsh Harriers.

Clearly further experimental work is needed to tease apart the relative importance of environmental predictability, and the interaction between initial food resources and later predictability, which will differ for small versus large species. Tropical versus temperate species matched for body size and phylogenetic similarity may be the best route to gaining a greater understanding of the role of resource predictability in egg-size hierarchies.

In conclusion, these results for African harriers suggest that patterns of increasing or decreasing egg size with laying order are adaptive shifts dependent more on the initial food resources and their predictability than either species-specific or body-size related response, or an energetic constraint imposed by food at the time of laying. It would seem that the body size trends found by Slagsvold *et al.* (1984) may arise because larger species are less able to predict future resources than small species simply because the time interval between clutch planning and brood rearing is greater in larger species.

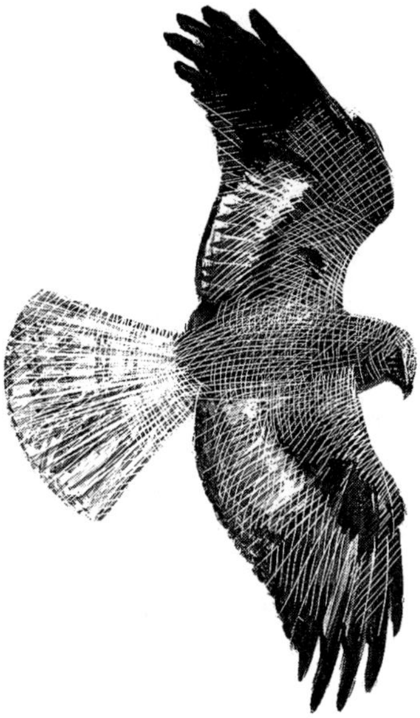

8 Food and reproduction in the tropics

A traditional view of both food resources and reproduction in tropical regions is that the environment is aseasonal, and food such as seeds, insects, or rodents do not peak and trough in the same boom and bust manner as found in the temperate climes. The aseasonal nature of tropical environments is certainly true, and as a northerner accustomed to long cold winters followed by a predictable burst of activity in the boreal spring, it came as a surprise to me to find a general lack of seasons in subtropical Africa. It was replaced by a cycle based on rainfall, not temperature. For example, large parts of Africa experience two distinct rain seasons each year, which are followed by a flush of resources as large as, and in the case of seeding plants and flying insects, often larger than, the flush of resources found in temperate regions. Thus while aseasonality may be true of the tropics it is not true to say that resources cannot peak and trough—they do, but in a less predictable manner. Why is this? In temperate regions, temperature change drives seasonality, and that in turn is driven by a predictable interaction between sun and earth. Thus temperate seasons are more forecastable in timing and extent than seasonality driven by unpredictable weather systems, and zones of warm and cool air causing rain events. Since many birds, both temperate and tropical, are timing their breeding to coincide with the flush of food (Perrins and Birkhead 1993), we predict, and find, a certain rhythm to temperate breeding schedules, and a far more chaotic pattern to breeding in tropical areas.

One classic example is that of Darwin's Finches *Geospiza* spp. on the Galapagos. On the equator, there is little detectable change in day length and maximum monthly temperature varies from about 30° to 38° C across the whole year (Grant and Grant 1989). Rainfall generally occurs in the period January to April, with dry conditions thereafter. However, ocean warming via the El Niño phenomenon created extensive rainfall in 1983 when 200 times the normal rain fell in eight months. Two years later there was no rain at all. In the high rain year some plants seeded five times, and produced seeds that themselves germinated and seeded; this provided enormous food resources for the finches (Grant and Grant 1989). The breeding response of finches was directly related to rainfall—while birds do not lay *larger* clutches in good rains, they laid more—up to seven clutches in the high rainfall year, two to three in the more typical years, and none in the drought year. Adult survival was also affected by the size and hardness of the seeds produced during these different rainfall periods (Grant 1986). The Grants' studies clearly shows how rainfall influences the most critical aspects of breeding and survival of these tropical finches through its effect on the quantity and quality of their food resources. In this chapter we will explore how these and other factors influence reproduction in harriers, particularly in the tropics, and on what cues birds time their breeding.

The influence of food on egg size and clutch size

At the proximate level several factors can influence the size of eggs, clutches, and broods. Among these are the body size of the breeding female, the experience and age of the breeding pair (Coulson 1966; Ollason and Dunnet 1978; Newton *et al.* 1981; Poole 1985), and the local food conditions at the time the eggs are being formed (Lack 1947; Klomp 1970; Nisbet 1973; Winkler and Walters 1985). While these factors may not necessarily dictate the evolution of clutch size, they can account for the variation found within a population. Raptorial birds (falconiforms) are particularly suitable for examining the role of ecological variables in clutch and egg size determination because provisioning (by males to dependent females) is easily assessed, as is the prey base of small mammals taken by prey specialists such as harriers.

Food resources have been shown to play a major role in the breeding of raptors based on evidence from

(1) correlational studies of cyclic variations in prey abundance and clutch size (Cavé 1968; Hagen 1969; Hamerstrom 1979; Korpimäki 1984; Simmons *et al.* 1986b; Village 1990);

(2) correlations between poor- and good-food habitats and clutch size (Newton 1976b; Newton 1986);

(3) correlations between natural provisioning rates and clutch size (Simmons *et al.* 1986b; Donazar *et al.* 1992);

(4) experimental provisioning of extra food and increases in clutch size (Newton and Marquiss 1981; Dijkstra *et al.* 1982; Meijer *et al.* 1988; Korpimäki 1989).

The close correlation between the number of voles in the environment in North America and the number of nesting Northern Harriers is considered in Chapter 5.

Evidence linking prey in the environment to the reproduction of harriers is provided by significant correlations between male food provisioning rates and clutch size for the same Northern Harrier population in New Brunswick. In 1980 and 1981, I combined the pre-hatch and post-hatch provisioning rates of nine males to compute an overall provisioning effort and found significant ($P <$ 0.02) correlations for both clutches and number of young reared in the successful nests (Fig. 8.1a, b). Realizing the importance of pre-hatch provisioning to clutch formation I then separated pre- and post-hatch provisioning effort in 1982, a year of generally poor food resources, and found similar correlations (Fig. 8.1c). Clearly, male provisioning effort for Northern Harriers had a major influence on their reproduction and success.

However, experimental manipulation of food resources provides the strongest support for a causal relationship between food and reproduction. None had been carried out on harriers and not all species respond to extra food by increasing their clutch size (Powell 1983; Poole 1985; Arcese and Smith 1988; Simmons 1993). Moreover, none of these experimental studies, bar the latter on African Wahlberg's Eagles *Aquila wahlbergi*, were conducted in tropical raptors and thus the generality of food's influence on various parameters of reproduction has yet to be shown. Below I will describe some of the experiments undertaken on African Marsh Harriers but first we will compare the seasonality of food and reproduction between temperate and tropical harriers.

Seasonal declines in productivity

It is well documented that clutch sizes in temperate latitudes decline with breeding season for many species (reviewed by Klomp 1970; Murphy and Haukioja 1986), and raptorial birds (Daan and Dijkstra 1982; Newton and Marquiss 1984) are no exception. Among harriers this is well documented for the Northern and European Marsh Harriers (Altenburg *et al.* 1982; Barnard *et al.* 1987). For

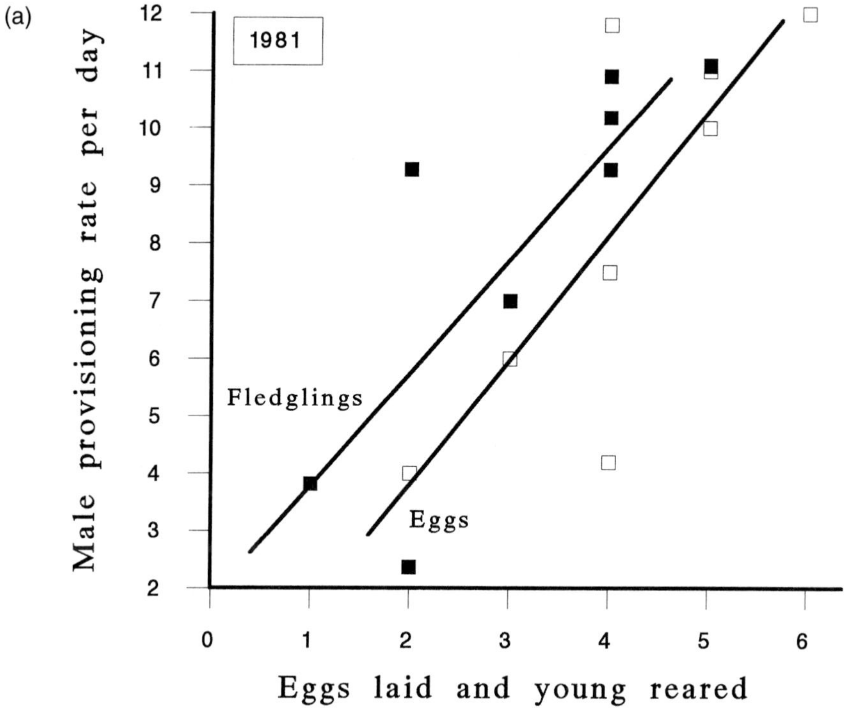

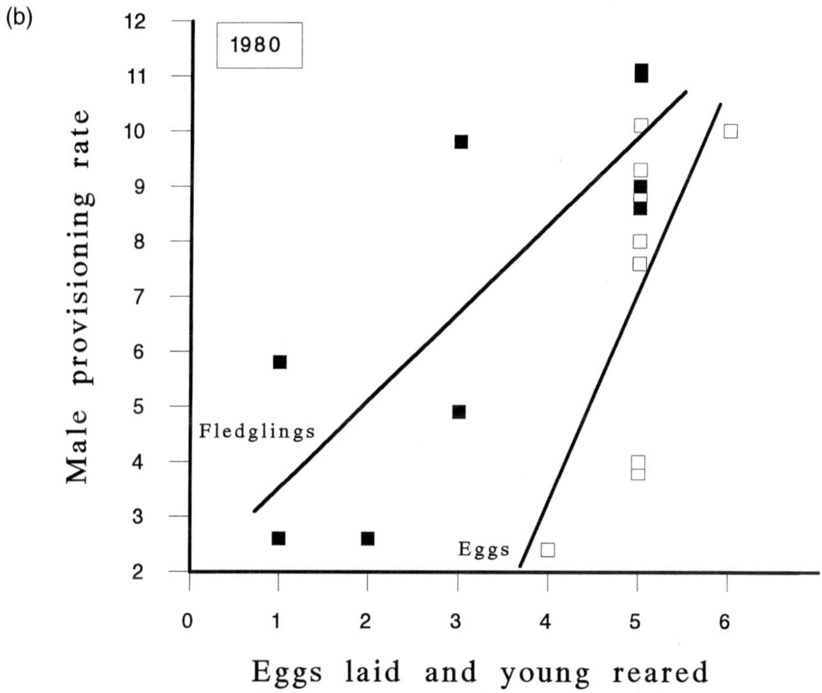

Fig. 8.1 Caption (see overleaf)

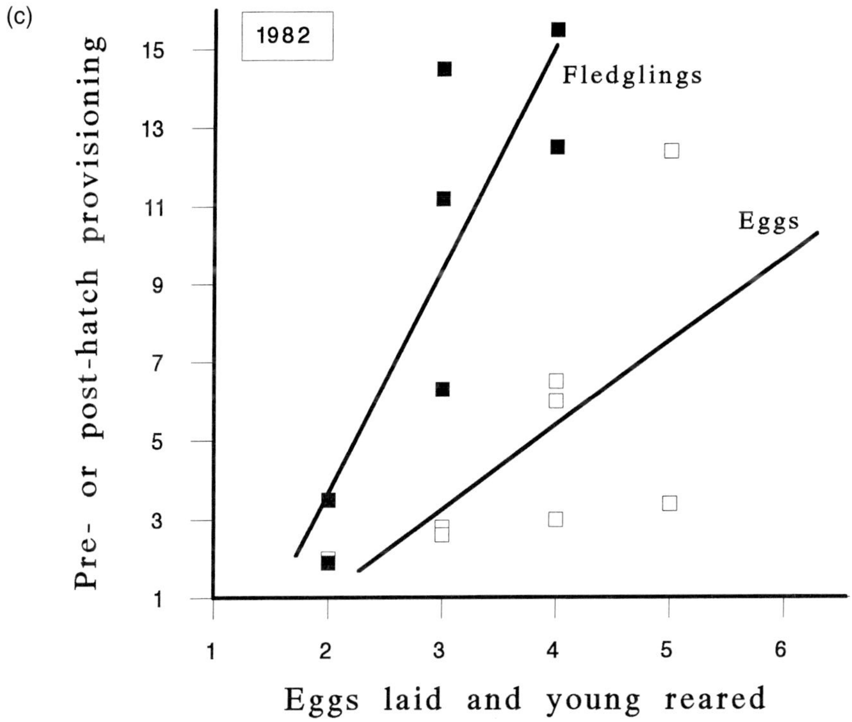

Fig. 8.1 Food provisioning performance (items delivered to nests/14-h day) in relation to clutch size and brood size of Northern Harriers, New Brunswick. Graph (a) depicts data from 1980, a food-rich year and (b) a slightly poorer year; (c) is an atypical year in that spring vole abundance (associated with clutch size) was very low, but rose steeply .through July/August as young were fledging. Lighter squares represent clutch size, dark squares fledglings. All correlations were significant.

example, among European Marsh Harriers breeding in north-western Holland, females laying eggs at the beginning of April produce clutches of 5 eggs; by mid-May average clutch size had dropped by one egg to a clutch of 4.0, and females laying thereafter lay clutches averaging about 3.6 eggs. In Canada, the seasonal decline in clutch size for Northern Harriers varied depending on the level of prey in the environment—in food-rich years, clutches declined slowly from just over 5 eggs to 4 eggs in one month (similar to European Marsh Harriers). However, when food was poor, the decline amounted to 2 eggs per clutch in a month (from 5.5 to 3.5) and in the poorest year the decline amounted to 2.8 eggs in one month (Fig. 8.2). Very similar declines are known from the Tengmalm's Owl, which also responds functionally and numerically to fluctuating food (vole) abundance (Korpimäki and Hakkarainen 1991). Clutch size declines have also been reported for Montagu's Harriers in western France and central Spain (Arroyo 1995; Arroyo *et al.* 1999), and while annual food abundance also affected the average clutch size, seasonal variations in food resources could not explain the seasonal

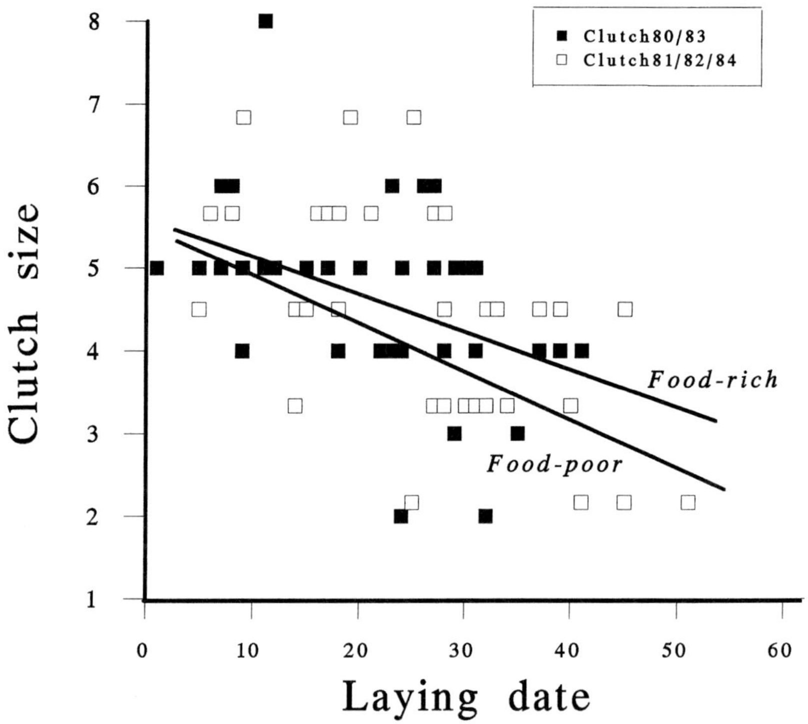

Fig. 8.2 Seasonal declines in clutch sizes of Northern Harriers in food-rich and food-poor years. Note the steeper decline in food-poor years. 1 = May 1.

decline in clutch size (Arroyo *et al.* 1999). Thus temperate-breeding harriers exhibit steeply declining clutch sizes.

What is curious about these declines is that prey levels, particularly of small mammals, generally *increase* as late small clutches are laid; that is, young prey animals are coming into the environment, increasing rather than decreasing the prey base (Daan *et al.* 1989). Thus, debate has centred on whether the declines are adaptive shifts by 'prudent' parents in response to declining fitness returns (Daan and Dijkstra 1982; Meijer *et al.* 1988; Daan *et al.* 1989; Rowe *et al.* 1999), or whether late nesting individuals are incapable of gathering more food. For this to be true, they would either have to be poorer quality foragers or confined to poorer territories, and therefore constrained to laying smaller clutches (Newton and Marquiss 1984; Murphy 1986; Nilsson 1991). Seasonal declines are apparently unstudied in subtropical regions, possibly because of the aseasonality of the climate. However, they do occur, at least among harriers. For African Marsh Harriers the clutch size decline over the season (Fig. 8.3) had a much shallower slope and egg laying occurred over a season that extended for 120 days, or three months. This is in contrast to the typically 50-day egg-laying period exhibited by European Marsh and Montagu's Harriers (Altenburg *et al.* 1982; Arroyo 1995) in western Europe and the 55-day season of Northern

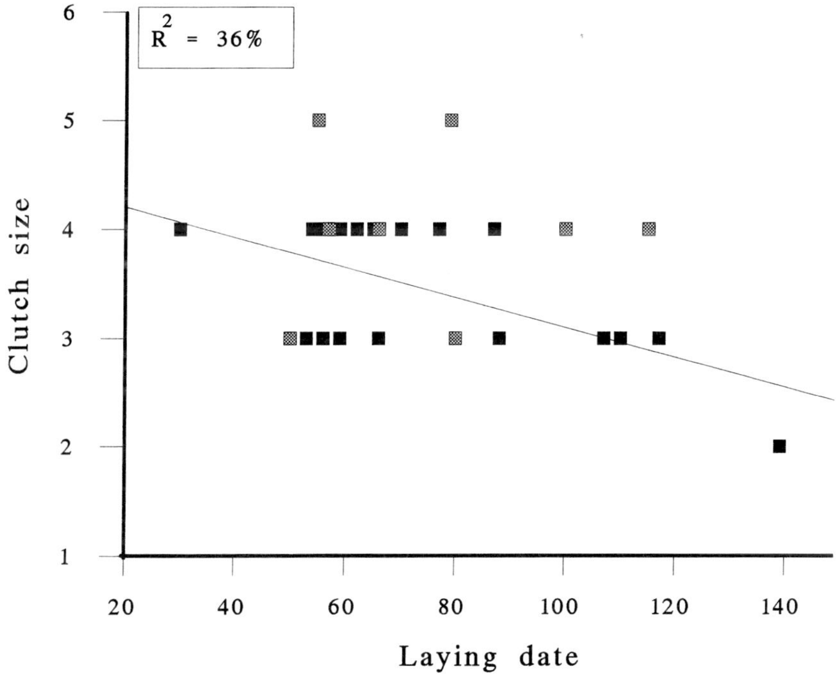

Fig. 8.3 Seasonal decline in clutch size of African Marsh Harriers. Note the very long laying period (120 days) relative to temperate Northern Harriers (previous figure: 55 days).

Harriers in New Brunswick (Fig. 8.2). Recent unpublished data on the Rock Kestrel (*Falco tinnunculus rupicolus*), a subspecies of the Eurasian Kestrel in southern Africa, suggests that seasonal declines may also occur in other raptors, but they have only been documented for fledging success, not clutch per se (van Zyl 1993). This species also laid eggs over a three-month laying period. These longer egg-laying seasons are one of the major differences between tropical and temperate species and reflect the much longer breeding seasons in tropical environments found by Wyndham (1986) for many other birds (Fig. 8.4).

Measurement of provisioning rate and biomass

To measure food supply, considered potentially the single most important influence on reproduction (Newton 1979; Martin 1987), I monitored the food provisioning performance of males at 19 harrier nests over three years (described further in Chapter 1). Because I needed to determine how much food females received while they were actually forming their eggs, I carefully observed them for signs of the egg-laying lethargy stage—when egg formation is assumed to begin. In observing them closely I realized that these lethargic females became visibly pregnant (Fig. 8.5). Whereas the deepest part of a harrier is typically its pectoral muscle mass, for a pregnant female the belly and vent

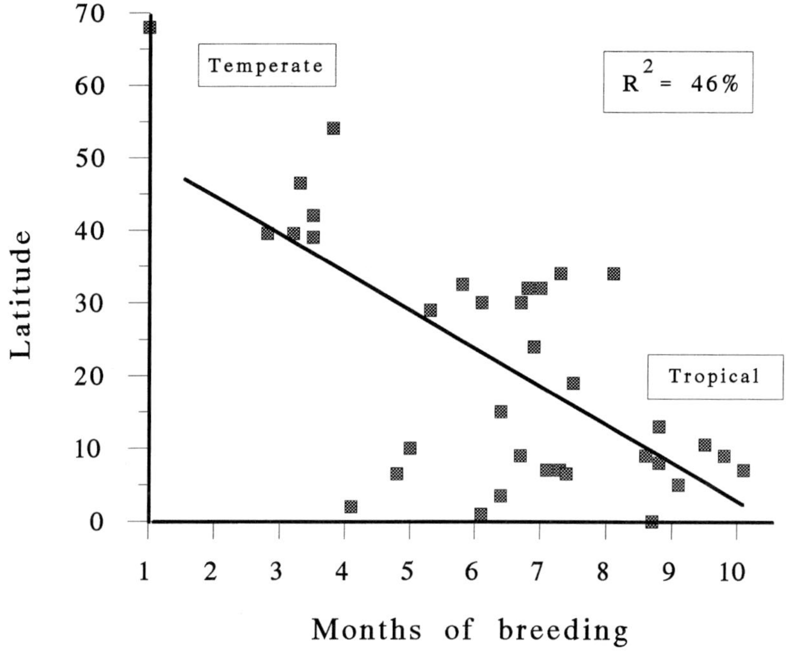

Fig. 8.4 Differences in the length of breeding seasons for birds at different latitudes. A significant increase is apparent towards the equator, where up to ten months is available for breeding. For every 11° decrease in latitude, breeding seasons increase by one month. Adapted from Wyndham (1986).

Fig. 8.5 Female harriers become visibly pregnant just before egg laying as shown by the distended belly and vent region of the lower bird.

regions visibly distend and become the deepest part, changing the bird's flight profile. Thus, I could visit any nest and observe the female for signs of pregnancy. Egg-forming was taken as the period from seven days before the first egg appeared until the clutch was complete. This is corroborated by the work of Meijer *et al.* (1989) who examined egg-formation energetics in kestrels.

Provisioning performance: rate relative to stage of breeding

Male African Marsh Harriers provided prey to their mates and offspring mainly between the hours of 07h00 and 17h00. Rates were relatively constant at 0.3–0.4 prey offerings per hour, and did not differ significantly from even hourly rate ($X^2 = 13.7$, 10 df, $P > 0.2$) with all years combined. During the 12-h days at the start of breeding in August, rates of provisioning were therefore 3–5 items daily.

Between years, seasonal patterns of prey provisioning were similar (Fig. 8.6). Males began provisioning their mates an average 0.7 to 1.6 items/day before egg formation, increased mean delivery rates to 4.5 to 6.4 items per day when females were forming eggs, decreased to 3.7 to 4.9 items during incubation, and increased substantially thereafter to levels about twice as high just after hatch.

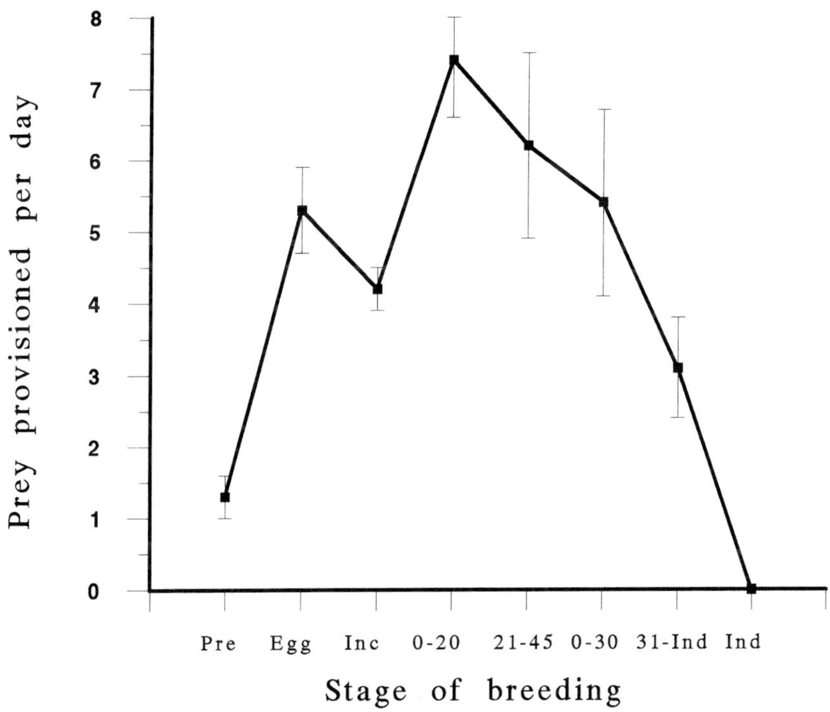

Fig. 8.6 Differences in provisioning performances (items/day ± 1 s.e.) of African Marsh Harriers relative to stages of the breeding cycle; all males combined, 1984 to 1986. Note the higher levels during egg formation relative to incubation feeding.

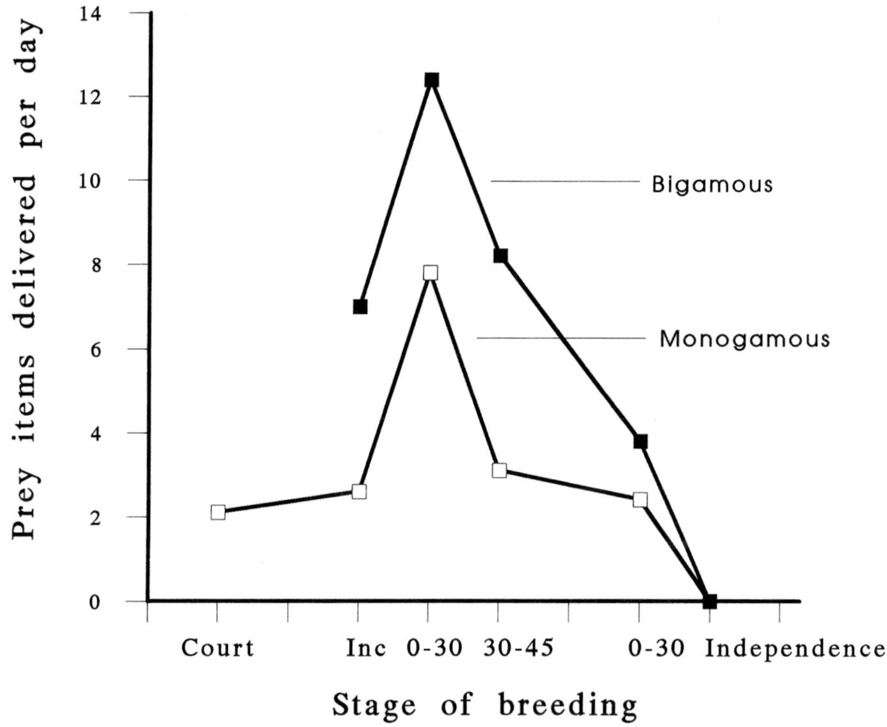

Fig. 8.7 Male provisioning performance (items/day) of European Marsh Harrier males from courtship through to the late fledging stage (adapted from Altenburg *et al.* 1982). The upper graph depicts the provisioning rate of a bigamous male in relation to monogamous males. A trigamous male whose provisioning rate was even higher could not be shown.

Rates decreased once young were flying and reached their lowest average levels in the period just before independence. Similar results are apparent for subtropical/tropical *Accipiters* (Simmons 1986a; Delannoy and Cruz 1988) and were apparent for temperate European Marsh Harriers in Holland. Provisioning there began before egg laying, increased to highest levels of about eight items per day, and decreased suddenly through the nestling and fledging stages (Altenburg *et al.* 1982). The decrease during incubation found for African birds was not apparent (and possibly masked) in the European data (Fig. 8.7) because Altenburg's group did not distinguish the egg-forming period from other periods. While the overall trend was similar, the amount of food provisioned to nests was very much greater for the polygynous males that they observed even though these males foraged over the same general terrain as the monogamous males. For example, six bigamous males provided on average over three times as many prey items to the nest during incubation stage as did monogamous males, and their rate was 50% higher at peak provisioning times during the nestling stage (~7.8 vs. 12.4 items/day: Fig. 8.7). One trigamous male provided even higher numbers of prey (~13.3 items/day), and continued this trend in the post-nestling stage when other males were declining in the provisioning performance (Altenburg *et al.* 1982).

One of the important points to emerge here is that, at least in terms of items, *monogamous* African and European birds provisioned roughly equal numbers of food parcels, yet males that became polygynous in Europe exhibited much higher rates than African birds. However, because such comparisons take no account of the biomass provisioned, for African birds I used the method of ranking the size of items in a bird's foot and equating it with biomass from known items (Chapter 1).

The biomass provisioned (Fig. 8.8) showed similar trends as prey items, although the average sizes of prey items differed with respect to the stage of breeding. In 1984, largest prey were delivered before egg formation ($\bar{x}$ = 54 g), during egg formation in 1986 ($\bar{x}$ = 40 g), but after hatch in 1985 ($\bar{x}$ = 42 g). Combining years, prey size did not differ significantly with respect to month (P = 0.08) or the stage of breeding (P = 0.06), even though there was a steady increase towards largest prey types in the most stressful period of breeding, the

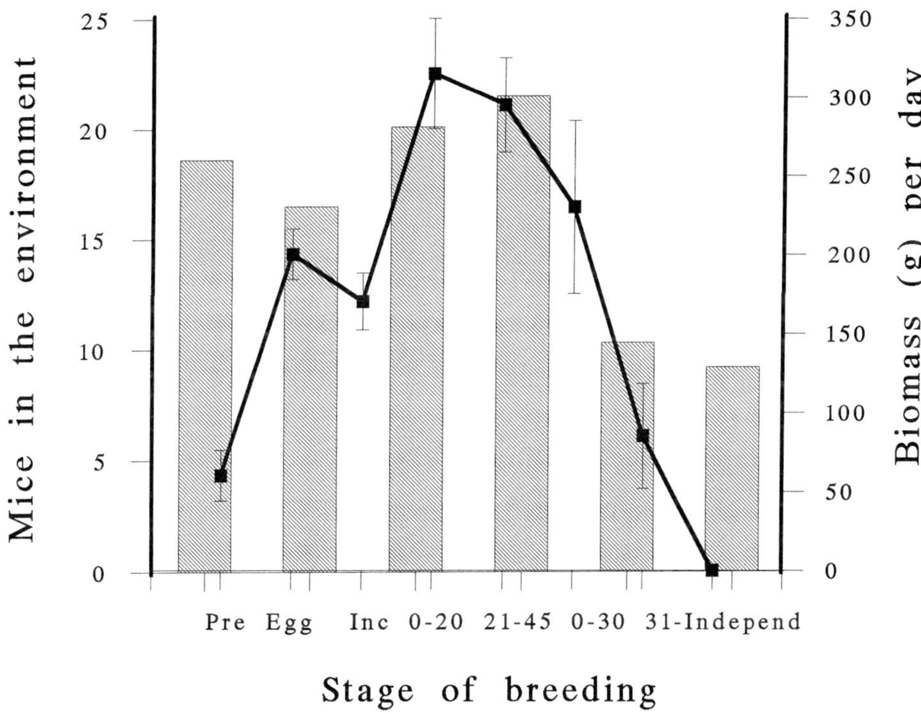

Fig. 8.8 Male African Marsh Harrier provisioning performance (biomass/day) in relation to seasonal differences in mice found in the environment (mice captures/100 trap days). The close association between the two measures (except during egg forming) suggests prey-capture return of harriers is mainly dictated by the abundance and availability of mice.

Table 8.1 Average daily biomass deliveries by male African Marsh Harriers with respect to stage of breeding; years combined. Means are given with standard errors, sample size, and the number of hours observed.

	Pre-breeding	Egg forming	Incub	Hatch: 0–20 d	Nestling: 21–45 d	Fly: 0–30 d	Fly: 31–Indep
Biomass (grams/day)	58 ± 15	199 ± 16	164 ± 16	313 ± 34	296 ± 30	231 ± 53	86 ± 24
Deliveries (items/day)	1.4 ± 1.7	5.3 ± 1.5	4.2 ± 1.8	7.7 ± 3.2	7.1 ± 2.2	5.9 ± 3.6	2.4 ± 1.6
Number of males	23	11	19	15	9	7	6
Hours	496	205	426	337	113	173	56

last half of the nestling stage. By combining average prey biomass and the rate of delivery, the biomass delivered per day can be computed (Table 8.1). A sharp (3-fold) increase in male provisioning effort was again evident just as eggs were being formed. Because this may have an important bearing on clutch and egg sizes, we will look at this period in detail.

Provisioning before breeding

Some males, particularly those exhibiting seasonally low rates, fed their mates well before egg formation. The earliest such provisioning occurred 36 days before the first egg. However, the average was 8.5 ± 12 days (n = 18) before egg laying; close to the time at which females are assumed to start forming eggs (7 days). Rates averaged about 60 g per day (Table 8.1), the equivalent of one large mouse, but varied widely, from no prey items from 35% of the males, to 176 g per day from the best provider. It is important to note that six males did not start to feed their mates at all until they had started egg laying; of these, four were newly formed inexperienced pairs, (the males were known to be young based on iris colour and plumage) and two were older, established males whose diets were being experimentally supplemented. It is known in several species that males may become 'lazy' when fed and merely wait by their feeding posts in anticipation of the forthcoming meal. This occurred with these African harriers and may explain the behaviour of the experienced males not feeding their mates. Following initial prey passes, biomass provisioning increased threefold as females became sedentary shortly before becoming visibly pregnant and egg laying. Excluding supplemented pairs, provisioning performance was variable between males during this critical period (Table 8.1). The poorest provisioner providing 174 g per day (4.0 items) in 1984, and the highest 306 g per day (7.8 items) for a replacement clutch in 1986.

Provisioning performance and clutch size

If we exclude the supplemented pairs for now, we find no correlation between clutch size and the biomass provisioned by males during egg formation (r = 0.02, n = 11, n.s.), nor between clutch size and the frequency of delivered items (r = 0.23, n = 11, n.s.). However, on including all food supplemented birds and

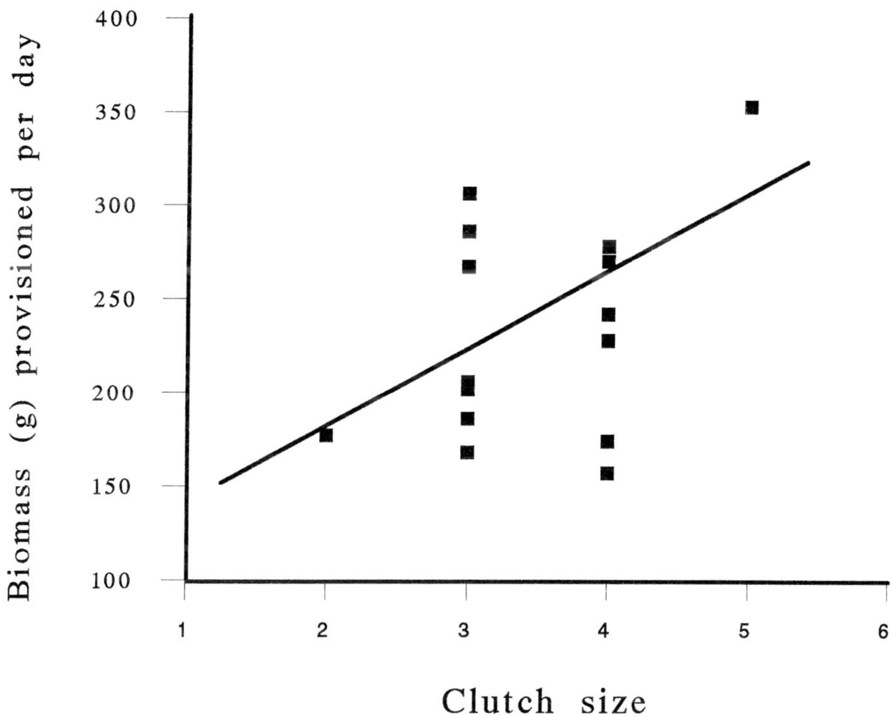

Fig. 8.9 Clutch size of African Marsh Harriers in relation to provisioning performance of males during egg forming. While the correlations are weaker than those for Northern Harriers (Fig. 8.1), the relationship is significant ($P < 0.05$).

calculating their biomass rate as the sum of that artificially and naturally provisioned, significant correlations were apparent between clutch size and provisioned biomass ($R^2 = 18.5\%$, $P < 0.05$: Fig. 8.9). This combination is necessary because the 'laziness' induced in supplemented males by the act of feeding them artificially reduces their apparent prey provisioning rate to their mate and thus lowers the apparent intake of provisioning food to a female who is in fact receiving more food than normal.

David Lack (1947) originally proposed that clutch size should conform to the number of offspring that a pair could adequately rear. One way in which a female might assess the number of young she can rear is to assess male provisioning performance early in the season, and use this as a guide to provisioning later in the season (Niebuhr 1981; Wiggins and Morris 1986). This is only likely for species such as the raptors—and some sea birds—in which females are highly reliant on their mates for food; otherwise an assessment of food in the environment might be more important. To allow such a male-based assessment, correlations must exist between egg-formation provisioning and that provided during the nestling phase. Weak correlations did indeed exist between provisioning performance during egg formation and that during the nestling period (r = 0.68, P = 0.09), although samples were severely reduced (n = 7) due to several nests failing prior to hatch and the need to exclude supplemented pairs. Nevertheless, the correlations suggest that females could use early provisioning at least as a guide to future performance and potentially base their clutch and brood size on future provisioning. Females in stable pair bonds could also use previous provisioning performances of their mates to assess future breeding prospects—but only if provisioning remained relatively constant from year to year. I could check the constancy of provisioning performance during the nestling stage for only three males between years, and they showed identical (5.2, 5.3 items/day), similar (9.3, 7.9), and dissimilar but increasing (4.7, 10.4) values between years.

Clutch size, body size, and laying date

To distinguish the possible effects of other factors, I introduced female size, average egg volume, the laying date, the female herself, and the total biomass provided into a stepwise regression analysis, with clutch as the dependent variable. Biomass provisioned remained an important factor influencing clutch size, and together with laying date explained a highly significant proportion of the variation (Table 8.2). The other variables played no significant role. This result has recently been confirmed in another harrier. In western France, Beatriz Arroyo and co-workers examined clutches in an impressive 387 nests of the colonial Montagu's Harrier and found that clutch size also varied significantly with laying date. Although they did not determine provisioning rates, they did note a decline in average clutch size with generally declining food abundance from year to year. They also found that clutch size did not vary for females of different ages (from 2–10 years old). This is an important result because it relates to one of the hypotheses suggested as a reason for declines in clutch size with seasons (below). For African birds, to determine whether declines in male provisioning performance of late layers could explain decreased clutch sizes (Fig. 8.2), I correlated provisioning rate with laying date.

Table 8.2 Factors explaining significant amounts of the variation in clutch size of African Marsh Harriers. Analyses used stepwise regression procedures on 16 nests. Egg volume, the female herself, and female body mass played no role.

Independent variables	Proportion of total variation explained (R^2)	Significance level (P)
Laying date	28%	0.0022
Provisioned biomass during egg formation	27%	0.0038
Date + biomass	55%	<0.0001

Table 8.3 Provisioning performances (items/day) of young or newly paired male African Marsh Harriers relative to old, experienced males, 1984–86. Means are given with standard deviation, range, and sample size.

Breeding stage	Young/newly paired males	Old/established males
Provisioning during egg formation	6.7 ± 2.7 4.0–10.0 (n = 4)	5.6 ± 1.4 4.0–7.7 (n = 6)
Provisioning during the nestling stage	8.9 ± 4.0 5.0–13.9 (n = 5)	6.6 ± 2.0 4.2–10.4 (n = 10)
Mean prey sizes	40 ± 24 g	39 ± 24 g

There was no significant trend for earlier birds to exhibit higher rates ($R^2 = 9\%$, $n = 16$, $P = 0.23$).

Provisioning and male experience

The effect of experience on provisioning effort by males was determined by dividing pairs into experienced (paired for >1 year), and inexperienced pairs. While some young males did not start provisioning until their mates began laying, there were no significant differences in provisioning performance between young and old birds either during egg formation or the most demanding period of the year, the nestling stage (Table 8.3). It was apparent, however, that newly mated males invariably exhibited the highest recorded rates at each stage. Since average prey size was almost identical for the two groups (Table 8.3), age and experience played some role in the provisioning performances of different males. However, these differences for 13 males explained no variation at all in clutch size when included in the stepwise regression procedures. This is in contrast to that found in temperate zone raptors (Newton 1986; Poole 1989), including European Marsh Harriers, in which older males give rise to significantly earlier and larger clutches than young males (Altenburg *et al.* 1987).

Factors influencing egg size

Large bodied females (based on wing chord measurements) produced significantly smaller eggs than small females. This was apparent either by taking females as independent data points in each year that they laid ($r = -0.86$, $n = 22$, $P < 0.0001$: Fig. 8.10) or by averaging egg sizes for each female across all years ($r = -0.910$, $n = 6$ females, $P < 0.01$: Table 8.4). Total clutch volume was also significantly negatively related to female body size ($r = -0.84$, $n = 6$, $P < 0.025$): the smallest females thus not only laid large eggs but put greater resources into

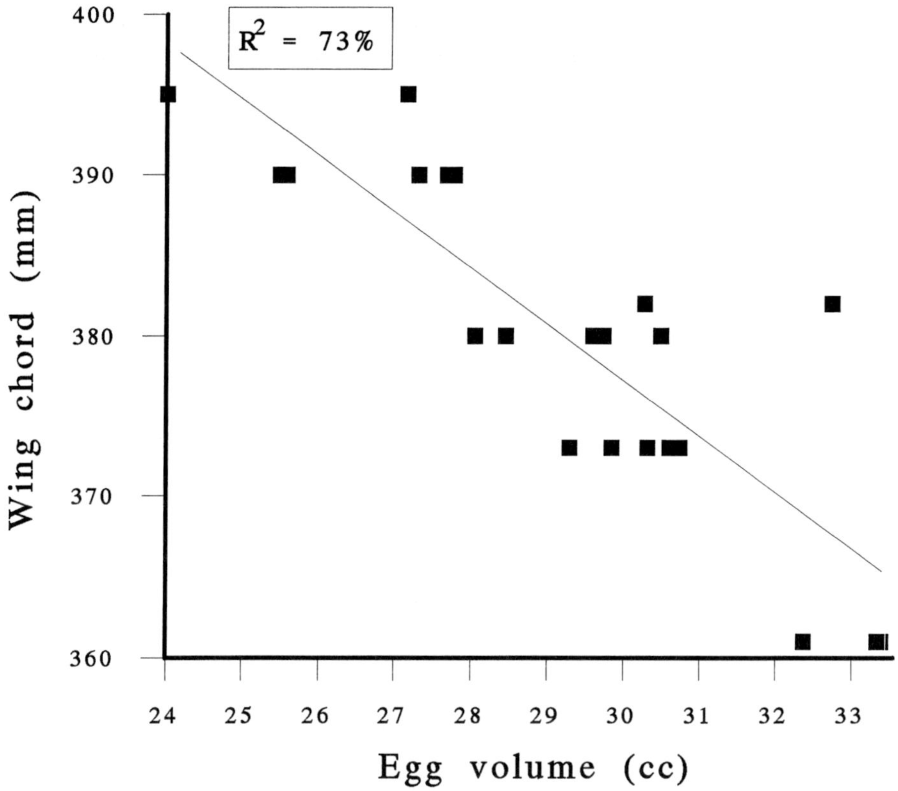

Fig. 8.10 Egg volume of African Marsh Harriers in relation to wing chord of the female. Smaller females laid larger eggs than large females; a highly significant relationship ($P < 0.0001$).

Table 8.4 Egg and clutch sizes of African Marsh Harriers in relation to female harrier body proportions. Where two or more measurements were taken on females caught at different times, the average was used; similarly, egg volumes were averaged from up to four years' data, 1984 to 1987.

Female	Mass (g)	Wing chord (mm)	Mean clutch size	Egg volumes (cc)				
				1984	1985	1986	1987	Mean
EV1	560	380	3.25	27.57	30.01	29.11	29.25	28.99
EV2	588	373	4.50	30.25	28.81	30.12	—	29.73
EV3	639	390	3.75	27.20	27.30	26.83	25.10	26.60
GV2	580	382	3.00	32.25	29.78	—	—	31.02
RONDV	490	361	4.00	—	33.08	32.86	—	32.97
R-SWRT	550	395	3.50	—	23.53	26.68	—	25.11

Table 8.5 Factors explaining variations in the average egg volume of African Marsh Harriers. The analysis employed stepwise regression procedures on 16 clutches, using one (averaged) egg volume for each clutch.

Independent variables	Proportion of total variation explained (R^2)	Significance level (P)
Female body size	63%	< 0.0001
Female	15%	0.0009
Laying date	9%	0.01
Size + female + date	87%	< 0.0001

their total clutch. Tarsus measurements, sometimes used as an index of body size, were unrelated to any measure of reproductive investment. Female body size was not the only factor that significantly affected egg volume. Stepwise inclusion of the provisioned biomass, laying date, and the female herself indicated that female size, the female, and the laying date explained a significant proportion of the variation (Table 8.5). There was no evidence, however, that small females laid earlier than larger ones ($R^2 = 0.64\%$, $n = 13$).

Gathering these results together gives us a picture of what factors are influencing the productivity of female African Marsh Harriers. The main findings were that male marsh harriers provisioned their mates at low rates up to a month before eggs were laid, and substantially increased rates of biomass provisioning at the time eggs were being formed. Some inexperienced males, however, only started provisioning as their mates began to lay, although thereafter their provisioning rates were typically higher than older males'. Clutch size was significantly related to the amount of biomass provided as well as the laying date, with these variables explaining 55% of the variation. Since early and later provisioning rates of the same male were correlated ($P = 0.09$), albeit weakly, females could use their mates' early provisioning as a guide to their future brood size. Egg size, on the other hand, was unrelated to provisioning performance, but was best explained by allometric relationships with female body size, an intrinsic aspect of the female herself, and the date on which she laid the eggs. Thus, since clutch size was related to food supply and males provide all of the food for females at this time, male performance seems critical for female reproduction. However, several other ideas have been proposed for why males feed their mates.

Why do males feed their mates before egg laying?

The most prominent alternative explanation for pre-egg laying feeding by males is one of courtship and pair bonding (Lack 1940). As with other species (Royama 1966; Tasker and Mills 1981; Donazar *et al.* 1992), this is unlikely for marsh harriers because it starts in new pairs after the pair bond has formed, and is found in pairs established for over three years. Similar to gulls (Niebuhr 1981;

Tasker and Mills 1981) and kestrels (Donazar *et al.* 1992), pre-breeding feeding for harriers starts in earnest about nine days before females lay their first eggs, and is three times more frequent than earlier deliveries. For harriers this initial burst of feeding is also approximately 20% higher than that found during incubation (Figs 8.6 and 8.8), similar to trends recently found for temperate breeding Ospreys (Birkhead and Lessells 1988) and European Marsh Harriers (J. Underhill-Day *in litt.*). This strongly suggests that males work particularly hard to bring their mates larger than normal amounts of food during egg forming.

However, before we can claim such increased effort for male harriers we must check that their increased rate is not simply a result of increased food in the environment. There are several ways to do this:

(i) determine predation success in the months corresponding to the diferent stages of breeding (July–August for pre-breeding), late August to early September for eggs, September for incubation, and so on);

(ii) determine mouse captures per month;

(iii) determine the weight of pellets which are known to be positively correlated with food intake (Tarboton 1978; Mendelsohn 1982) in other mouse specialists in Africa, and which indeed were 38% heavier for the food-supplemented harriers (see below).

Since all measures showed the same trend, I have depicted the most direct—the mice present in the environment (Fig. 8.8). The trends indicate that food abundance cannot explain the increased provisioning of females by male harriers: at the time their provisioning was increasing—when eggs were being formed (August)—the number of mice in the environment was at its lowest level for this part of the year (Fig. 8.8). Thereafter both mouse populations and provisioning rates increased to the most stressful breeding period—the mid- to late-nestling period. There seems to be little doubt, then, that males are working extra hard to provide their mates with more food during this critical period.

Nevertheless, 6 of 17 males began provisioning only after egg laying had commenced. Although two were supplemented birds, four were inexperienced males who began provisioning as their mates began laying. Since these females did not lay smaller than typical clutches it suggests that food *during* laying is more important than that supplied before laying commences. Thus we can assume that increased provisioning during egg formation has an important function in readying the female for egg laying, and allows her to lay slightly larger clutches if supplies are high. This conclusion is supported by the correlations between provisioned biomass and clutch size (Table 8.2), as well as increased clutch sizes in experimentally supplemented birds (Chapter 9).

The possibility that early provisioning functions as a secondary mate choice cue is supported for one temperate harrier from associations between order of settling and provisioning performance (Chapter 5). This is difficult to check for African birds because of the longer-term pair bonds, but it was noticeable that on average males in newly formed pairs were better provisioners than other males in the egg-formation period. As males stabilized pair bonds, their rates decreased (Table 8.3). This would be expected if females choose the best provisioning males as mates when making initial choices. Female African Marsh Harriers could use early provisioning as a rough guide to future food supplies from the male, and the small data set on between-year provisioning by males suggests that a female could also use past performance of her mate to assess his future potential. For other temperate species there is also some evidence that females dependent on their mates for food can use their initial performance as a guide to future performance (Wiggins and Morris 1986; Donazar *et al.* 1992).

In other raptors, notably Ospreys and Lesser Kestrels, 'courtship' feeding was found to play little or no role in pair bonding. For kestrels, but not Ospreys, it also played a role in clutch size determination (Donazar *et al.* 1992; Poole 1985). As an alternative Poole (1985, 1989) suggested that 'courtship' provisioning facilitates copulation; some males with low rates experienced more unsuccessful copulations and low mate fidelity. This is supported by Tasker and Mills (1981) for gulls, although it is unlikely for solitary nesting Scottish Ospreys (Birkhead and Lessells 1988). Even though infidelity and extra-pair copulations are rare among raptors (Simmons 1990; Møller and Birkhead 1993b), twice as many African Marsh Harrier copulations were unsuccessful when food was not present than when it was present (Chapter 6), and food and early copulations are significantly related in other harriers (Pandolfi *et al.* 1998; Arroyo 1999). Hence there is some evidence that pre-breeding food enhances the chances of successful copulations for harriers as it does in gulls and some Osprey populations. In summary, pre-egg laying food for a female African Marsh Harrier probably functions to ready her for egg laying and explains a significant but small proportion of the variation in clutch sizes. This is supported by results from supplementary feeding experiments in harriers reported in the next chapter, and several other raptors (Newton and Marquiss 1981; Meijer *et al.* 1988; Korpimäki and Hakkarainen 1991). Secondarily, the presence of food enhances copulation success (but not frequency) and may be used in mate assessment for Montagu's Harriers. A pair-bonding role is unlikely.

Role of laying date

Both clutch size (R^2 = 28%) and egg size (R^2 = 9%) were partially explained by laying date, but are these adaptive decreases, in the sense used by Daan *et al.* (1989), or could they be explained more simply by declining food provisioning performance of males starting later? For African Marsh Harriers there was no significant decline in provisioning performance with season which might explain the significant seasonal decline in clutch size. I was able to check this result for Northern Harriers where provisioning performance was also closely monitored prior to hatch. While clutch size (which varied from two to five eggs) was significantly correlated with male provisioning performance in terms of items per day ($P < 0.05$, $n = 8$), there was only a very weak ($r = -0.29$) and non-significant correlation for the same males and their laying date. In other words, like African Marsh Harriers, Northern Harrier clutches could be partly explained by provisioning performance, but the decline with season could not be explained by provisioning performance. This somewhat surprising result makes intuitive sense when we remember that even late-laying females are provided with large quantities of food because their mates appear to work particularly hard when eggs are being formed (Fig. 8.8), and for the latest egg-forming females, mouse abundance and their vulnerability to harriers is increasing (Chapter 4).

Nevertheless, when examining the provisioning rates of Hen Harrier nests in northern Britain, Redpath and Thirgood (1993) did find a significant decline in provisioning rates of males at later nests ($R^2 = 0.55$, $P = 0.01$, $n = 9$). Males hatching young earliest, at the end of May, provided over 15 items/14-h day, while males hatching chicks a month later provided only about one-third of this total (~5.6 items/day). Since they, too, found seasonal declines in fledging success their results suggest the most likely reason for the decline is reduced foraging returns of late-starting males. We do not know from their study whether this was due to inferior males, or declining food in the environment.

Is it possible that small female African Marsh Harriers (which laid large eggs) also laid earlier than large females. There was no correlation between female size and laying date. Since neither ecological nor allometric factors affecting clutch size could explain the seasonal decline, laying date did appear to play an intrinsic role in breeding performance. In other words if there exists a decreasing likelihood that young can be reared the later in the season the clutch is started, then it would be in the female's interests to lay a smaller clutch. But does fledging

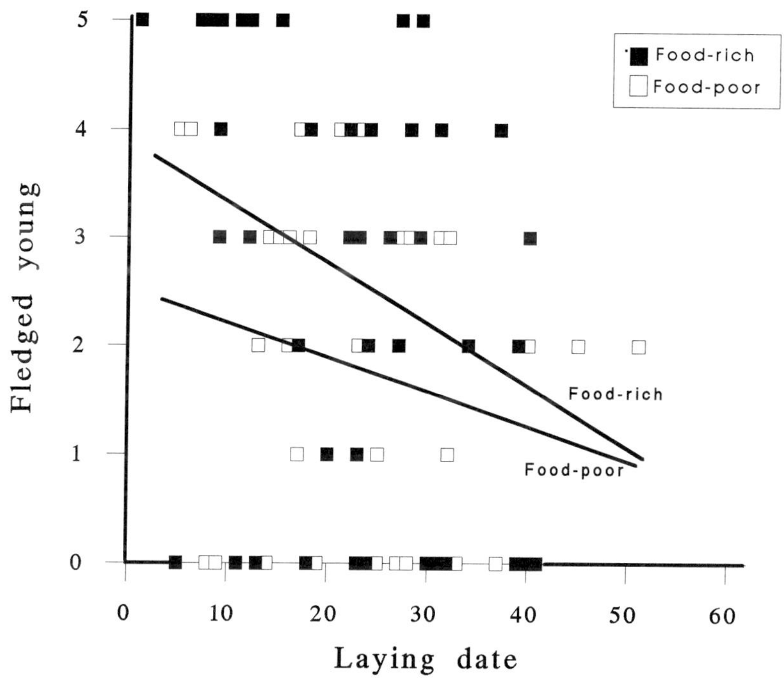

Fig. 8.11 The seasonal decline in fledging success of Northern Harriers. Declines were steeper in food-rich years, mainly because more young were reared by females that started earlier.

success (when harriers first begin to fly) decline with season? This is well documented in Northern Harriers in which the number of fledged young in New Brunswick drops from an average of almost four young (in food-rich years) to just over one young 50 days later (Fig. 8.11). The decline is less steep in food-poor years for the simple reason that the number of young fledged at the start of the season averages only 2.5, dropping to one youngster 50 days later. Such declines are also apparent in the relatively aseasonal climes of southern Africa where breeding success showed a non-significant decrease with season (Fig. 8.12). Moreover, we know from measuring the harriers main food resource—mice—and their availability as judged by a measure of vegetation depth (chapter 4) that both decline as the season progresses. Thus harriers should find it increasingly difficult to provide large broods with sufficient food as the food level in the environment and its vulnerability decline. Thus, I tentatively conclude that seasonal declines in clutch size may be an adaptive response to declining fitness returns by both African Marsh Harriers and Northern Harriers, and the declines cannot be explained by constraints at the time of egg-forming.

Do other harriers fit this picture? The only other species for which some clues are available is the Montagu's Harrier. As reported earlier, Arroyo *et al.* (1999), found that food abundance in the environment was associated with decreasing clutch sizes from year to year. They also found that the seasonal decline was

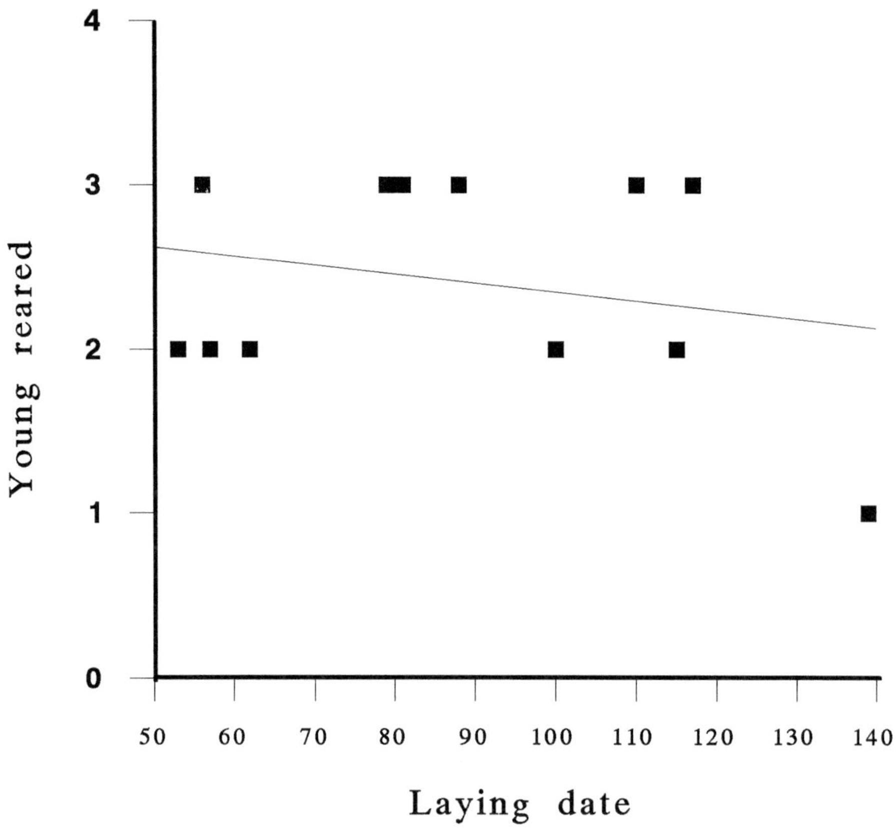

Fig. 8.12 The non-significant decline in fledging success of African Marsh Harriers. Pairs hatching young earliest in the season reared about one young more than pairs hatching young 100 days later.

unlikely to be related to younger females laying later because there was no difference in clutch size for females aged 2–10 years. While first-year birds did show lower clutches they made up such a small proportion of the breeding birds (28% in peak years, 5% in medium years and 0% in low vole years; in total 15% of 378 females), that it is unlikely that female age could have accounted for seasonal declines. This suggests that seasonal declines were not age related and may also be an adaptive response. However, Arroyo *et al.* suggested that poorer quality birds were laying towards the end of the season. They did so based on the finding that the weight of the females laying later clutches declined with laying date. In other words those females that laid clutches later were in poorer condition than females laying earlier. This in itself does not suggest that birds were constrained to laying smaller clutches, for two reasons. First, a poor condition female may still find it adaptive to reduce her clutch size because she is not being provisioned well by her mate. Her poor condition may not necessarily mean she is a poor quality bird, as argued by Arroyo *et al.* Instead, it may simply be a reflection of male performance or more competition for the food source from other males and their young as the season progresses. It is known from their work that food *abundance* in the environment increases, even though food *availability* decreases in some habitats (Arroyo *in litt.*). This suggests that later birds should do better, but this was not the case. Even if females were planning smaller clutches to account for their (or their mate's) poorer abilities in harvesting the increased food supply, the fact that females reared a lower proportion of the eggs laid (Arroyo unpublished) does suggest poorer abilities. Ideally, experimental work, as undertaken by Dijkstra's group (feeding females laying mid-season), could elucidate the exact mechanics of adaptive vs. constrained clutch size declines. For harriers, therefore, there is some controversy over the reasons for seasonal declines in clutch size. I expect that when further evidence is accrued, both camps will probably be correct—females may begin later because their mates do not provision them well enough to start earlier (suggested by Redpath and Thirgood's data) and thus lay large clutches, but having started later it is then adaptive for them to reduce their clutches because of the reduced likelihood of raising a larger brood.

Timing of breeding

If early-settling females are reproducing more successfully than late-settling ones, what cues are they using to time their reproduction to seasonal fluctuations? This may be particularly difficult in birds in aseasonal environments where day length and temperature changes are less pronounced than in temperate regions. I started this chapter by arguing that temperature changes were more important in temperate regions, and rainfall was probably of most importance in subtropical regions. What evidence is there for this idea?

In boreal regions Emperor Penguins are the most extreme example of species that start breeding well before optimal conditions prevail. Adult females lay a single large egg on the featureless inland Antarctic ice sheets when temperatures hover around –40° C. The egg is then incubated, precariously balanced on the bird's feet, and shortly thereafter transferred to the male who faces severe winter storms, and slowly starves for the next three months while incubating! Indeed, when the breeding grounds were first discovered and the first egg brought back for western science, the discoverer was ignored because his claims were too outlandish to believe (Bond *et al.* 1988)! The penguins undertake this incredible hardship for one main reason: to time their breeding such that their nestling's principle growth period coincides with the main food source of krill.

But there are other options open to raptors in timing their breeding and we assume most are linked to prey availability. Thus birds might start breeding (a) when prey first becomes easier to catch (Newton and Marquiss 1982); (b) so that maximum prey availability coincides with maximum food demands of their young, mid-way through the nestling period (Lack 1954); or (c) well before maximum prey yields exist such that fledglings start to self-forage when maximum prey availability occurs (Daan and Dijkstra 1982; Daan *et al.* 1989). Alternatively, birds may time the period of maximum food demand to a change in the composition of the diet (Barnard *et al.* 1987), or in temperate or boreal zones initiate breeding as early as possible to enhance fledgling survival (Perrins 1970; Cooke *et al.* 1984). In a review of nine temperate-zone studies, Daan *et al.* (1989) concluded that young fledge either as prey abundance begins to increase (five species) or as it starts to decrease (four species). They pointed out, however, that few studies attempt to estimate the availability of prey, relying instead on measures of prey abundance, a poor substitute for assessing the difficulty in prey acquisition for parents planning a brood.

For African birds, the influence of food on the timing of breeding is even less clear, and its importance as a major factor has been questioned by Brown and Brown (1984). These authors argued that some species breed at all times of the year, and egg laying is not obviously timed to any particular climatic fluctuations (see Dittami 1987 for a specific example). Similarly, breeding in East African vultures is irregularly timed in relation to presumed peaks in carrion availability (Houston 1990). However, for other tropical birds, food was found to be a better predictor of egg laying than day length, temperature, or rainfall (Maclean 1976), even though some African birds fledge young well before maximum prey peaks occur (Earlé 1981). Other than perhaps vultures (Brown 1988; Houston 1990), *Elanus* kites (Mendelsohn 1982), and kestrels (Kemp 1990) little is known about the timing of breeding in relation to food in African predators.

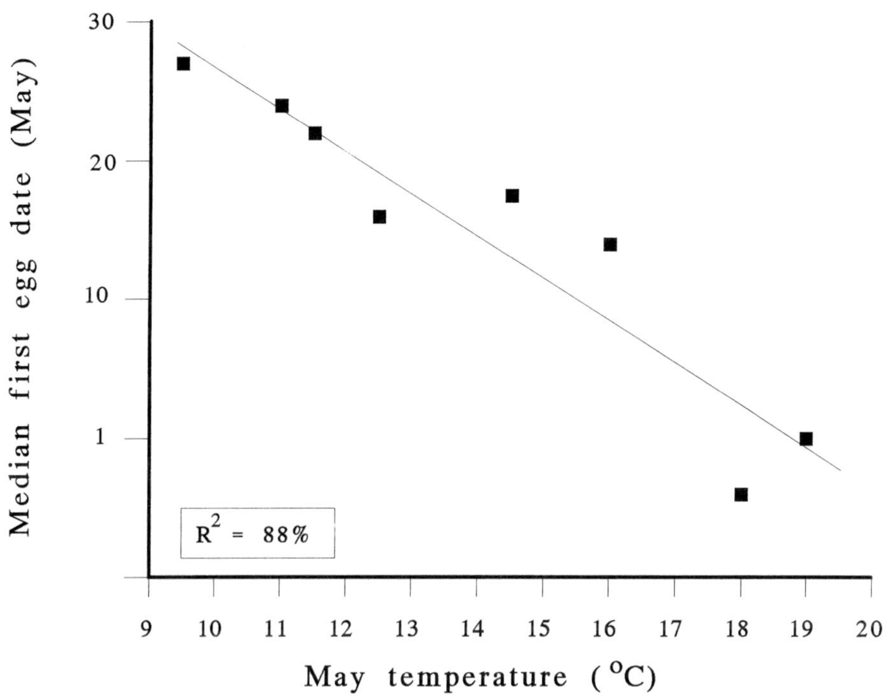

Fig. 8.13 The significant trend for Montagu's Harriers to begin breeding later in regions where average May temperatures are lower (adapted from Schipper 1978).

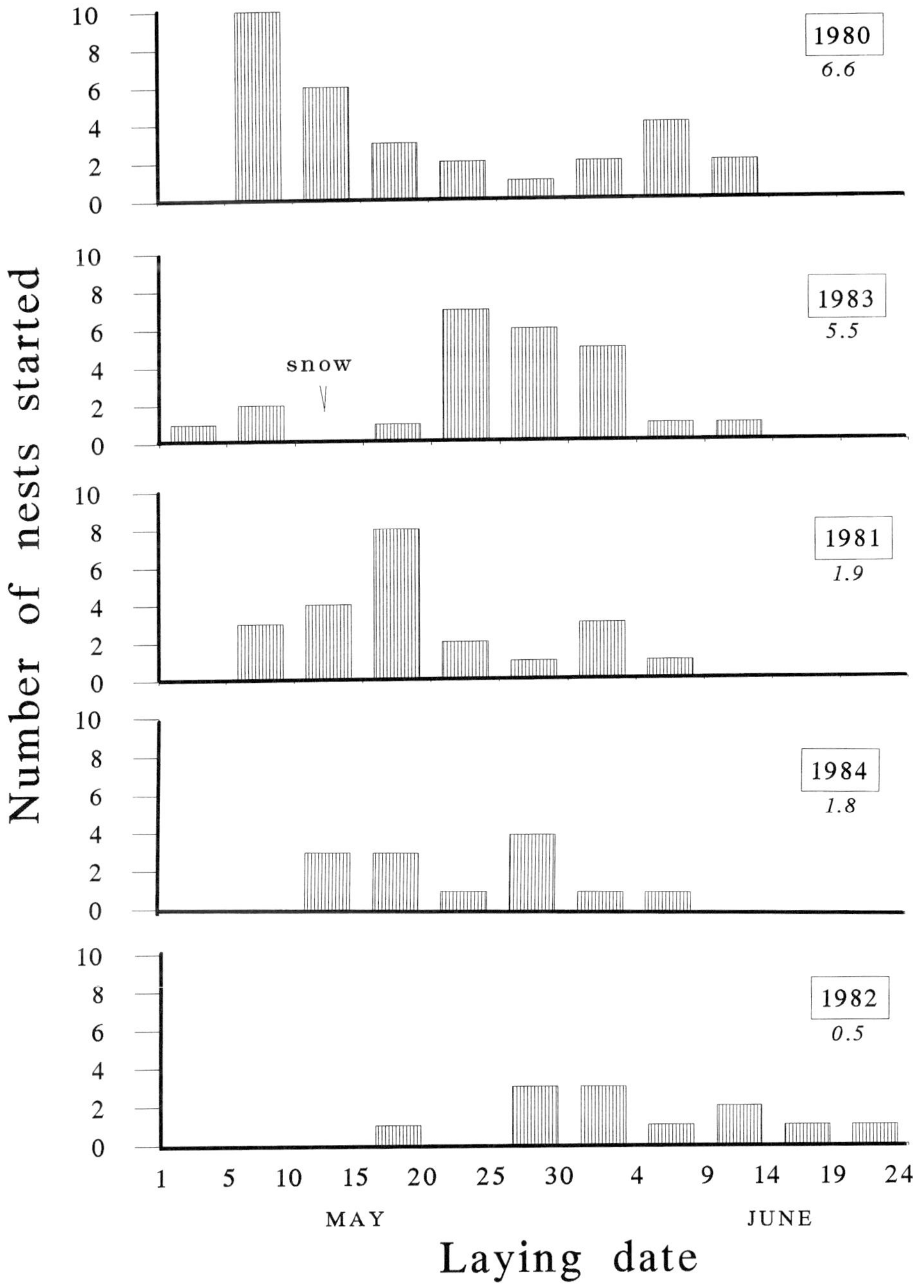

Fig. 8.14 The change in nest initiation dates represented by five-day egg-laying peaks for Northern Harriers in New Brunswick in relation to decreasing vole abundance (given under the year box). Note the complete cessation of egg laying during a cold snap when snow fell over the study site in 1983.

Harrier breeding in relation to climatic factors

The evidence that breeding is temperature influenced in temperate harriers is fairly good. Wibe Schipper and colleagues assessed the initiation of breeding of three European harriers in relation to spring temperatures in May. For each species they found a close correlation between median first egg date in May and the typical May temperature, across study sites from Denmark to southern France (Schipper 1978). The largest number of study sites (eight) were for Montagu's Harriers and the relationship between temperature and first egg date was highly significant (Fig. 8.13). For every 5° C drop in temperature in May, egg dates were retarded by about 16 days. The same was apparent in New Brunswick where laying dates were assessed in relation to vole abundance (Fig. 8.14). While there was a strong trend for first eggs to be laid progressively earlier with increasing vole abundance, in 1983 a cold spell with unseasonal snow during early egg laying, caused all birds to stop laying and only resume about ten days later (Simmons *et al.* 1986b). Thus it seems that food abundance and temperature combined dictate laying in temperate-breeding harriers.

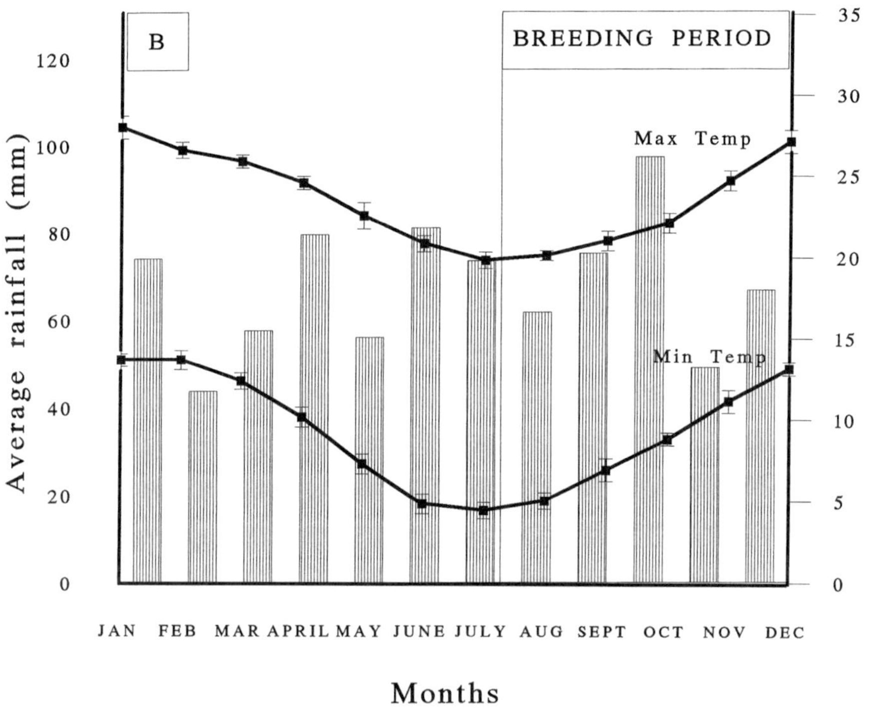

Fig. 8.15 Breeding of African Marsh Harriers in relation to monthly maxima and minima in temperature and rainfall. Rainfall throughout the year in this southern Cape area precludes timing in relation to the beginning of any rain season, and a rise in minimum temperature appears to be the main cue to start egg laying in July. Further north rains are less even (see Fig. 8.17) and breeding is timed differently (Fig. 8.16).

We can gauge the different abiotic factors (presumed to be rainfall) in more tropical regions by determining the egg-laying patterns of harriers in a rainfall transition zone in Africa. This zone occurs in southern Africa from the southern Cape northwards. Furthest south (below about 30° S) so-called 'winter-rains' result in rainfall in each month (Fig. 8.15), similar to temperate climes. Above 30° S, winter rain falls away and rain is highly seasonal with a long dry period separating the strong rains of the austral spring and summer from the rest of the year. African Marsh Harriers below 30° S began nesting just as the average minimum temperatures had passed their lowest point and started to rise again (Fig. 8.15). Indeed in the low-lying areas of marshland used by these birds, frost was not uncommon as the first eggs appeared. Thus breeding was initiated as temperatures began rising, despite rains both before and after this. It may have been fortuitous that rainfall peaked in October, about one month after the peak egg-laying month. This coincided with maximum prey provisioning by males (Fig. 8.8) and peak numbers of harrier nestlings (Fig. 8.16). Since one of their principal prey species, the Striped Mouse *Rhabdomys pumilio*, breeds in response to rain (Taylor and Green 1976) harriers may be timing their breeding to coincide with the first temperature rise then the first rains which produce a flush of vulnerable young mammals. Many passerine birds in Africa also breed in response to rain (Harrison *et al.* 1997), which means at least two newly available and vulnerable

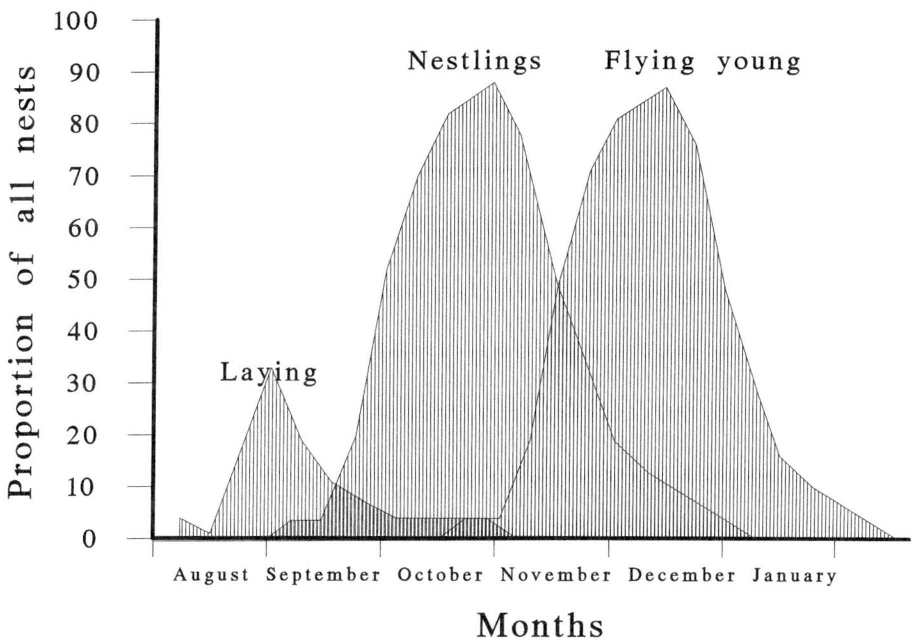

Fig. 8.16 The spread of breeding from egg laying, through hatching, to fledging for African Marsh Harriers in the southern Cape Province (34° S). Note that egg laying extends over three months in this subtropical species, whereas for temperate congeners the equivalent period is 50–55 days. The total breeding spread covers more than six months.

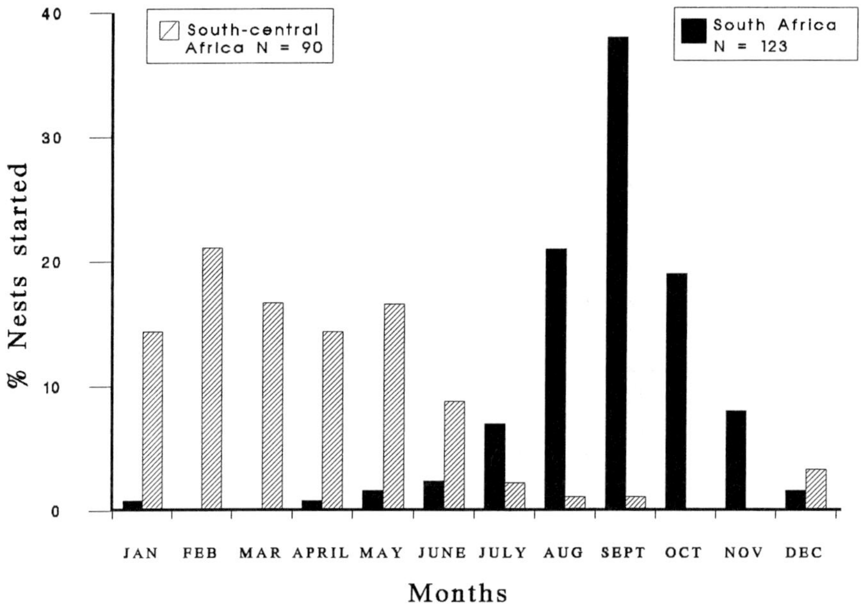

Fig. 8.17 Differences in the egg-laying months for African Marsh Harriers in south-central Africa (peak = February) and South Africa (peak = September).

prey types can be exploited by African Marsh Harriers. The September peak in nesting was not peculiar to just my study area (at 34° S) but is apparent from 123 nest records from throughout South Africa (Fig. 8.17). However, farther north breeding peaks about six months later in south-central Africa (Fig. 8.17). Interestingly, this change cannot be explained by differences in seasonal temperature slopes in south-central Africa (e.g. Zimbabwe, Zambia, Malawi) because temperature exhibits the same patterns as those depicted in Fig. 8.15 (Harrison *et al.* 1997). As expected, rainfall does not follow the same pattern but has a pronounced peak in February (mid-summer), followed by a long five-month dry season. Harriers appear to time their breeding to this peak in rainfall (Fig. 8.18). Note that unlike harriers in South Africa, which begin before the peak, birds do not start egg laying until the rains have begun in south-central Africa. This may arise because there is no additional food (for breeding) whatsoever in these regions prior to the rains, so even if pairs would find it adaptive to breed early they may not be able do so because of food constraint. This would tally with the idea that food resources closer to the equator are of lower amplitude (Ashmole 1963; Ricklefs 1980), giving rise to the smaller clutches in tropical regions (Skutch 1949; Chapter 9).

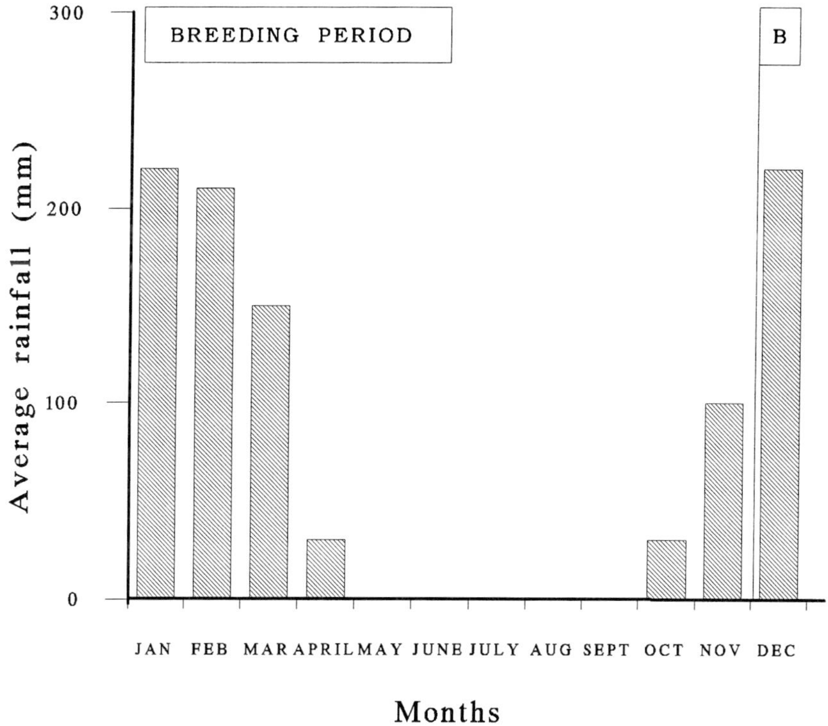

Fig. 8.18 Initiation of breeding of African Marsh Harriers (shown at top) north of 26° S in relation the onset of highly seasonal rains in south-central Africa (Zimbabwe/Zambia). Rainfall patterns from Harrison *et al.* (1997), breeding data from Benson and Benson (1977), Tarboton and Allan (1984).

Breeding in relation to prey populations

Having determined some of the abiotic factors associated with harrier breeding, we can now look in further detail at how African Marsh Harriers time their breeding to food quality and availability. Seasonal changes in prey capture success and diet were documented prior to, and throughout the breeding period. Dry pellet weight, an index of biomass ingestion (see below), was used as a monthly index of the biomass consumed by harriers. In addition, prey abundance and changes in availability were drawn from live trapping programmes and a measure of vegetation depth (Chapter 4). This approach allowed an appraisal of the timing of breeding of marsh harriers in relation to seasonal variations in (i) diet composition, (ii) prey abundance, (iii) prey availability, (iv) prey capture success, and (v) biomass intake. We will then compare our finding with temperate breeding birds to determine if there are any answers common to all birds.

Pellets were collected by providing breeding and non-breeding birds with 1–2-m high posts in their territories. These were accepted quickly and subsequently a small wire basket was wound around the base—high enough so that mammalian predators could not reach the collections, but low enough for harriers to perch unimpeded. Pellets were gathered as soon as I arrived in the study area (winter pellets) and monthly thereafter for six months. Each pellet was dated, air dried, weighed to the nearest 0.01 g on a Mettler PE 1600 electronic balance, and the contents later identified by experts (Margaret and Graham Avery) from collections at the South African Museum. Both mammalian and avian prey could be identified to genera and often to species and the results were arranged by month, species, and weight. In two years (1984 and 1986), 251 prey items from 156 pellets were identified (Simmons *et al.* 1991). For other mammal-eating raptors, notably Black-shouldered Kites, dry pellet mass is known to decrease in parallel with decreases in both biomass ingestion (Tarboton 1978) and the trappability of the bird's rodent prey (Mendelsohn 1982). Thus pellet mass may be used as an index of prey intake by mammal-eating raptors. The relationship was verified for harriers by comparing the dry weight of pellets cast by food-supplemented harriers to those cast simultaneously by unfed birds (Simmons 1989). Birds food-supplemented on average 164 g per day for about

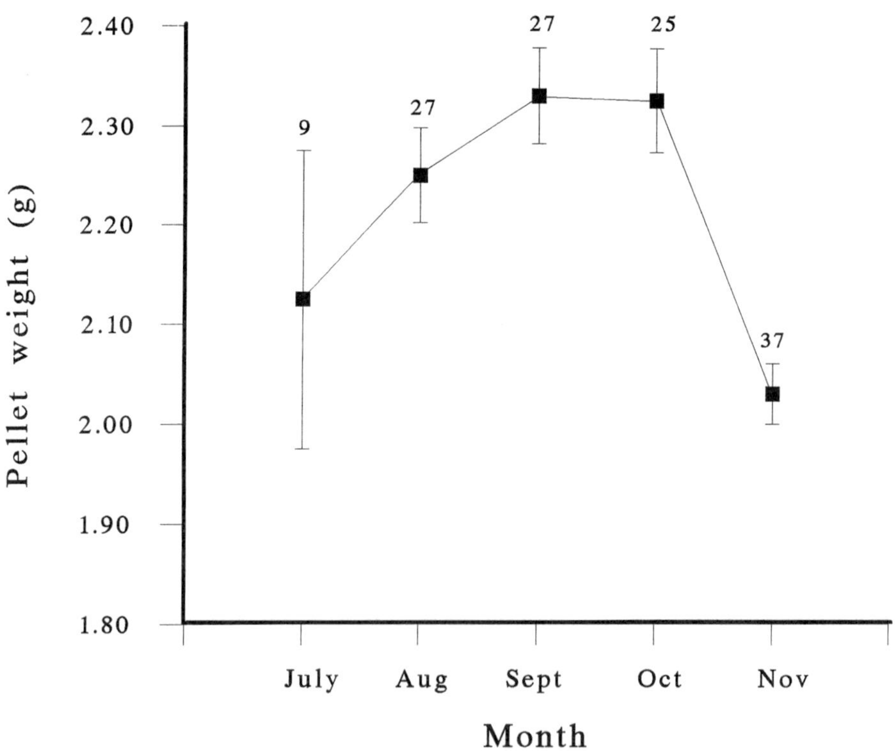

Fig. 8.19 Monthly changes in average pellet weight of African Marsh Harriers from the southern Cape Province from 1984. Since pellet weight changes with biomass intake, these weights reflect seasonal differences in biomass intake of adult birds. Number of pellets are given above s.e. bars.

one month cast pellets that were 37% (August) and 39% (September) heavier than pellets simultaneously cast by unfed birds. Seasonally, pellet weights in the best represented year (1984) increased slowly from July, peaked in September/October and declined in November (Fig. 8.19). While the differences

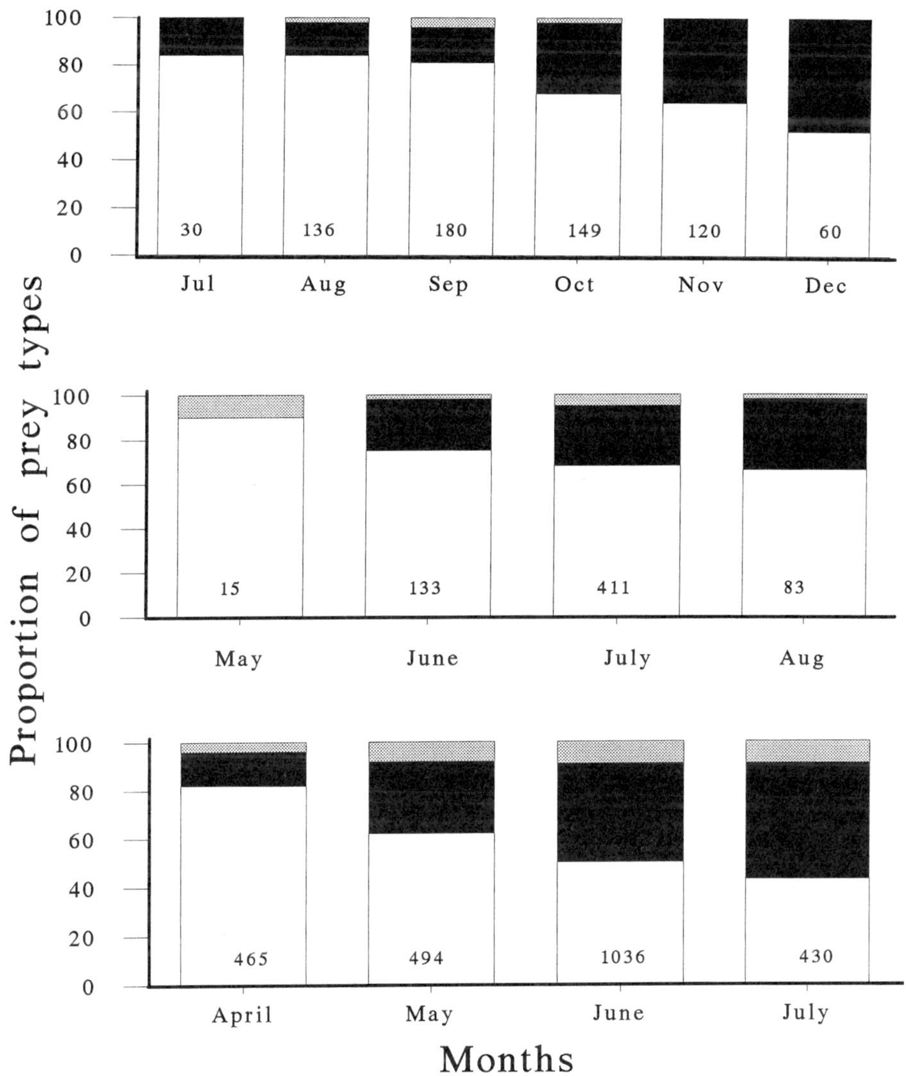

Fig. 8.20 Increasing dietary shift from mammals to birds for three species of harriers: African Marsh (top); Northern (middle), and Montagu's (bottom). For each species, mammals (white bars) predominate during courtship and egg laying, and birds (black bars) become increasingly common as young harriers begin to appear. Only for African Marsh Harriers do birds ever exceed mammals in frequency. Other prey (snakes, frogs, insects) are shown at the top of each bar. Principal mammals are mice for African birds, voles for Northern harriers, and young rabbits for Montagu's Harriers in Spain. (From this study, Barnard *et al.* 1987; Arroyo 1997.)

were not statistically significant (ANOVA $F_{4,\,124} = 1.17$), they reflected provisioning performance and mice abundance in the environment (Fig. 8.8), and independently determined mouse availability judged from vegetation density profiles (Chapter 4). Thus, harriers appeared to consume most food in September/October when they were feeding hatching and half-grown young (Fig. 8.16), and least when their young were reaching independence. It is noteworthy that pellet mass also partly reflected the proportion of birds in the diet, implying that lighter feathers may account for lighter pellets. However, seasonally, bird prey increased through October into December (Chapter 4; Fig. 8.20), while pellets peaked, then decreased (Fig. 8.19). Such trends cannot be explained by the lighter feather component expected in pellets.

Direct observations at nests over 2000 hours identified 374 prey deliveries: 74% were micro-mammals (rats, mice, and shrews), 23% were passerines and waterbirds, and the remainder comprised small frogs (2%) and fish (1%). The proportion of micro-mammal prey differed over the three years: in 1984 harriers delivered mainly (90%) micro-mammals, declining to 63% in 1985, and 54% in 1986. The dietary intake of more difficult to catch birds simultaneously increased from 8% in 1984 to 32% (1985) and 42% in 1986. Hence increased predation on birds occurred between years, as micro-mammal abundance decreased.

Seasonal changes in diet

Seasonal differences in prey composition were also evident from direct observations. The proportion of micro-mammals was highest in August and September (81% and 84%), started declining in October and November, and reached a trough in December when young harriers were gaining independence (Fig. 8.20). This analysis combined years, but between-year variation also existed in prey shifts. Thus in 1984, a year of heavy predation on micro-mammals, an almost constant supply of small mammals was delivered in each month (92%, 94%, 93%) declining only in November and December to 79% and 75%, respectively. By contrast, in 1986 (a low in micro-mammal predation) maximum contributions occurred in August (75% of all prey), and declined steadily to 43% in December. The one consistent pattern was for the maximum micro-mammal contributions to occur in August/September before and during the start of harrier breeding, and the minimum in December, when they finished (Fig. 8.20). This was confirmed by trends in pellet composition: few birds were found in July (12%) and October (3%), but harriers switched to greater numbers in November (29%) and December (33%), in parallel with that found by direct observations (Fig. 8.20). The small pre-breeding season sample of pellets ($n = 33$) indicated that harriers took mainly micro-mammals (91%), most of which (56%) were *Otomys*.

This pattern of a mainly small-mammal diet during critical courtship and egg-laying times, is not peculiar to African birds but is found in two temperate-breeding harriers also. Figure 8.20 shows dietary data for the African Marsh Harrier (top), Northern Harriers (middle), derived mainly from direct observations (Barnard *et al.* 1987), and Montagu's Harrier (bottom) pairs in Spain (Arroyo 1997), derived from a combination of direct observation, pellets, and remains. In the latter population the small mammals were young rabbits, and because their size was larger than the birds taken, the data is presented as biomass, not proportion of each prey type. The first month shown is the one in which courtship/egg laying predominates, and ends when young are nearing independence. In each case the harriers are relying on small mammals to the extent that 80% or more of their prey is mammalian. For each species it is also obvious that birds, and mainly passerine birds, become more and more important as breeding progresses. By the end of breeding, birds comprise about 50% of the diet, but it is not known when mammals then begin to predominate again for any species. Since birds are generally difficult to catch it probable that they are the newly available source of new fledglings. If they are nestling birds (see below), are harriers timing their breeding to coincide with this newly available food source? By plotting the hatching patterns of a Northern Harrier population against the sudden increase in bird prey (as a ratio of the birds to mammals), Barnard *et al.* (1987) found a close correlation ($r = 0.66$, $P < 0.05$, $n = 10$ weeks) between the peak in the weekly hatching patterns of the harriers and increasing numbers of birds in the diet (Fig. 8.21).

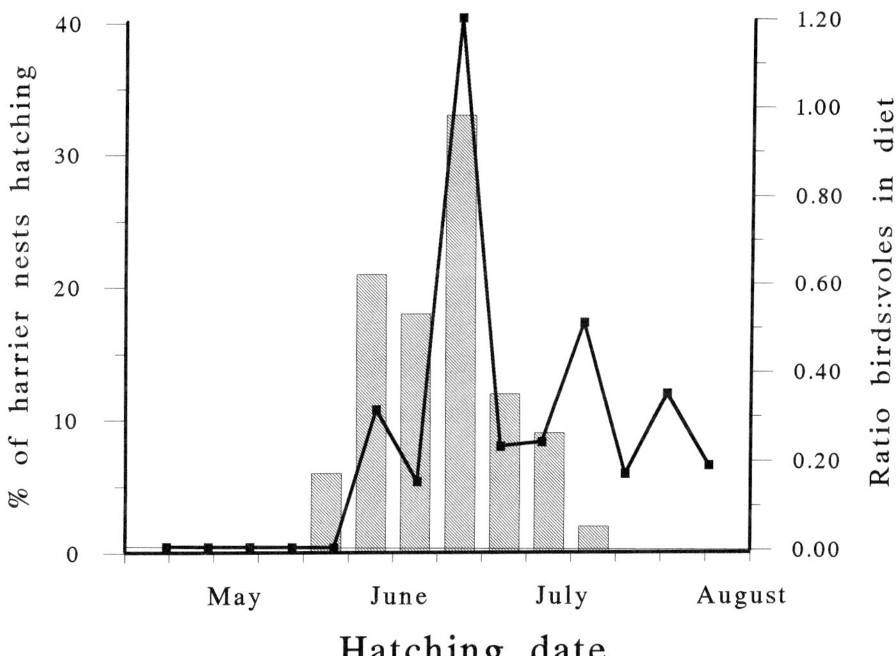

Fig. 8.21 The coincidence of peak Northern Harrier hatching patterns (bars) with the increases in bird prey in the diet during weekly intervals in New Brunswick (after Barnard *et al.* 1987).

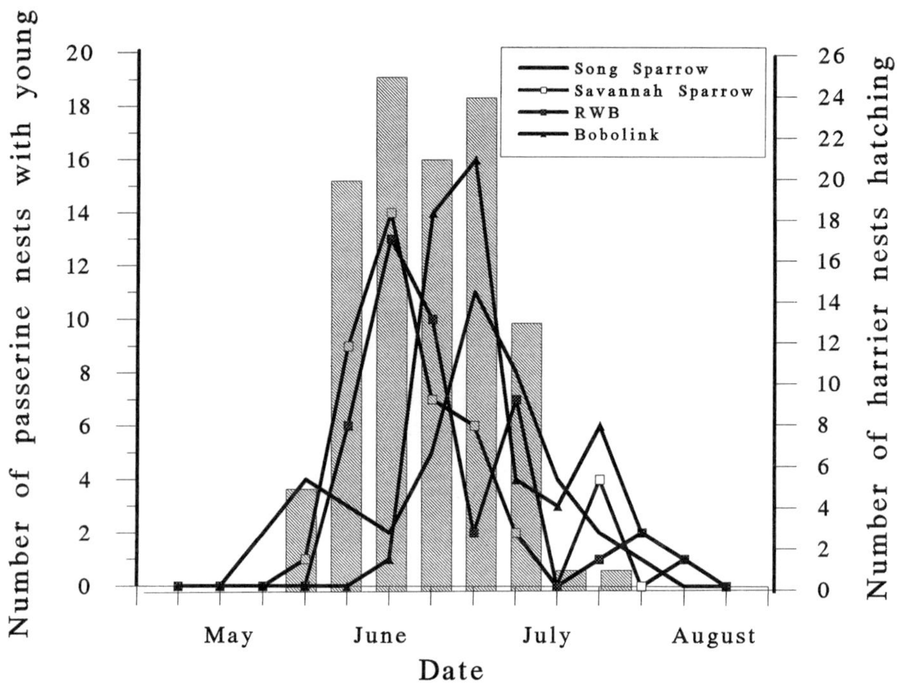

Fig. 8.22 The coincidence of hatching patterns of Northern Harrier young (bars) in New Brunswick with the presence of nestling passerines (lines) in the environment. Passerine data derived from (179) nest record cards in southern New Brunswick; includes only the four principal prey species—two sparrows (85 cards), the Redwinged Blackbird (RWB: 42 cards), and the Bobolink (52 cards).

Birds in the diet stayed high thereafter suggesting that certain harriers were timing the hatching of their young to the new flush of birds. However, this alone does not show that the harriers were exploiting newly available young birds. So Barnard *et al.* (1987) assessed the general hatching patterns of the four main bird prey species (Bobolink, Redwinged Blackbird, Savannah Sparrow, and Song Sparrow in southern New Brunswick) from 179 nest record cards collected since 1923, by the Canadian Wildlife Service. By plotting the number of nests with nestling passerines against the hatching patterns of the harriers, it became clear that there was a close association between the newly available nestling passerines and the hatching of the harrier chicks (Fig. 8.22). The first harrier nests to hatch did so only as the first nestlings of their main bird prey appeared in the nests. This makes it unlikely that the harriers were simply trying to catch more birds, adult or young, but rather that a new source had become available for them. Did this make any difference to their breeding success? Barnard and workers gave two points that suggested that it may have done: first, most Northern Harriers initiated their clutches only as the first clutches were approaching hatching (implying a strategic timing), and second, nests started during the midseason 'laying peaks' were significantly more successful than early nests in 1980 and 1981. However, this effect was swamped out when further years were added; then those nests started earliest (1–14 May) fledged significantly more young than those begun in the laying peaks (15 May–3 June) ($t = 3.07$, 59 df, $P < 0.05$). The strategy for Northern Harriers breeding in New Brunswick therefore may best be described as: breed early in the season, but if this is not possible, time breeding for the middle–end of the month and exploit the newly available food source of birds that occurs as harriers hatch.

Seasonality in micro-mammal weights

The September/October peak in biomass ingestion for African birds suggested by the change in pellet mass, may have arisen from either (a) harriers catching prey more frequently, or (b) the average prey biomass peaking in these months. Did micro-mammals—the most frequently caught prey—increase in biomass? Parallel trapping studies of *Rhabdomys* and *Otomys* indicated that the mean mass of neither species increased in a predictable seasonal fashion. *Rhabdomys* varied by a maximum of only 12 g in any one year, and *Otomys* varied by 18 g. Both species were below their average weights in October.

Seasonal changes in predatory success

The increased biomass intake in September/October was best ascribed to seasonal variations in successful predation attempts by breeding adult harriers. The winter months ($n = 25$ attempts) saw a median success rate of 56%, followed by a *decrease* in the next two months, when harriers initiated breeding, to 41% ($n = 49$). This low was followed by a sharp increase in success in October, to 74% ($n = 31$), when harrier nestlings were being fed. The sudden change coincided with more young birds entering the diet, and a simultaneous peak in availability indices of mice (Chapter 4). Success dropped in November and December to 50% ($n = 13$) when young harriers were reaching independence. In turn, these predation differences coincided with increasing then decreasing abundance of micro-mammals trapped on the study area each month (see Chapter 4).

To summarize these findings:

(a) African Marsh Harriers initiated nests in August/September, hatched and fed growing young in September/October, and brought young to independence in November/December;

(b) they fed mainly on micro-mammals which slowly declined as provisioned prey, reaching a trough as young harriers reached independence;

(c) biomass ingestion for harriers peaked when they were feeding young;

(d) this was due to increased capture success at this time, not to an increase in micro-mammal or bird prey mass.

These findings suggest that winter (July) and summer (December) were times of relative food scarcity, while spring (September/October) was a time of abundance. Therefore, harriers laid eggs at a time when micro-mammals were common in the diet, but predation efficiency and biomass intake were both low. Hence, African Marsh Harriers cannot be said to commence breeding to coincide with a change in diet (micro-mammals were prominent in the winter diet), a high in predatory success or biomass intake, or even an influx of juvenile micro-mammals. Similarly, and contrary to the majority of temperate-zone studies reviewed by Daan *et al.* (1988), both predation success and biomass intake were near, or at their lowest for the season (Fig. 8.8) as young were reaching independence. Instead, for the earliest nests, the period of maximum nestling demand (50 to 60% through the nestling period: Collopy 1986; MacWhirter 1994), coincided with maximum predation success, maximum biomass intake, and a shift in diet composition from mainly micro-mammals to birds. While birds are normally difficult prey to catch (Temeles 1985), those chased and caught by harriers in October were increasingly juveniles. This situation parallels that for Northern Harriers, which also switch to a newly available food source of nestling passerines as their own eggs hatch (Figs 8.20 and 8.21). Moreover, African Marsh Harriers enjoyed a double benefit at the most demanding time of year because maximum *Rhabdomys* abundance (Fig. 8.8) *and* availability (as measured by vegetation height) occurred in October (Chapter 4). That some benefit is apparent to harriers starting earlier was apparent from the higher success of early relative to late breeders (1.3 vs. 1.0 fledglings; n.s.), based on a small sample ($n = 25$ nests)

Differences in breeding schedules of temperate and tropical species

Why should African harriers differ from the temperate species reviewed by Daan *et al.* (1989) in bringing young to independence at times of poor food availability? African Marsh Harriers do have significantly longer breeding schedules than temperate congeners (e.g. 120 vs. 50 days for egg-laying spread), suggesting that the sheer length of the breeding process may play some role. A second possibility is that the principal prey of the African birds, *Rhabdomys pumilio*, typically exhibit breeding seasons as short as one month (Henschel *et al.* 1982) and in this population juveniles never exceeded 25% of the total population (Simmons 1989). A short breeding window and a relatively small pulse of newly available juvenile micro-mammals could both contribute to these African harriers' inability to time the fledging of their young to coincide with peak prey availability and abundance as do some temperate zone species. However, other factors may be important for longer-lived tropical birds (Snow and Lill 1974; Fry 1980). These birds may make the 'best of a bad job' by superimposing the greatest period of stress (foraging for completely dependent young) with that of maximum prey yield, thereby minimizing the impact of such stress on each adult's residual reproductive value. Birds doing otherwise may survive less well, a major factor determining lifetime success in many species (Newton 1989). In this view, low prey yields may coincide with the independence of harrier young as an unavoidable *constraint* of the short optimal 'window' in which harriers breed. This may explain why Northern Harriers also coincide the hatching of their own young with the flush of new passerines, which have largely disappeared by the time young become independent in August (see Fig. 8.22). Price *et al.* (1988) give further discussions of the selection pressures that may be involved in the timing of breeding in different species.

The idea (Perrins 1970) that birds merely start breeding as soon as conditions improve enough for individuals to lay down reserves for egg laying appears improbable for African harriers. In July and early August, when eggs are being formed, biomass intake and predation efficiency are at their lowest for the season. Furthermore, weather conditions can be severe, with temperatures of –1 to –2° C, in the low lying marshes used for nesting. Indeed, harriers, like Tengmalm's Owls and Emperor Penguins (Stonehouse 1969; Korpimaki 1987), could not start breeding at a poorer time of year. The implication is that harriers are timing their breeding for a future beneficial purpose. In this case, so that maximum prey yields coincide with maximum nestling demands.

Is clutch size determined by female condition or male feeding frequency?

Several researchers have used female body condition (breast muscle size, weight, or fat indices) at the time she lays eggs as a measure of food availability (Drent and Daan 1980; Poole 1985; Newton 1986; Meijer *et al.* 1988; Arroyo *et al.* 1999), and hence the number of eggs she is likely to lay. Female raptors may not need to be in good condition per se, but instead perhaps rely on the provisioning by their mates as a guide to the number of eggs to be laid. This is the 'income' model of Drent and Daan (1980). Provisioning frequency may be important if the pair use it as a guide to future food supply, and base their clutch size on the number of young the future food supply will then support. Alternatively, each female may require a large amount of extra protein a few days before she actually forms an egg, some of which may be stored in the breast muscle by some non-provisioning species (Jones and Ward 1976) or stored as fat by arctic-breeding geese and penguins that do not feed at all while laying (Ankney and McInnes 1978). It seems that larger species are more likely to employ the fat deposition strategy than smaller species (Perrins and Birkhead 1983).

Raptor provisioning frequency and its relationship with clutch and brood sizes have been documented for several species (Drent and Daan 1980; Poole 1985; Erickson 1986; Simmons *et al.* 1986b; Donazar *et al.* 1992), but most showed parallel increases in female condition with provisioning frequency, making it difficult to differentiate between previous supply or 'capital' (stored as fat) and current 'income' models. The implication that females need to put on weight to breed is based on studies of weight increase by breeding, but not by non-laying sparrowhawks and owls in Europe (Newton *et al.* 1983; Hirons *et al.* 1984; Dijkstra *et al.* 1988). This view has been eroded somewhat by the finding that for starlings and kestrels most of the weight increase or decrease is attributable to reproductive tissues and the weight of the developing egg itself (Ricklefs and Hussell 1984; Meijer *et al.* 1989). A further dent was put in this traditional view when Dijkstra *et al.* (1988) and Meijer *et al.* (1988) experimentally showed that female Kestrels increase in mass but that it has little bearing on clutch size when laying date is taken into account. For larger Ospreys, females put on almost no extra weight before breeding (Poole 1985). So, what do harriers tell us about this unresolved problem of the importance of *income* versus *capital* strategies?

In this study, some young males did not provision their mates at all until egg laying began and then provided at high rates: since their mates laid clutches no smaller than females provisioned prior to laying, this is inconsistent with the idea that females must acquire good condition *before* laying. It offers strong support for the income model of Drent and Daan (1980). In their studies of the diminutive Tengmalm's Owls, Korpimaki and Hakkarainen (1991) found that birds lose only 10–15 g in the period from egg laying through to incubation yet they are laying a clutch of 5–6 eggs that amounts to 60–75 g. They suggest that a small owl such as this is incapable of storing such reserves and must rely on provisioning at the time of laying. Other indices of owl reproduction also supported this 'income' model in their study. Thus while females obviously require

extra protein to form eggs, I suggest that fat increase may be a fortuitous by-product of increased feeding or perhaps important for short-term food stress during incubation. Hence correlations between good body condition and clutch size may simply reflect feeding frequency of the male at the time the eggs are formed. Ian Newton tested the importance of male feeding performance by removing three male sparrowhawks from their nests shortly after their mates had laid the first egg. Each female *immediately* stopped laying despite staying around the nest and having gained enough weight to lay 4–6 eggs. This is expected for the income model and contrary to the good condition or capital model of breeding. As such, it highlights the importance of a motivated and competent male for a female during this critical period of breeding. Experiments contrasting the clutches of females experimentally supplemented only until they begin laying, relative to females supplemented only during egg forming and laying, can clarify this aspect of avian breeding biology. Two studies have attempted this and found that the extra energy provided only after laying began was put into increased clutch size (Nilsson 1991), offering further support for the income model of Drent and Daan (1980). When clutch size was not increased there was some evidence that egg-size hierarchies within the clutch were altered (Chapter 7; Nilsson and Svensson 1993b). Possibly females that were unable to advance their laying date and thus the reproductive value of the egg put the extra food into increasing offspring quality (via increased egg size) rather than quantity; or altered the sex ratio of the eggs. Both possibilities are in need of further investigation.

9 *Clutch size and latitude*

One of the most widespread ecological phenomena for which there is yet to be a single satisfying explanation is why the number of eggs in a clutch declines with latitude. This is one area where research on raptorial birds can make an impact because several studies have now compared food resources between northern temperate and southern subtropical regions and arrived at similar conclusions. These global trends in avian reproduction are well known and were first analysed by Moreau (1944) and Lack (1947); they found that avian reproductive output, particularly clutch size, decreases progressively towards the equator for both passerines and non-passerines. This has been confirmed many times for both passerines (e.g. Klomp 1970; Kulesza 1990) and falconiforms (Donazar 1990; Jenkins 1991), yet the selection pressures on small clutches in the tropics are still widely debated (Skutch 1949; Lack 1954; Ashmole 1963; Hussell 1985; Murray 1985; Kulesza 1990; Martin 1996). The uncertainty in exactly which selection pressures and suite of life-history traits (Stearns 1992) are of greatest importance arises partly for three reasons:

(1) different selection pressures may operate on different species;

(2) most analyses have concentrated on across-species comparisons;

(3) almost no studies have specifically sought answers in tropical/subtropical regions.

Most approaches have been to correlate environmental variables with wide-scale differences in breeding output to reach broad conclusions. Such studies have concluded that either food supplies or predation levels have influenced the small clutches. However, correlational studies cannot pinpoint the causes. One way to avoid such pitfalls is to undertake brood-enlargement experiments that test a parent's ability to rear more than the usual number of young. These can simultaneously test if predation levels increase as predicted by Alexander Skutch's nest-predation hypothesis. As important, supplementary feeding of birds prior to breeding will determine if females are simply constrained by food early in the season, and thus constrained to lay small clutches. Even with this approach different results may arise if environmental conditions change, putting birds under different stresses. Of course weather patterns are highly variable, and it thus hardly surprising that experimental studies spanning two or more years reveal costs to parents that single-year studies do not always uncover (vander Werfe 1992). As suggested above, few such experiments are undertaken in subtropical areas so that selection pressures on birds laying the smaller clutches are not directly assessed. For example, in 42 brood-enlargement studies reviewed by vander Werfe, only 17% were within the tropics or subtropics (23° N or S). Most of the exceptions include early brood-enlargement experiments with sea birds which provided mixed results (Ydenberg and Bertram 1989). Food supplementation experiments in subtropical areas are even rarer since only 9% of 32 studies have been undertaken there (see Table 9.15). One of this chapter's main aims is to redress this imbalance.

I will begin by summarizing our current knowledge of clutch-size declines in harriers, review the current ideas seeking to explain reasons for decreasing output in equatorial regions, and then report on experiments with African Marsh Harriers to test some of these ideas. This is the second time that a combination of brood-manipulation experiments (to determine if parents can rear more than the planned brood size), and food-supplementation experiments (to determine if females are limited by early-season food resources), have been reported for any African species, and the second time for any tropical species. African Wahlberg's Eagles were previously supplementary fed and brood sizes manipulated (Simmons 1993a, 1997b), while others have exclusively reported brood-manipulation experiments from tropical settings (e.g. vander Werfe 1992; Young 1996). Nevertheless, results for tropical species are beginning to form some clear answers to the clutch size—latitude problem.

In an effort to reduce the plethora of ideas explaining small clutches in tropical birds to facilitate field testing, I again employ the approach that clutch-size hypotheses can be categorized as either *constraint* (e.g. phylogenetic, physiological, ecological) or *adaptation* (e.g. strategic balancing of present output against future reproductive potential). Hence the low reproductive output of African harriers may be constrained by:

(1) the low availability of pre-breeding food preventing birds from laying larger clutches;

(2) inadequate food later in the breeding season disallowing them to nourish more than the future brood size. Alternatively, clutch size may be strategic in that

(3) larger clutches would increase present parental investment and thereby limit each adult's lifetime reproduction, or

(4) larger broods would necessitate lower investment per nestling, decreasing nestling growth and fledging weight, ultimately impairing each nestling's future survival.

Hence a small clutch may be a trade-off between the quantity and quality (survival) of present and future offspring. One can experimentally test the idea of constraint by: (1) artificially feeding pre-breeding pairs to alleviate pre-breeding food limitation and (2) increasing brood size to determine the ability of parents to rear more than their average brood size. To determine the adaptiveness of harrier clutch sizes one can: (1) determine the impact on adult survival of increased brood sizes, and (2) assess the growth and fledging weights of increased broods relative to controls, as a guide to the future recruitment of young from different sizes broods.

Such brood manipulations have been reported eight times previously for terrestrial tropical species (Ward 1965; Lill 1974; Mader 1982; Simmons 1986a; Tarburton 1986; Poiani 1993a; Simmons 1993a, 1997b; Young 1996), and they allow us to make comparisons with results from the present study and thus there general applicability to tropical species. If one does find that clutches are limited by the ability of parents to provide food, the next question becomes: why? Are parents simply not working as hard, and not living long enough to increase their lifetime output, or alternatively, are they simply unable to gather the same food resources because levels in the environment are lower? One way of prising apart these adaptive and constraint arguments is to compare foraging activity and yield between harrier species in temperate and tropical settings to assess relative reproductive effort and reward. This is possible with harriers because foraging yield has been studied in two European congeners (the Hen and European Marsh Harriers) by Schipper (1977), Serge Daan's group, and Redpath (1992). Direct comparisons of effort and reward are therefore possible for the first time. First we must establish whether harriers show the same latitudinal declines that are well documented in other species.

Clutch size declines in harriers

Despite latitudinal clutch declines in non-passerines not being as pronounced as those in passerines (Moreau 1944), at least four species of harriers exhibit latitudinal declines. First, from the small sample of available data, *ranivorus* clutches declined by 0.45 from an average clutch of 3.48 eggs in South Africa to 3.0 in Malawi approximately 18° further north (Fig. 9.1). The decline was significant ($r = 0.97$, $n = 4$, $P < 0.05$), and averaged 0.2 eggs for each 10° of latitude, and is similar to that found in African non-passerines (Moreau 1944) and Australian passerines (0.27 eggs/10°: Yom-Tov 1987), but much less than the steep declines found over similar latitudinal gradients in temperate zone species (Klomp 1970; Yom-Tov *et al.* 1994), including harriers.

The average clutch size of 3.48 ± 0.13 eggs (Table 9.1), for female harriers that weighed on average 570 g, means African Marsh Harriers lay relatively small clutches for both their size and in comparison with temperate congeners. This clutch is one egg less than the typical average for temperate harriers of similar size. For example, the 520-g Hen Harrier lays clutches averaging 4.5 eggs in western Europe, and 4.4 in North America (MacWhirter and Bildstein 1996). Other comparisons are given below but it is important to note that since we are comparing species within a single genus (*Circus*), and of similar size, both phylogeny and allometric differences are effectively held constant. This may help reveal ecological or lifespan influences on clutch size variation.

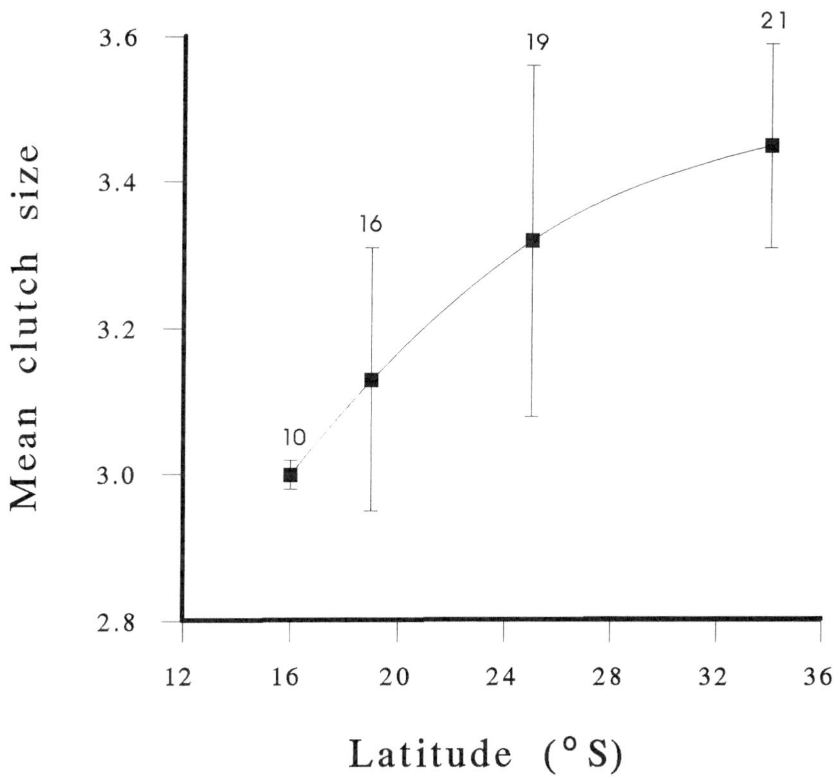

Fig. 9.1 Latitudinal decline in clutch size of African Marsh Harriers through south-central Africa.

Table 9.1 Latitudinal declines in mean clutch sizes of African Marsh Harriers through south-central Africa, and Australasian Swamp Harriers in south-central Australia.

Country	Latitude	Average clutch ($x \pm$ SD)	n	Source
African Marsh				
Malawi	16° S	3.00 ± 0.00	10	Benson and Benson (1977)
Zimbabwe	19° S	3.12 ± 0.72	16	Benson *et al.* (1964)
N South Africa	25° S	3.32 ± 1.05	19	Tarboton and Allan (1984)
S South Africa	34° S	3.48 ± 0.64	21	This study
Australasian Swamp				
Mildura	34° S	3.5 ± 0.6	4	Cupper and Cupper (1981)
Melbourne	38° S	3.6 ± 0.6	5	Baker-Gabb (1982)
Tasmania	42° S	3.8	9	N. Mooney, Tasmania

In Europe, Wibe Schipper, famous for his studies of predation and breeding in three species of harrier, selected six study sites in a virtual north–south line from the northern Netherlands to the Camargue in southern France. Curiously he never reported latitudinal trends in clutch size, even though he reported significantly earlier laying in study sites with higher temperatures in May (Schipper 1978). We can use some of his data in conjunction with more recent data to assess these trends. While Corbacho *et al.* (1997) have already reported a significant decline in clutch size with latitude for the Montagu's Harrier, Arroyo *et al.* (1999) challenged this finding, reporting that their reanalysis using only studies with two or more years' data did not support a significant decline. They used only studies of more than two years because of the wide year-to-year variation in clutches of Montagu's Harriers (e.g. Butet and Leroux 1993). This meant that single year studies may have reported erroneously large or small clutches. Even with these one-year studies included, their data did not show any clutch size decline ($r^2 = 4.8\%$, $n = 19$). Why should this be?

I reanalysed all available data for the Montagu's Harrier and reduced extraneous sources of variation introduced by known east–west (i.e. longitudinal) effects. East to west variation is known to influence clutch size of various species across Europe and North America (Klomp 1970; Perrins and Birkhead 1983), with species in eastern Europe exhibiting larger clutches than those to the west. Thus, I analysed only those Montagu's Harrier clutches from a band 10° west and 10° east of the Greenwich meridian (0° longitude Fig. 9.2). This excluded five studies of the 19 reported by Arroyo and co-workers: three from Italy, one from Poland, and one from Belarus. I added the studies of Schipper (1978) to the data set (Table 9.2). The correlation (Fig. 9.2) between clutch size and latitude ($r^2 = 33\%$, $n = 16$) with longitude essentially held constant was significant ($P = 0.02$, two-tailed). Similar (but non-significant) declines were apparent when this method was applied to both Hen and European Marsh Harriers (Fig. 9.3). The decline amounted to a decrease of 0.3 to 0.4 eggs for every 10° decrease in latitude. This is twice the equivalent decrease found in southern harrier clutches as they approached equatorial regions (above). The

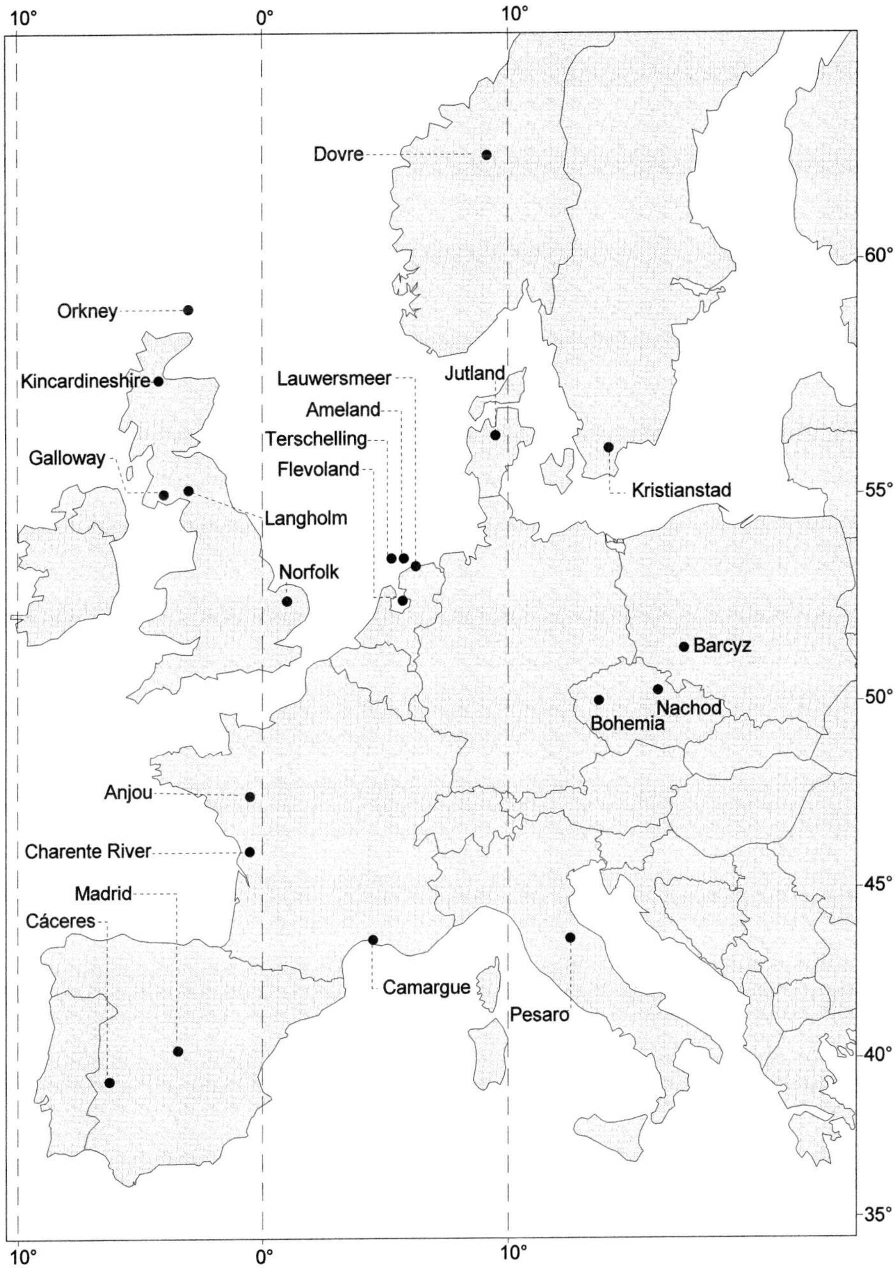

Fig. 9.2 Some of the major study sites where research on Hen, Montagu's and European Marsh Harriers has been undertaken. Those outside the 10° line of longitude were omitted from clutch size–latitude analyses.

Table 9.2 Clutch sizes of three species of European harrier from 43 studies. Coastal sites (C) are also distinguished from inland (I) sites for some analyses.

	Region	Latitude	(C/I)	Clutch size ($\bar{x}\pm$ 1 SD)	*n*	Source
Montagu's Harrier						
The Netherlands	Terschelling	53.5	C	4.17 ± 0.7	24	Schipper 1978
	Ameland	53.5	C	4.07 ± 0.73	14	Schipper 1978
	Flevoland	53.0	C	4.37 ± 0.96	30	Schipper 1978
England		52.3	C	4.02 ± 0.07	227	Underhill-Day 1990
France	Morbihan	47.4	C	4.27 ± 0.14	29	Hays 1971
	Milly	47.1	I	3.69 ± 0.19	26	Cormier 1985
	Noirmoutier	47.0	C	3.61 ± 0.17	23	Cormier 1985
	L' Aiguillon	46.2	C	3.75 ± 0.80	101	Arroyo *et al.* 1999
	Deux Sevres	46.1	C	4.28 ± 1.09	143	Arroyo *et al.* 1999
	Saintongeais marshes	46.0	C	3.64 ± 0.4	162	Butet and Leroux 1993
	Rochefort	45.6	C	4.00 ± 1.04	133	Arroyo *et al.* 1999
Spain	Madrid	40.4	I	3.93 ± 0.84	215	Arroyo *et al.* 1999
	Caceres	39.3	I	3.31 ± 0.65	19	*in* Arroyo *et al.* 1999
	Madrigalejo	39.1	I	3.48 ± 0.99	98	Corbacho *et al.* 1997
	Ciudad Real	38.4	I	4.30 ± 1.00	120	Castano 1997
Portugal	Castro Verde	37.4	C	2.82 ± 0.40	11	Onofre 1992
[a] Belarus		53.0		3.42	12	*in* Arroyo *et al.* 1999
[a] Poland		53.0		3.90–4.40	125	*in* Arroyo *et al.* 1999
[a] Italy	Marche	43.4		3.80	37	Pandolfi and Giacchini 1991
[a]	N. Apennine	43.3		3.32	31	Faralli 1994
[a]	Emilia-Romana	44.3		3.70 ± 0.80	61	Martelli and Sandri 1991
Hen Harrier						
Norway	Dovre	62.2	I	4.55	29	Hagen 1969
Scotland	Orkney	59.0	C	4.74 ± 0.92	300	Picozzi 1984
	Kincardineshire	58.0	C	4.70 ± 1.25	27	Picozzi 1978
	Galloway	55.0	C	4.78		Watson 1977
England	Norfolk	52.5	C	4.61		Underhill-Day 1990
The Netherlands	Terschelling	53.5	C	4.53 ± 1.14	43	Schipper 1978
	Ameland	53.5	C	4.68 ± 1.01	34	Schipper 1978

Table 9.2 Continued

	Region	Latitude	(C/I)	Clutch size ($\bar{x} \pm$ 1 SD)	*n*	Source
	Flevoland	52.5	C	4.46 ± 1.02	24	Schipper 1978
France	Charante River	47.0	C	3.92 ± 1.13	26	Cormier 1983
Spain	Madrid	40.5	I	4.55 ± 1.29	49	B. Arroyo *in litt.*
European Marsh Harrier						
Sweden	Kristianstad	56.0	C	5.07 ± 1.22	43	Bengtson 1967
Germany	Schleswig-Holstein	54	I	4.35	123	Bock 1979
England	Norfolk +	52.5	C/I	4.61 ± 1.47	114	Underhill-Day 1984
The Netherlands	Terschelling + Ameland	53.5	C	5.70 ± 1.84	10	Schipper 1978
	Lauwersmeer	53.0	C	4.8 ± 0.9	153	Altenburg *et al.* 1987
	Flevoland	52.5	C	4.8 ± 1.0	288	Altenburg *et al.* 1982
Czechoslovkia	Nachod	50.4	I	4.72 ± 0.99	171	Divis 1982
	Bohemia	50.0	I	4.45	58	Divis 1986
Poland	Barcyz Valley	51.5	I	4.86 ± 0.74	118	Witkowski 1989
France	Charente-Maritime	45.6	C	4.1 ± 1.0	495	Bavoux *et al.* 1989
	Camargue	43.5	C	4.62	21	Thiollay 1970
Spain	Madrid	39.0	I	4.61 ± 0.14	49	Gonzales 1992

[a] Omitted from analyses of latitude and clutch to reduce variation imposed by longitudinal effects.

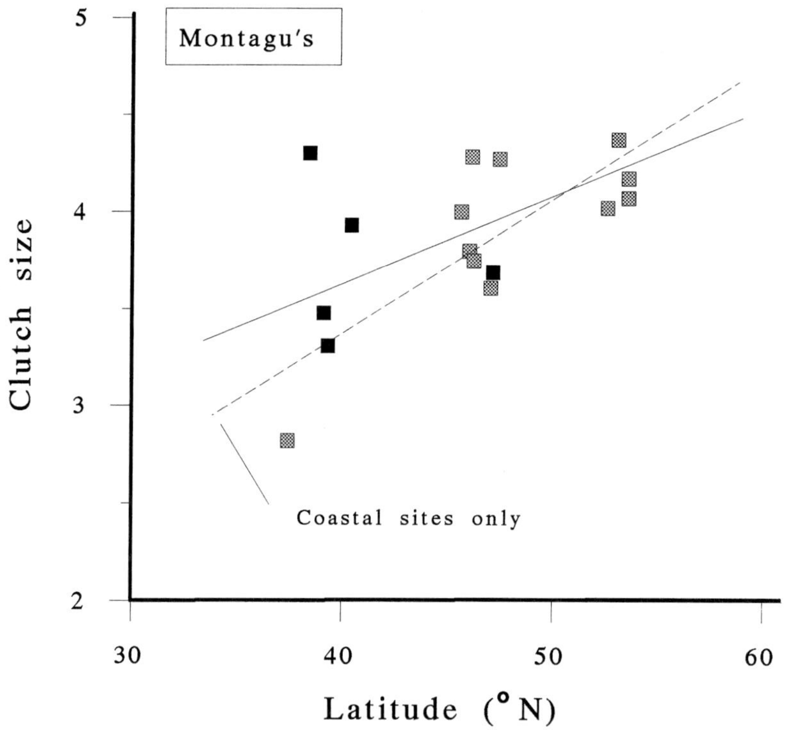

Fig. 9.3 Latitudinal decline in clutch size for Montagu's Harriers though western Europe coastal sites (light squares) are distinguished from other sites because of closer correlations.

significant variation of clutch size with latitude, with longitude essentially held constant, may explain why MacWhirter and Bildstein (1996) found no correlation between clutch size and latitude in a review of 16 studies of the Northern Harrier, spread east to west across North America—they did not account for longitude.

While all European species showed declining clutches, these general latitudinal correlations masked another trend found on inspection of the data for each species. Within the longitude band chosen, most study sites were coastal and a few were 'inland' sites more than 100 km from the coast (only two of ten Hen Harrier study areas were inland sites). On excluding these inland sites, latitudinal correlations increased for each species. For Montagu's Harriers, clutch size at 11 coastal sites declined even more steeply (0.7 young/10°) and were more highly significantly ($P < 0.01$) related with latitude (Fig. 9.3). For Hen Harriers, clutch size decline became significant and occurred at almost the same rate as for Montagu's Harriers (Fig. 9.4). European Marsh Harrier declines also increased but using coastal sites alone did not enhance significance levels (Table 9.3). In a similar analysis Witkowski (1989) also found no significant decline in

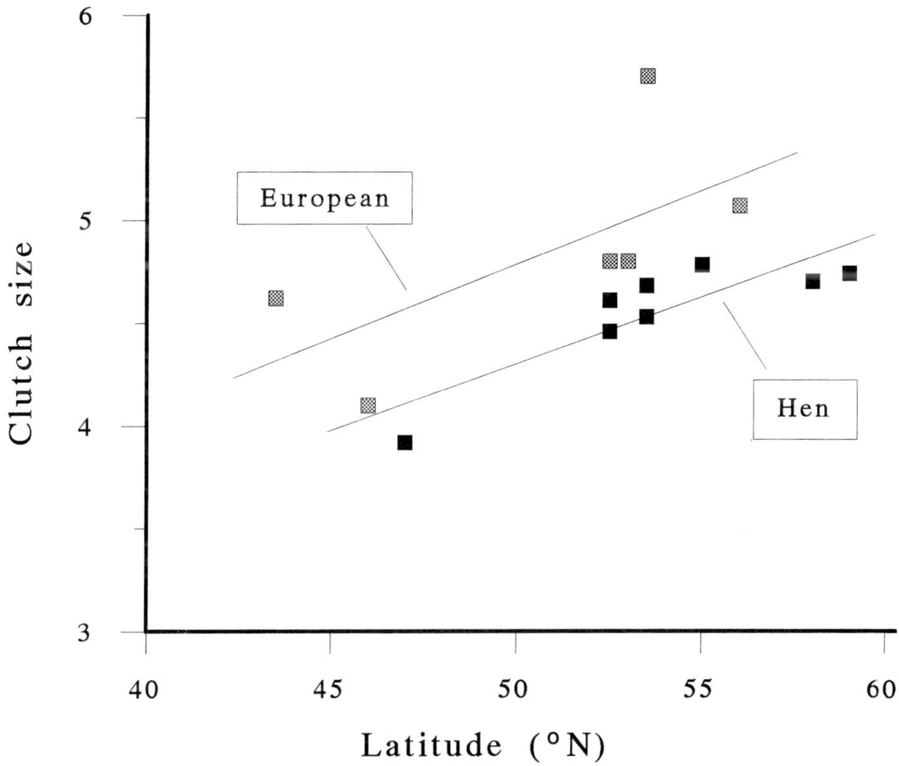

Fig. 9.4 Latitudinal declines in cluth size for two European harriers: the European Marsh Harrier (top) and Hen Harrier (bottom). Only birds breeding in study sites from western Europe (Scotland to Spain) are used to reduce variation caused by longitudinal variation. See Table 9.3 for regression equations.

Table 9.3 Summary statistics for clutch size declines with latitude for three European harriers. The table compares declines for all study sites relative to coastal sites alone in western Europe.

Clutch size statistics	Montagu's (345 g)	Hen (520 g)	European (670 g)
All study sites	$r^2 = 0.33$, $n = 16$, $P = 0.02$	$r^2 = 0.21$, $n = 10$, n.s.	$r^2 = 0.33$, $n = 11$, n.s.
Coastal sites	$r^2 = 0.59$, $n = 11$, $P < 0.01$	$r^2 = 0.74$, $n = 8$, $P < 0.01$	$r^2 = 0.51$, $n = 6$, n.s.
No. of eggs =	0.6 + 0.07 (latitude)	1.08 + 0.064 (latitude)	1.20 + 0.072 (latitude)

clutch size for this species. That each species showed virtually identical declines, however, suggests that similar selection pressures influence them across the latitudinal gradient. The only difference between these three harriers was evident in the initial size of clutches—the largest species (European Marsh) showed the largest clutch size for any given latitude (Table 9.3), and the smallest (Montagu's) exhibited the smallest clutch size.

Clutch size differences are also evident for two subtropical harriers, although samples are small. The Australasian Swamp Harrier, the largest of the harriers at 890g, was found by the Cupper's (1981) study to average 3.5 eggs at 34° S while David Baker-Gabb's extensive studies found average clutch size to be 3.6 eggs at 38° S and 3.8 at 42° S in Tasmania (Table 9.1). He also noted that average clutch size was significantly larger in New Zealand (4.6 eggs, n = 25) than Australia. However, much of New Zealand's land mass lies at the same latitude as the southern part of Australia from which the comparisons were made. Therefore, the one-egg difference does not follow the usual latitudinal trends. That pre-breeding carrion is readily available to New Zealand harriers but not Australian birds, and that New Zealand birds have no competitors in contrast to those in Australia, may explain these differences (Baker-Gabb 1983). Other explanations are explored below, but the situation would repay experimental testing.

In South America, the Longwinged Harrier breeds both north and south of the equator, and is the only harrier to do this. In Trinidad, at 9° N, the bird lays a clutch of only two eggs (the smallest for any harrier) while in Argentina (~ 35° S) it lays 3 to 4 eggs (del Hoyo *et al.* 1994) with an average of 3.4 (n = 6 Bó MS).

In summary, each harrier studied showed some latitudinal decline in clutch size, the significance of these declines increasing when potentially confounding variables such as longitude and proximity to coast were controlled for, and the decrease in number of eggs (0.6–0.7) per 10° of latitude was virtually identical for each species of European harrier. Thus of six harriers for which data is available, five show latitudinal declines, three significantly so, conforming to trends in many other species. For other raptors, strong latitudinal trends in numbers of young fledged have been recorded for Peregrine Falcons breeding throughout the African continent (Jenkins 1991).

Using these relationships between clutch size and latitude, it is possible in a crude way to predict the clutch size that an African Marsh Harrier breeding at 20–34° S might be laying. Of the three European harriers, the Hen Harrier is closest in both body size and prey type to the African Marsh Harrier. A bird of this size breeding at 20° S is predicted to lay 2.4 eggs and in fact lays 3.1 (Benson *et al.* 1964), while birds laying at 34° S on my study area should produce 3.3 eggs, and indeed averaged 3.5. Black Harriers, breeding at 33° S, also averaged 3.6 eggs per clutch (n = 5; Chadwick 1997) and for a bird of just over 500 g, their clutch, too, is relatively accurately predicted (3.2 eggs) by equations derived from the Hen Harrier. Thus, remarkably (given all the factors that may influence reproduction), clutch sizes of similar sized African harriers can be roughly gauged from clutch size declines of their northern counterparts. Do these correlations also apply to Australasian and South American birds breeding at southern latitudes? Baker-Gabb's studies (1982, 1983) of Australasian Swamp Harriers in south-eastern Australia indicate a mean clutch varying between 3.6 and 3.8 eggs at an average latitude of 40° S. Clutch size equations for the largest temperate harrier (*aeruginosus*), suggest their clutch size should be 4.1 eggs. Birds on the islands of New Zealand, with a central latitude of the same degree, lay clutches averaging 4.6 eggs, a slightly poorer agreement with the prediction. However, South American Cinereous Harriers with a mean body mass of 496 g breeding at 51° S would be expected to lay a clutch of about 4.3 eggs. Their clutches were smaller, averaging 3.6 ± 1.0 (n = 14; Jimenez and Jaksic 1988). These comparisons suggest that the relationships are good guides to clutch size but not universally applicable and other factors such as local conditions and island versus mainland sites are important. Nevertheless, factors controlling the clutch size of harriers in northern regions may be the same as those in southern regions of similar longitude—a useful start to our investigation.

So what might these factors be and which ones are critical in determining clutch sizes? To analyse this it is necessary to first review the plethora of ideas that have been forwarded. To keep things in perspective, it is sufficient to say that three of the factors most often invoked to explain limits to clutches in tropical regions are: (i) food resources; (ii) high predation levels; and (iii) reduced day length.

Clutch size hypotheses reviewed

Density-dependent (Ashmole)

Philip Ashmole's (1963) hypothesis is one of the earliest and most prominent ideas to emerge from research on latitudinal trends in clutch size (Ricklefs 1980). It posits that clutch size variation is dependent upon two factors: (i) seasonality of resources and (ii) resultant population levels. In temperate areas, where great seasonal variation in food resources exist, winter mortality and migration decrease population density to its lowest level until just before breeding. Since seasonal resources then rapidly increase (Ricklefs 1980), surviving breeders enjoy enhanced access to relatively high resource levels. Because the reverse is true in the tropics—stable, sedentary populations exploit less pronounced amplitudes in resource levels—surviving breeders will enjoy relatively lower resource levels. Consequently breeders are expected to reproduce relatively poorly. It should be noted that even if the rain-driven seasonal pulse of resources in tropical regions is as high as that in temperate regions, the population levels of (tropical) birds exploiting them is likely to be relatively higher because their numbers are not likely to be reduced to the same degree from adverse weather and migration from temperate climes. If significant mortality in tropical species was the case, one would probably see more migratory movement. This is essentially an easily testable (food) constraint hypothesis, and is tested below using experimental manipulations and direct comparisons as follows:

Test 1

Comparisons of limiting resources gathered per unit time by ecologically similar temperate and tropical species should highlight whether tropical species are indeed food-limited relative to temperate counterparts. This has only recently been evaluated in other species and all results (below) show parallels with those found for harriers.

Test 2

If breeders are constrained by limiting food resources, food supplements to pre-breeding birds should increase their clutch sizes above that of unfed controls. Clutches should indeed equal that of temperate congeners unless other limiting factors or adaptive restraint occurs. If so, clutches may increase up to a ceiling below that of temperate congeners. No such tests exist for subtropical/tropical species.

Test 3

If populations in temperate regions are reduced by over-winter mortality and migration, one might expect that population densities in temperate regions should be *lower* than those of harriers in regions closer to the equator (Ricklefs 1980). This is a counter-intuitive prediction because the higher prey populations thought to exist in temperate regions are expected to support higher density populations of their predators.

How good is clutch size as a measure of reproductive investment?

For larger species such as harriers, clutch size is fortuitously a good measure of annual commitment to breeding. For smaller species, however, this may not be so for the following reason: while there exists a fairly low amplitude in food resources in tropical areas it is often accompanied by a relatively lengthy breeding season. For subtropical harriers and kestrels this was evident in egg laying dates that extended over 120 days, compared with the 50 days for temperate kestrels and harriers (Chapter 8). For passerines with a shorter turn-over time, both temperate and tropical species may lay second clutches, but for some tropical and south temperate species, they may simply continue while resources remain sufficient (Earle 1981; Grant and Grant 1989; Maclean 1990; Rowley and Russell 1991). Thus while individual clutches may be smaller than those of temperate congeners, total egg production may be enhanced by the length of the breeding window. That this can occur even in larger species when conditions are favourable is exemplified by the Bat Hawk *Macheiramphus alcinus*, which lays a single egg under both good and bad food conditions. In one study where food was in constant supply, a pair simply re-laid once the previous chick had successfully fledged (or failed), so breeding was continuous over a nine-month cycle (Hartley and Hustler 1993). The same phenomenon has recently been reported for supplemented Black Eagles in South Africa (R. Davies pers. comm.). Continuous breeding is commonly found in oceanic birds in aseasonal equatorial environments (Perrins and Birkhead 1983), and has been recorded in good food conditions among moorhens *Gallinula chloropus* in South Africa (Siegfried and Frost 1975). The prediction is that where food resources remain low but predictable for an extended period, clutch size may be lowered in a trade-off with future clutches. A similar point was made by Martin (1995).

Table 9.4 Clutch size variation (cv) between harrier species breeding in subtropical (**in bold**) and temperate regions.

Species	♀ mass (g)	Area	Mean clutch size (n)	Range and cv (%)		Source
Circus ranivorus	570	Malawi +	3.08 (26)	3–3	18	Benson and Benson 1977
		Zimbabwe		2–4		Benson *et al.* 1964
		N + S South Africa	3.38 (39)	2–5	25	Tarboton and Allan 1984; this study
C. assimilis	725	SE Australia	3.11 (80)	2–4	18	Baker-Gabb 1982
C. approximans	890	SE Australia	3.60 (5)	3–4	15	Baker-Gabb 1982
C. aeruginosus	720	Poland	4.86 (118)	2–6	15	Witkowski 1989[a]
		Finland	4.66 (79)	3–7	18	Hilden and Kalinainen *in* Cramp and Simmons 1980
		Czechoslovakia	4.72 (171)	3–10	21	Divis 1982
		N Holland	5.21 (71)	3–8	21	Schipper 1978
		N Holland	4.80 (288)	3–8	21	Altenburg *et al.* 1982
		England	4.60 (114)	2–8	32	Underhill-Day 1985
C. cyaneus	510	Orkney	4.70 (700)	3–8	20	Picozzi 1984a
		N Holland	4.56 (101)	2–9	23	Schipper 1978
		Scotland	4.70 (27)		27	Picozzi 1978
C. hudsonius	546	E Canada	4.30 (80)	2–8	29	Simmons *et al.* 1986b
C. pygargus	345	Spain	3.93 (215)		21	Arroyo *et al.* 1999
		W France	4.28 (143)		25	Arroyo *et al.* 1999
C. cinereous	496	S Chile	4.3 (14)	1–5	28	Jimenez and Jaksic 1988

[a]From DDT-free period only.

Marginal habitats (O'Connor)

As populations expand, some pairs are forced into sub-optimal breeding habitats in order to breed at all; one example is given for European Kestrels *Falco tinnunculus* by O'Connor (1982). Thus in subtropical habitats, in which habitat saturation is more typical, proportionately more pairs may be breed in sub-optimal habitats, and thus be constrained to small clutches. The overall effect would be a lowering of *average* clutch size relative to temperate counterparts. One would expect to see similarly high clutches in both populations but a greater variation and skew towards low clutches in tropical regions. This idea is also a density-dependent hypothesis and was tested by comparing the variation and range of tropical clutch sizes with temperate congeners. The data (Table 9.4) do not support this idea. The largest clutches among African Marsh Harrier populations were consistently lower than those for temperate species and the coefficient of variation in clutch size varied from 18.2% to 25.0% from Malawi to South Africa. This variation was generally lower than that exhibited by temperate-breeding congeners. Intuitively, this hypothesis is also untenable, at least for marsh harriers because they display much narrower habitat tolerance than the adaptable Kestrel which spawned the hypothesis. Nests are generally confined to marshy habitat both north and south, and consequently the probability that a wide range of sub-optimal breeding habitats would be occupied is remote.

Day length (Lack)

On first analysing latitudinal clutch size trends, Lack (1947) developed an existing idea that (i) clutch size is adapted to the number of young that parents can adequately feed, and (ii) latitudinal variations arise because day length constrains foraging time and thereby the acquisition of food necessary for feeding nestlings (cf. Hesse *et al.* 1937). Because day length decreases uniformly as latitude decreases, this is a simple and elegant explanation of uniform clutch declines. The hypothesis was challenged by Skutch (1949) who showed that a Central American tananger (*Ramphocelus passerinii*) was capable of increasing its foraging effort to nourish more young than it normally reared (violating Lack's first assumption), and by Rowan (1966) who pointed out that Cape Sparrow *Passer capensis* clutches increased 30% between Johannesburg and Cape Town (8°), while day length increased by a mere three minutes. Also, nocturnal owls (Strigiformes) should follow the reverse trend, being nocturnal food gatherers. They do not, as Lack (1947) and others (Murray 1976; Owen 1977; Donazar 1990) have noted. This hypothesis is therefore unsupported and generally discarded in modern treatments.

Adaptive lifetime response (Williams)

In response to Lack's general hypothesis that parental birds should lay as many eggs as they can adequately rear (and hence the commonest clutch size should be the most productive), Williams' (1966) proposed that lifetime benefits may reduce clutch size below the annual optima (see also Charnov and Krebs 1974).

Thus, as Klomp (1970) and others have shown, the commonest clutch may be smaller than that which parents can adequately rear, and smaller than that producing the most surviving offspring (Boyce and Perrins 1987). Parents (especially longer-lived species), by rearing fewer young, may live longer and enjoy increased lifetime output (Murray 1985; Stearns 1992). Tests of this require data on the costs of increased reproductive effort on future survival and reproduction (Askenmo 1977; Högstedt 1981a; Nur 1988a). This can be done by determining lifetime reproductive output of individuals that differ in their breeding strategies. A simpler method is to determine the costs in terms of decreased future adult survival or depressed future reproduction for birds attempting to rear experimentally enlarged broods (e.g. Roskaft 1985; Lindén and Møller 1989). Provision must be made for the probability that individuals will naturally vary in their ability to rear young (Nur 1988b; Pettifor *et al.* 1988), and that long-lived species may respond differently to short-lived passerines because of the greater premium on longevity (Saether 1988; Lindén and Møller 1989).

Optimal allocation of energy (Cody)

Cody (1966, 1971) surmised that each individual has limited time and energy for survival and reproduction, so birds that optimize perfectly should compromise between (i) predator avoidance, (ii) competition for resources (cf. Ashmole above), and (iii) reproductive effort. Since predation and competition for resources in the tropics is assumed to be high, energy for reproduction will be decreased. This idea is worded rather generally, so precise predictions and tests are not possible. However, it can act as a broad framework within which the idea of trade-offs between predator avoidance and resource competition on the one hand, and reproductive investment on the other, can be viewed.

Predation avoidance and risk-spreading (Skutch–Snow)

Skutch (1949, 1985) and Snow (1976) interpreted the low clutches characteristic of neotropical passerines as adaptations to (i) decrease the number of provisioning trips to reduce the chances of predators cuing into parental activity. Alternatively, small clutches may (ii) decrease the period in which the eggs are in the nest (assuming asynchronous hatching is typical) or (iii) allow fast replacement of the clutch. These ideas arise from the extreme predation rates found in their studies of New World passerines and among many African passerines (Maclean 1990). These hypotheses are testable in the same manner as Ashmole's hypothesis: supplementing the diet of pre-breeding birds should not increase clutch size if predation or risk-spreading are the chief selection pressures limiting larger clutches. Correlations are also expected between brood size and predation rate if larger broods occurring in larger nests require greater provisioning frequency or call louder or longer than smaller broods. A more direct test, increasing brood sizes of selected nests and monitoring predation rates relative to unmanipulated controls, was not possible in this study but has been reported twice before in subtropical regions. As we shall see, the

results do not support the Skutch hypothesis. Reduced nest size itself may be an adaptation to decrease the probability of predator detection (Snow 1976, 1978), and small size may then constrain clutch size (Slagsvold 1982). There is both supportive across-species (Kulesza 1990), and contradictory within-species, evidence for this in New World passerines (Willis *et al.* 1978; Skutch 1985). Direct experimental tests are few, but small nest size in open-nesting species does limit clutch and brood size (Møller 1982; Slagsvold 1982), and large nests appear to be more susceptible to predation, at least in temperate regions (Møller 1990a).

Thomas Martin (1996) has pointed out that a paradox exists between the presumed selection for small broods in order to reduce the length of time that young are in the nest, and the slow growth rates found in tropical species. If predation exerted a strong influence on clutch size, it is likely that growth rates would be accelerated (Bosque and Bosque 1995). Yet growth rates are slow in both large and small tropical birds (Lack 1968; Bortolotti 1986; Simmons MSa). Furthermore, intense predation rates would be expected to shorten incubation periods which are also longer in tropical species than in same-sized temperate ones (Martin 1996). Thus while predation rates may be high, they do not influence either incubation or growth rates in the expected manner. Experimentation may be fruitful.

Ectoparasitism

With the rise in appreciation of the major role that parasites play in both reproductive success and viability (Hamilton and Zuk 1982; Møller 1993), it is inevitable that, when their levels are high, they may be seen to play a role in the small clutches in the tropics. Parasites may decrease nestling mass and ultimately their survival; the result may be that larger broods are less successful than small broods (Moss and Camin 1970). Parasites may affect the adults directly, reducing their reproductive capacity and ability to lay eggs. In a test of this idea in which North American, Russian, and Australasian passerines with differing degrees of parasitism were examined, results were equivocal. Australian species with more contagious ectoparasites had smaller clutches than other species (Poiani 1993b). However, no trends were apparent within samples of North American or Russian passerines (Gregory *et al.* 1991; Poulin 1991). Nevertheless, experimentally increased parasite loads do have some effects on clutches within species (Møller 1990b; Korpimäki and Hakkarainen 1992). Thus the role of parasites in the evolution of small clutch size in the tropics remains ambiguous and no such experiments have been undertaken with harriers.

High quality offspring (Nelson–Simmons)

I propose two additional hypotheses: high quality young are of greater worth in more stable (tropical) environments because of the longer lifespan (Snow and Lill 1974) and presumed greater competition for territories in saturated habitats (Nelson 1978; Simmons 1993b). Hence smaller clutches may have evolved because females laying fewer eggs can (i) lay larger eggs with concomitant advantages of enhanced survival (Williams 1993), or (ii) rear fewer but better-nourished offspring, which fledge at higher weights and are cared for in a prolonged post-nestling stage. In doing so, offspring may experience increased survival, and a greater chance of recruitment to the breeding population.

Test 1

Determine the ingestion rate per offspring of subtropical breeders relative to those of temperate birds, concurrently taking account of the lower metabolic rate of tropical species (Ricklefs 1976; Wasser 1986). Subtropical birds should ingest equal or greater amounts of biomass per chick than temperate species if selection for offspring quality has been favoured. Note that if food in tropical environments is lower, ingestion rates equal to those in temperate areas probably indicate that parents are working hard to supply sufficient food, supporting this idea.

Test 2

Assuming survival is related to fledging weight, tropical birds are predicted to fledge at higher weights than temperate counterparts, allowing for allometric differences in adult body mass. Alternatively, parental care following fledging

Table 9.5 A comparison of biomass received per nestling per 14-h day during maximum provisioning periods, for subtropical and temperate harriers. Biomass estimates were calculated from observations at nests and known relationships between prey size rank and biomass.

African Marsh Harriers[a]				Northern Harriers[b]		
Nest	Total biomass delivered (g)	No. of nestlings	Biomass / nestling (g)/day	Nest	No. of nestlings	Biomass / nestling (g)
EV1	259	2	130	B5	2	118
EV2 Renest	342	2	171	B7	3	150
EV3	516	3	172	B17	2	139
EV1	543	2	272	H1	3	113
EV2	397	4	99	H2	5	106
EV3	567	3	189	H5	4	165
RONDV Re	160	3	53	H11	2	193
FYNBOS	588	3	196	H12	2	121
GV1	213	3	71	H13	2	136
GV2	268	2	134			
EV1	418	3	139			
Y1	399	3	133			
FYNBOS	314	4	100			
RS	164	2	82			
B1	189	2	94			
Means $n = 15$	356 ± 149	2.7	**136 ± 57**		2.7 $n = 9$	**138 ± 28**

[a] Data for 1984–86.
[b] Data from MacWhirter 1994 for New Brunswick, Canada, 1983 and 1991.
Note: correcting for body mass differences (570 vs. 510 g) suggests that Northern Harriers would be provisioning an average of about 155 g/chick/day. See text for details.

may be longer to enhance survival since this is the period when mortality is greatest (Perrins and Birkhead 1983).

Test 3

Egg volume of similar sized congeners might be larger in tropical than temperate species since egg size and survival are positively related within species (review in Williams 1993). Large eggs are known to produce better quality chicks within species (e.g. Thomas 1983; Williams 1993), a result verified for both Montagu's Harriers (Arroyo *et al.* in press) and one subtropical species (Simmons 1997b). This is only likely, however, if egg size is not constrained by body size.

We can now test some of these ideas by comparing species from temperate and tropical regions.

North–south comparisons: nestling feeding rates, egg size, and dependence periods

During maximum prey provisioning periods (when nestlings are 21–45 days old), African nestlings received from 53 to 272 g per chick each day, averaging 136 ± 57 g per chick for years combined (Table 9.5). According to Schipper (1973: Fig. 4), nestling European Marsh Harriers in the northern Netherlands each received about 260 to 452 g per 14-hour day ($\bar{x}$ = 354 g/day). In a Polish study Witkowski (1989) found that young birds received 129–164 g/day. Allowing for the difference in body mass between the two species (1 : 1.26), the corrected daily intake for African birds (i.e. 136 g × 1.26), at 171 g, is approximately half that received by European birds in The Netherlands and similar to that received in Poland. How do these compare with *similar sized* temperate harriers? In a detailed two-year study of biomass provisioning to sons and daughters in Northern Harrier broods in Canada, MacWhirter (1994) corrected for indigestible parts such as viscera and bones (a 9% reduction for small mammals). For comparison I have added these back into his estimates. During 14-hour days he found that biomass ingested per nestling per day averaged 138 ± 28 g, remarkably similar to that found for African birds (136 g: Table 9.5). However, since Northern Harriers are marginally smaller than African Marsh Harriers (female body mass ratio, 0.89 : 1), Northern Harriers were supplying their chicks with slightly more energy per unit body mass than African birds. This amounted to approximately 155 g/chick/day (1/0.89 × 138 g). Thus, two of the three comparisons suggest a slightly higher intake for temperate harriers. However, this comparison takes no account of lower temperatures and the likelihood of higher maintenance costs in temperate regions. This is inconsistent with the idea that African Marsh Harriers are reducing clutch and brood size to enhance offspring survival through increased biomass intake. Instead, African birds appear to be simply limited in the amount of food they can deliver to their young.

The second possibility promoting enhanced quality—weight at fledging—was not directly measured in this study, but the lower biomass provisioned to individual African birds suggests that it is unlikely. Moreover, young African birds stay in their parents' care for between 74 and 90 days after hatch, when they receive no more food, and then disperse up to 1000 km (Simmons 1986c). Virtually identical post-nestling periods are reported for European Marsh Harriers in Poland (80–90 days: Witkowski 1989) and The Netherlands (> 74 days: Altenburg *et al.* 1982). However, European birds are larger and fledglings should take longer to mature and complete feather growth. A more accurate comparison is with the Northern Harrier, where fledglings in two areas reached independence at between 45 and 56 days in Wisconsin and between 52 and 66 days in New Brunswick (MacWhirter and Bildstein 1996). The median date of 56 days is 32% shorter than the median date of 82 days for African birds. How can we interpret this? On first sight it would appear that because parental care in African birds is longer, greater investment is put into this stage to enhance quality relative to temperate birds. However, the alternative, which we cannot reject, is that African birds grow slower and require extended parental care until their feather growth is complete. Many examples exist of tropical raptors growing more slowly than temperate ones of similar size, including eagles (Bortolotti 1986; Simmons MSa), and sparrowhawks (Simmons 1986a). Thus harriers are no different in this regard and the extended post-nestling period probably simply reflects the slower growth and maturation. In a comparison of House Wrens breeding in temperate and subtropical regions, Young (1996) concluded differently: he argued that because tropical birds devote more time to rearing their young (64% longer) than temperate wrens, they were investing more which may enhance offspring quality. He did not consider that this may arise from constraints imposed by food in the environment.

The third possibility promoting increased quality of young in stable environments—increased egg volume—was also not evident. Relative to the similar-sized *C. cyaneus*, African females laid eggs that were, on average, no different in volume ($\bar{x}$ = 29.6 vs. 29.2 cc: this study; Cramp and Simmons 1980). In summary, it seems that African Marsh Harriers are not adaptively altering brood size, egg size, or their period of parental care after fledging to enhance offspring quality. Hence small clutches cannot be seen as a direct result of selection for offspring quality.

Allometric consequences of long lifespan

Per unit body mass, tropical species are longer lived than temperate species (Snow and Lill 1974; Fry 1980; Brown and Pomeroy 1984; Simmons 1989; Rowley and Russell 1991). Doubts concerning such a relationship for neotropical passerines were raised by Karr *et al.* (1990) but their findings have been questioned by Martin (1996) as probably spurious for methodological reasons. Assuming tropical species are longer lived, correlates of greater longevity include decreased clutch size, longer incubation periods, and slower growth rates (e.g. Lack 1968; Stearns 1992; Ricklefs 1993). Hence a long lifespan may pleiotropically constrain tropical species to decreased reproductive output. Physiologically this may arise because tropical raptors exhibit lower basal metabolic rates than temperate raptors (Wasser 1986). Empirical lab studies have additionally demonstrated a genetic basis to inverse correlations between survival and reproduction (e.g. Rose and Charlesworth 1980; Partridge and Farquar 1981), and the possibility remains that small clutches may be nothing more than pleiotropic effects of increased lifespan. Research should thus address the question: why are tropical species longer-lived? Adaptive scenarios suggesting that smaller clutches and broods have evolved to enhance lifespan and thereby lifetime reproductive output (Newton 1989) have been suggested by Moreau (1944) and Murray (1985). This is simply the alternative of the cause–effect relationship between lifespan and reproductive output, and is best tested by looking for adaptive restraint in reproductive output.

To evaluate this idea I marked a small sample of African birds and resighted birds over four years. Based on a sample of 45 adult birds older than one year that were either patagially marked (91%) or clearly recognizable (9%), annual survival rate was 82%, with little variation between years (75% $n = 8$; 85% $n = 20$; 82% $n = 17$). This gives an estimated mean lifespan of adult birds as 6.4 years. Survival estimates for adult Hen Harriers from Picozzi (1984b) were 85%, with 90% for female *cyaneus* and 72% for males. While these data are similar, the different methods used may have clouded the comparisons. Survival data based on other studies reported by Newton (1979, table 49) show that the mean expectation of further life was 2.8 years for two populations of *C. cyaneus*. Mean expectation of life for African Marsh Harriers was 5.1 years, based on the same formula used by Newton ($[2 - m]/2m$, where m = annual mortality). The latter comparisons, indicating increased longevity of the smaller subtropical harrier, is in agreement with the differences in lifespan found for temperate vs. tropical eagles (Simmons MSa). These findings indirectly support the idea that clutch size could be smaller due to inverse correlations between lifespan and reproduction.

However, does their longer lifespan allometrically result in low reproduction, or is long lifespan a consequence of low reproduction? If results (presented below) had indicated that harriers were not constrained by food-gathering factors, the possibility would have existed that they were withholding reproductive effort, thereby increasing lifetime success through increased survival. Several lines of evidence suggest otherwise: (1) African birds spend just as long foraging for food as European Marsh Harriers, but are less successful in their return (below). (2) Food availability was shown to limit clutch and brood sizes (Chapter 8), and thus a direct link can be seen between food-finding capabilities and reproductive output. Thus, tropical raptors' low reproductive output, in concert with their lower metabolic rates (Wasser 1986), stable environments, and typically sedentary nature, may explain increased longevity. Lifespan may be decreased if parents attempt to rear more young than they plan. This is well documented in temperate species (review in Lindén and Møller 1989; Stearns

Table 9.6 Summary statistics of the effect of supplementary food on initiation date, egg size, clutch size, total clutch volume, and hatching success of African Marsh Harrier nests from 1984 to 1987. Those comparisons marked * differed significantly.

Dependent variable	Fed		Unfed	
	Mean ± s.e.	n	Mean ± s.e.	n
Initiation Date[a]	14 Sept ± 7 d	9	11 Sept ± 6 d	21
Egg size[b] (cc)	30.10 ± 0.67	9	29.24 ± 0.66	20
Clutch size[c]	4.00 ± 0.27	8*	3.47 ± 0.13	21
Total clutch volume (cc)	120.84 ± 9.57	8*	101.52 ± 5.16	20
Hatch success[d] (%)	87.5 ± 8.5	6	86.2 ± 5.2	12

[a] No significant difference ($F = 0.01$, df = 1,27, $P = 0.92$).
[b] No significant difference (Chapter 8).
[c] Significant difference ($F = 5.43$, df = 1,27 $P = 0.03$) between fed and unfed birds with clutch size corrected for laying date. Sample size is eight because one clutch failed before completion.
[d] No significant difference ($F = 0.09$, df = 1,15, $P = 0.77$). Samples reduced because hatching not established in 1987.

1992), and there is evidence for reduced breeding in subsequent years (Simmons 1997b) or immediate physiological costs (Poiani 1993b) for tropical species rearing experimentally increased broods. I tentatively conclude that small clutches are caused by limits imposed by the environment and that lower reproductive output adds to increased longevity and not vice versa. This conclusion may be checked by experimentally supplementing birds' broods, and manipulating food to test what these limits actually are.

Experimental food supplements

The natural diet of nine pre-breeding pairs of harriers were supplemented, as described in Chapter 1. Supplemented pairs were assessed for effects on (i) initiation (first egg) date, (ii) egg size, (iii) clutch size, (iv) total clutch volume, and (v) hatching success. For first nests, feeding started on average 30 days before clutch initiation, and for seven of the nine nests continued until clutch completion for an average of 40 ± 13 days (range 31–62 days).

Feeding programmes must ensure that food actually *supplements* the natural diet of the intended birds; adult male Ospreys and Kestrels, for example, are known to become lazy when supplemented (Poole 1985; Meijer *et al.* 1988). This also occurred with adult harriers. In 1985, supplemented males provisioned 1.8 items (60 g) less per day than unfed birds, while in 1986 this rose to 2.4 items (96 g) per day. I therefore provisioned large amounts, averaging 164 g per day, that increased each female's daily provisions by 104 g in 1985 and 68 g in 1986. What effect did this supplemental feeding have on clutches and initiation dates?

Effect on clutch initiation date

Natural clutches were initiated over a period of 120 days (chapter 8) and first clutches were initiated on average on 3 September, with a slight but non-significant trend for variation between years ($F = 3.23$, df = 3,14, $P = 0.055$). Combining all first nests, mean initiation date was similar for three of the years

Table 9.7 Natural clutch size variation in relation to year and nest status. Clutches of food supplemented birds are not included because they were significantly different.

	Mean	s.e.	*n*
Year[a]			
1984	3.75	0.25	4
1985	3.50	0.29	5
1986	3.14	0.26	7
1987	3.80	0.20	5
Nest status[b]			
First	3.61	0.12	18
Repeat	2.67	0.33	3

[a] Year ($F = 1.65$, df = 3,17, $P = 0.22$), no significant effect.

[b] Nest status ($F = 0.47$, df = 1,19, $P = 0.51$), no significant effect when corrected for seasonal declines.

(27 August) and did not differ with respect to the female producing the clutch ($F = 1.07$, df = 9,7, $P = 0.47$). Most importantly, supplementary feeding did not advance the timing of laying. With initiation date of repeat nests (birds relaying after failure of the first) as a covariate, nine supplemented birds initiated clutches three days later than 21 unfed controls, a non-significant difference (Table 9.6).

Effect on egg size

As reported in the Chapter 7, supplementary food had no marked effect on mean egg size (Table 9.6), but did change the way females apportioned the extra energy within the clutch: instead of eggs increasing in volume in the laying sequence (last eggs largest), females reversed the hierarchy and laid increasingly small eggs after the second egg (see Fig. 7.6).

Effect on clutch size

The average natural clutch size between 1984 and 1987 was 3.61 ± 0.12 eggs ($n = 18$) for first nests alone. First and repeat nests did differ significantly (Table 9.7), but when clutch size was corrected for seasonal declines (Fig. 8.3) the difference disappeared ($F = 0.474$, df = 1,19, $P = 0.51$). In other words, repeat nesting females laid clutches no different from that predicted from the seasonal decline. Thus all natural clutches were combined and averaged 3.48 ± 0.13 ($n = 21$).

Average natural clutch size did not differ significantly ($F = 0.55$, df = 20, $P = 0.82$) with respect to female (range 2.0–4.0), or year (Table 9.7), but did show significant declines with season (Fig. 8.3). Clutch size declined by 0.14 eggs every 10 d or one whole egg for clutches initiated 70 d apart. The mean clutch size of eight fed birds (one failed before completion) was 0.53 eggs larger than that of 21 controls and was significantly higher (Table 9.6). Identical results were obtained when clutch size was corrected for season ($F = 5.43$, df = 1,27, $P = 0.028$). This result is unexpected when coupled with the lack of advance in laying date. Meijer *et al.* (1988) in particular have shown that Kestrels do not respond to supplements later in the season, and they suggested that laying date alone dictates the clutch size of many birds (Daan *et al.* 1990). I checked the unusual result for these harriers by calculating the expected clutch size of supplemented birds on the basis of the date on which they were laid. The expected clutch size was then compared with that observed. Observed clutches of supplemented females ($\bar{x} = 4.0 \pm 0.8$) were larger than that predicted by their laying date ($\bar{x} = 3.4 \pm 0.3$), and they were significantly different (matched pair t-test, $t = 1.94$, 7 df, $P < 0.05$). Harrier clutches were therefore strongly affected by the food supply independent of laying date, confirming results reported for several raptorial birds including Tengmalm's Owls and European Sparrowhawks (Newton and Marquiss 1981; Korpimaki and Hakkarainen 1991).

Effect on total clutch volume

Because average egg volume differed between females, so too did the total volume of the clutch ($F = 3.09$, df = 11,16, $P = 0.02$). The average total volume for unfed birds was 101.52 ± 5.16 cc ($n = 20$). Total egg volume did not, however, differ between years ($P = 0.67$), season ($P = 0.08$), or with nest status ($P = 0.19$). Total egg volume in first nests (110.03 ± 5.10 cc, $n = 23$) was slightly larger than in repeat nests (93.26 ± 12.70, $n = 5$). Like clutch size, supplementary food had a significant effect on total clutch volume (Table 9.6); fed birds laid clutches with a total volume almost 10 cc larger than unfed birds. While these differences probably arise from the significant larger clutches of fed birds (Table 9.6), they are treated separately to ensure that females were not merely laying smaller eggs to produce larger clutches. The results show they were not: females laying large clutches also produced large total egg volumes.

Effect on hatching success

The proportion of successfully hatching eggs in each successfully hatching nest was not significantly related to year (73% vs. 90.3% vs. 90.6%; $P = 0.26$), female ($P = 0.85$), season ($P = 0.63$) or first nests (91%) relative to repeat (75%; $P = 0.10$). Supplemented birds also hatched no greater proportion of their eggs than unfed birds (Table 9.6); thus supplementary feeding did not effect hatching success of harrier eggs.

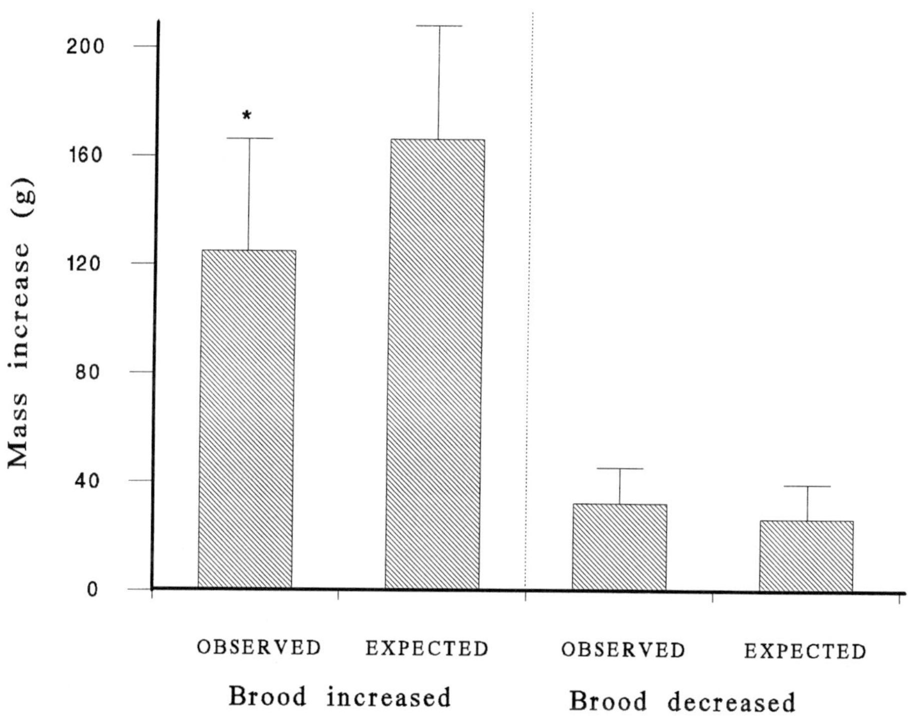

Fig. 9.5 Effects of brood size increases (left) on the increase in body mass of nestling African Marsh Harriers. A significant decrease in mass increment was apparent between observed and expected growth rates. However, little difference was apparent when broods were decreased (right) due mainly to depressed growth arising from heavy rain.

Effect on renests

Five female harriers produced replacement clutches, two of which were included in the feeding programme. While this sample is too small for statistical analysis, egg volume, clutch size, and total egg volume were all increased in fed birds. Egg volume increased 1.30 cc between fed and unfed birds, clutch size increased by 1.3 eggs, and total clutch volume by 41.81 cc; each difference for renesting birds is larger than that for first nest comparisons (above).

To summarize, supplementary fed harriers produced significantly larger clutches and larger total clutch volumes than unfed birds, but fed birds did not initiate clutches any earlier, produce significantly larger eggs, or enjoy enhanced hatching success relative to unfed controls. Despite small samples, feeding appeared to affect renests the most. Food did therefore appear to be limiting, supporting the second prediction of Ashmole's hypothesis. This is just one side of the equation—we now turn to the ability of African birds in rearing more than their planned size.

Experimental brood manipulations

To determine whether parents were capable of rearing more young, I manipulated four broods in 1985. Nestlings totalling 12 individuals were transferred between nests for periods of 7–8 days, to increase or decrease natural brood size by two or three chicks. Broods were selected such that introductions were less than four days older than the original chicks, and usually only a day older. Each individual's growth rate (mass and wing chord) was measured at the start and finish of each manipulation and the increase in growth was compared with that for unmanipulated nestlings of the same age. The observed differences for each group of nestlings were then averaged and compared to the average *expected* differences for the same group (Fig 9.5). I concurrently assessed the provisioning rates of the selected pairs to determine whether the parental effort changed in response to the manipulations.

Table 9.8 Provisioning responses of adult harriers to increased or decreased brood size. Typical rates of biomass provisioning (g/13-h day) are those when broods were unmanipulated, and are used as the 'expected' rates for comparison with the observed rates during experiments.

Nest	Brood size (g)			Rate increase or decrease (%)	X^2
	−2	Typical	+2		
GV1	192 g	208 g	–	− 8	1.23 n.s.
EV2	332 g	245 g	–	+ 36[a]	30.9 $P < 0.01$
EV2	435 g	452 g	–	− 4	0.6 n.s.
EV3	—	500 g	789 g	+ 58	167.0 $P < 0.01$
EV3	—	494 g	504 g	+ 2	0.2 n.s.
CV2	—	219 g	264 g	+ 21	9.0 $P < 0.01$

[a] Provisioning effort expected to decrease, not increase.

Table 9.9 Summary of brood manipulation experiments conducted on birds of prey. Sparrowhawks to eagles were assessed for the parents' ability to increase and decrease provisioning effort when brood size was varied. Nestlings in all but one species experienced increased food intake when broods were decreased (second last column).

Species	No. of broods manipulated	Brood size change	Adult provisioning effect when		Biomass change per individual when		Source
			Brood enlarged	Brood reduced	Brood enlarged	Brood reduced	
Cooper's Hawk	2	2–5–2	No change	No change	Decrease	Increase (260%)[a]	Snyder and Snyder 1973
European Sparrowhawk	2	2–4; 4–3	No change	No change	Decrease (n.s.)	Decreased (n.s.)	Geer 1981
Redbreasted Sparrowhawk	1	1–2–3–2–1	No change	No change	Decrease	Increase (172%)	Simmons 1986a
African Marsh Harrier	4	+2, –2	Increase	Decrease	Decrease*	Increase (33%)[a]*	This study
European Kestrel	50	+2, –2	Increase	Decrease	Decrease (n.s.)	Increase (33%)[a] *	Dijkstra 1988
American Kestrel	12	5–7; 5–2	Increase	No change	Decrease/increase[b]	Increase (n.s.)	Gard and Bird 1990
Rock Kestrel	7	+2, –2	Increase	No change	No change	Increase (179%)	van Zyl MS
Gyrfalcon	2	+3, –3	Increase	Decrease	Decrease*	Increase (39%)[a] *	Poole 1988
Snail Kite	12	+1, 2	Increase	—	Decrease	—	Beissinger 1990
Savanna Hawk	10	+1	No change	—	Decrease	—	Mader 1982
Osprey	6	+2	Increase	—	Decrease	—	Poole 1989
Wahlberg's Eagle	8	+1	No change	—	Decrease	—	Simmons 1993; 1997b

[a] The % increase in biomass (g)/chick relative to biomass consumption in unmanipulated broods.
[b] In one exceptional prey-rich year, prey deliveries per chick increased to enlarged broods. Not significant.
Note n.s. = not significant; * = significant $P < 0.05$ as reported by the author; those biomass changes per individual with no significance level were not reported by the author.

Brood size manipulation: effect on nestling growth

Manipulated nestlings were always accepted by their foster parents despite an increase of brood size from two to four. Expected growth in mass of nestlings in increased broods was consistently depressed by an average 40 g (n = 14 comparisons) relative to unmanipulated controls (Fig. 9.5). Combining all manipulated nestlings, the differences in observed and expected growth rates were significant (t = 1.74, 26 df, $P < 0.05$). The opposite trend—increased growth rates when brood size was decreased—was not as marked. The first reduction resulted in growth rates only 14 g more than expected (n.s.), while growth rates in the second manipulation were affected by heavy rains and strong winds. The combined results were, therefore, inconclusive for decreased broods. Nestlings differed less with respect to feather growth than mass. Nestlings in increased broods suffered depressed feather growth averaging 4.9 mm, 1.3 mm, and 2.0 mm (n = 14 comparisons) relative to unmanipulated controls. The differences were not significant (t-tests). Feather growth in decreased broods was rain-influenced (see above) and in fact grew more slowly than expected. The results indicate that parents could not rear good quality nestlings when given more than their planned brood size.

Brood size manipulation: effect on male provisioning effort

To determine if the depressed growth rates were due to unchanging provisioning effort (rather than, say, inter-sibling rivalry), male and female provisioning effort was compared for experimental and natural broods. Rates were recorded during 14 manipulations (decreasing or increasing broods and also for typical brood size) during a total 163 hours of observation at four nests. The biomass delivered during periods when the natural brood size was present was taken as the 'expected' rate and compared to the biomass delivered to manipulated broods, a few days later. Results (Table 9.8) indicate that on the three occasions when brood size was reduced, adults decreased their provisioning rate (non-significantly) on two occasions and increased it significantly on the third. In three experiments where brood size was increased, adults increased their provisioning rate significantly on the first two of three occasions. Thus, some parents reacted by increasing provisioning but later showed no response when their brood was again increased, while others showed no response or an inverse response. African Marsh Harriers thus demonstrated no clear trend in response to increased or decreased brood size. Similar results are apparent in at least one other subtropical raptor: Redbreasted Sparrowhawks breeding in Natal, South Africa showed no response to increased brood size when it was manipulated from one to three young (Simmons 1986a).

While both studies are among the first for subtropical species, sample sizes are small. Thus I checked them against other brood manipulations undertaken on raptors (Simmons MSb). Nine other studies showed identical results (Table 9.9): while some adults did increase their provisioning rates, in no case did the amount of biomass received by each chick stay at pre-manipulation levels, and in several cases biomass per individual declined by well over 30%. Similar results are appar-

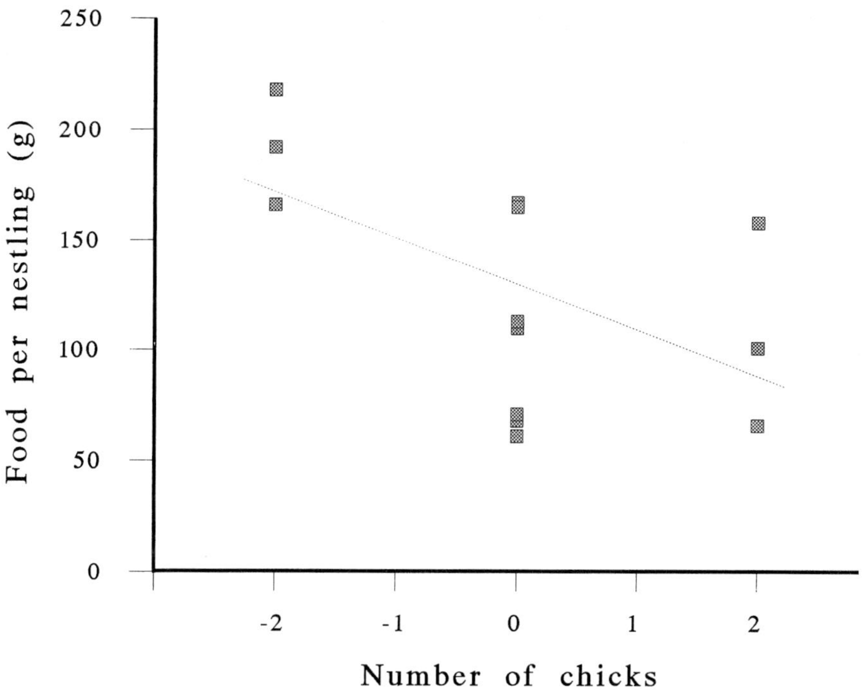

Fig. 9.6 Significant declines in provisioned biomass per nestling from increased, through unmanipulated, to decreased broods. Each rectangle represents two data points.

ent in many passerines (Nur 1988b), but not in certain siblicidal species (Mock and Parker 1997).

Thus the depressed growth rates among increased broods (Fig. 9.5) probably arose from reduced food biomass acquired per nestling during the manipulations. Indeed each nestling in the increased harrier broods received between 37 and 44 g less than in the unmanipulated state while nestlings in decreased broods enjoyed between 105 and 122 g more than in their unmanipulated state. Pooled, the results indicated that fed biomass decreased rapidly per chick from reduced to increased broods (Fig. 9.6). Even when parents did respond by delivering more biomass to increased broods (Table 9.8) the food received per nestling decreased in every case.

Egg quality

An additional possibility promoting increased quality of young in stable environments is an increase in egg volume: large eggs are known to produce better quality chicks within species (e.g. Thomas 1983; Williams 1993) a result verified for both Montagu's Harriers (Arroyo *et al.* in press) and one subtropical African eagle (Simmons 1997b). For African harriers, however, no such increase was evident; relative to the similar-sized *C. cyaneus*, African females laid eggs that

Table 9.10 Comparisons of provisioning between three intensively-studied harriers: *ranivorus, hudsonius* and *aeruginosus*. Provisioning effort is given in terms of items delivered/day, and live prey biomass delivered/day (**in bold**). Breeding stages are distinguished because of the differences in provisioning effort. (n = sample nests.)

Stage of breeding	African (*ranivorus*)	Northern[a] (*hudsonius*)	European (*aeruginosus*)
Pre-hatch	4.6	5.2	4.9[b]
	483 g	—	**657 g**[c]
	(15)	(8)	(4)
Nestling	7.5	9.4; 11.6	11.6
	790 g	—	**1550 g**
	(12)	(7); (7)	(3)
First flight – Independence	4.3	—	7.5
	452 g		**1005 g**
	(7)		(3)

[a] Mean biomass estimates not known.
[b] Probably artificially low because rate based mainly on provisioning during incubation (Altenburg *et al.* 1982, table 4) when harriers reduce provisioning rate.
[c] Biomass rate computed using mean prey weight from Schipper (1973) and rates from Altenburg *et al.* (1982), from similar study areas.
Sources: *ranivorus* this study; *hudsonius* MacWhirter 1985, Simmons *et al.* 1986a; Simmons *et al.* 1987; *aeruginosus*: Schipper 1973; Altenburg *et al.* 1982.

were, on average, no different in volume $\bar{x}$ = 29.6 vs. 29.2 cc: this study; Cramp and Simmons 1980).

It is curious that the two species whose eggs were larger than that predicted from their body size (Fig. 7.3) are both southern hemisphere harriers; the egg composition and breeding strategies of both the Spotted and Longwinged Harriers are worthy of further study.

Foraging yield: north–south comparisons

As mentioned previously, female raptors are often completely dependent on their mates for food to form eggs (Newton 1979) and as we have already seen (Chapter 8) there is good experimental and correlational evidence that provisioning rates affect clutch size in several species. Given this dependence, disparities in provisioning returns of temperate and tropical congeners may offer clues to latitudinal differences in clutch size. This has never been tested, despite the fact that it is a major prediction of Ashmole's hypothesis. I therefore compared *ranivorus* provisioning performance with known rates for the Northern/Hen Harriers, of similar body mass (510 g: Bildstein and Hamerstrom 1980), and one of larger body mass, the European Marsh Harrier (720 g). In this comparison I use the average biomass of *live* prey to calculate provisioned biomass because they are generally used by other authors (Schipper 1973; Barnard 1983).

As expected, the comparisons reveal that *ranivorus* harriers consistently provide fewer prey per day than temperate harriers (Table 9.10). This is most apparent when species are compared during the period of maximum prey returns—the nestling period. At this time *ranivorus* males provide, on average, four fewer items per day than *aeruginosus* males and two to four fewer items than *cyaneus* males. When biomass estimates are introduced, the discrepancies increase. Prey returns of *ranivorus* males in the pre-hatch period are then 74% that of *aeruginosus* males and as low as 51% during the nestling period. No comparisons are available for the critical egg-forming period.

While these comparisons are in the expected direction, taken alone such differences in provisioned food cannot explain why clutches of tropical harriers are smaller than those of temperate ones. As noted by Skutch (1985), lower provisioning may mean food is less abundant, perhaps harder to find, or that parents refrain from working harder for other reasons. To differentiate between these possibilities I collected data on provisioning effort and reward to determine how hard African marsh harriers work to procure each prey item. This approach avoids complications of relative laziness or constraints on foraging between species and is called *net hunting yield*: the rate of prey provisioning per hour of foraging (Altenburg *et al.* 1982).

While three males carried 14-g tail-mounted radios September 1984, I collected two-minute time budget data from them to determine how many hours they spent foraging. Simultaneously, I recorded the number of prey delivered to the nest to gauge prey delivery rate. By combining these two measures, the average time required to capture each prey (foraging yield) was computed. For the periods when males were lost to sight out of the territory, I ascribed their activity as follows: for each bird I calculated the proportion of each activity that occurred while the bird was in view but out of the territory (i.e. from 150–*c*.1000 m from the nest). These proportions then replaced the time lost and were added to the directly observed data to give a complete time budget for nesting harriers. This was necessary because for 31% of the 146 h that the males were monitored, they were lost out of the territory, and for at least 35% of that time they were likely to be foraging. Because of the importance of this analysis, I compared the results with a similar analysis of foraging yield for a slightly larger sample of African Marsh Harrier males at the same time of year. Sampled hours were longer (304.8 h), and while time lost was similar (33%), it was not incorporated into the time budgets. The first method may underestimate foraging yield since not all prey captured are returned to the nest, but it is a better measure of time spent foraging. The second method by contrast probably accurately represents foraging yield but under-estimates time spent foraging.

In up to 56.4 h of activity budget data per male, one renesting bird spent only 23% of the day foraging, and 42% perched in the territory. By contrast, two other males feeding incubating mates spent most time foraging or perching outside the territory (Table 9.11). The average time spent foraging by the three males was 29.3% of the 12-h day, or 3.6 h each day. Almost all foraging (99%) was outside the nest territory. In the second analysis (discounting time lost), foraging time was calculated as 24% of the day. In that time 84 prey were returned to nests, giving a hunting yield of 1.7 prey/h. The similarity between the time spent foraging by African and European Marsh Harriers is remarkable: three

Table 9.11 Net foraging yield and activity data for three radio-tagged males. Data collected in September 1984, with day length 12 hours.

Activity	EV2 (renesting)	EV1 (incub + hatch)	EV3 (incub + hatch)
Perched in territory	42%	14%	17%
Foraging	23%	31%	35%
Flap sailing	13%	6%	9%
Perched out of territory	9%	39%	29%
Soaring	7%	7%	7%
Perched on nest	3%	3%	1%
Territorial displays	2%	0%	2%
Undulating displays	1%	0%	0%
Hours observed	56	44	47
Hours foraging	12.9	13.6	16.5
Prey caught	13	12	12
Net foraging yield (prey/hour foraging)	1.0	0.9	0.7

aeruginosus males spent approximately 29% of a 16-h day foraging during incubation (Altenburg *et al.* 1982), compared with 24–29% by African males. Hence, at least during incubation, African harriers worked no less hard than temperate ones. Were they perhaps less successful foragers? Critically they were. The net hunting yield for the three African males during incubation varied between 1.0 and 0.7 prey items/h of foraging (x = 0.86 prey/h) in one analysis; in the second analysis which excluded both lost time and was less likely to be

Table 9.12 Net hunting yield comparisons for temperate and subtropical harriers. Studies were based on observations of focal birds and all studies discarded the time when birds were lost to sight. African birds exhibited a 51% lower yield than temperate birds.

Species	Mean hunting yield (items/h) (range)	Birds observed foraging (hours)	Principal prey
European Marsh[a]	1.7 (0.7–2.9)	3 (?)	Birds/mammals
European Marsh[b]	5.1 (4.5–7.5)	? (4.1)	Mammals/birds
Hen[c]	2.3 (1.7–3.6)	? (7.7)	Birds/mammals
Hen[d]	3.6 (–)	? (17.05)	Voles
Hen[e]	4.8 (–)	? (18.0)	Birds/mammals
Northern[f]	2.1 (–)	4 (14.4)	Small mammals
Mean:	**3.3 ± 1.5**		
African Marsh[g]	1.7 (1.3–2.1)	5 (49.1)	Small mammals

[a] Altenburg *et al.* (1982), breeding study, males only.
[b] Schipper (1977), breeding study, both sexes.
[c] Schipper (1977), breeding study, both sexes.
[d] Daan *et al.* (1982), winter study, both sexes, voles numerous.
[e] Redpath (1992), breeding study.
[f] Linner (1980), breeding study.
[g] This study, males only.

biased because of unseen prey capture, the yield was 1.7 prey/h. Since prey size was recorded, net biomass yield could be calculated. One male delivered 39 g/h, the most frequent deliverer brought 34 g/h, and the lowest delivery rate was 37 g/h. The average biomass yield for the three males was thus 36.5 g/h of foraging in the first analysis. In the second analysis, five males averaged 69 g/h because of the higher number of items caught per hour. Thus hunting yield is 37–69 g/h of foraging for African birds.

For temperate harriers foraging for mammals and birds (Table 9.12), only one of five studies recorded mean hunting yield as low as that for African Marsh Harriers. The crude mean hunting yield for temperate birds was 3.5 prey captures per hour of foraging—twice that of African birds. This is a critical result in our comparisons because the discrepancy suggests that African males are *constrained* to lower provisioning rates, and are not showing adaptive restraint by foraging less often. While this comparison takes no account of prey biomass, biomass per prey item is typically greater for *aeruginosus* males (Schipper 1973; Underhill-Day 1985), and this comparison is thus a conservative estimate of the disparity in total biomass procured and provisioned by African and European harriers.

Critics will argue that findings from one species do not prove a universal constraint imposed on food gathering abilities. So it is useful to compare these results with other comparisons made between temperate and subtropical species (Table 9.13). Five such comparisons are now available for passerines and raptors and each indicates a lower yield by subtropical species. In a pair-wise comparison of similar-sized sparrowhawks on three continents, South African birds provisioned 54% to 28% less biomass to each chick than did hawks in Britain and North America (Simmons 1986a). Comparing foraging Black shouldered Kites *Elanus caeruleus* across five continents, Mendelsohn and Jaksic (1989) determined that hovering birds in Chile and California spent less than half the time that South African birds did hunting for each kill. In real terms, assuming a similar prey capture success, a bird hover-hunting in southern Africa for a typical two hours per day would procure 2.8 rodents, 56% less than the 6.5 rodents procured by a bird in temperate regions in the Americas (Table 9.13). Comparing tropical Snail Kites in Venezuela with the same subtropical species in Florida, Beissinger (1990) found that tropical birds provisioned only 1 to 1.5 prey items per hour to their broods while Florida birds provisioned about three times this amount to the same-sized broods. In a comparison of the Southern African subspecies of the Kestrel *Falco tinnunculus rupicolus* with the European Kestrel, van Zyl (1993) showed that South African birds provisioned about 48% that provisioned to same-sized broods in Europe. Finally, Thiollay (1988) showed that foraging success was lower among forest birds in tropical French Guiana, than it was in forest birds in temperate France, even though foraging return to nests was not determined. These proportions are mirrored by the 51% lower foraging return for African versus European harriers (Table 9.12), and each comparison supports the notion that between 48% and 68% less food is procured per unit time in tropical or subtropical regions, and that this may act a major constraint on reproductive output.

Table 9.13 Summary of pair-wise comparisons of prey capture yield or provisioning differences between closely related temperate (Te) and subtropical (S) or tropical (**Tr**) species (**in bold**). The column 'Difference' expresses the reduced yield or provisioning of tropical and subtropical raptors relative to their temperate counterparts.

Species		Provisioning performance[a]	Difference (Te–S)/Te (%)	Locality	Latitude	Source
Redbreasted Sparrowhawk	(S)	**36 g/chick/day**	54	Natal, South Africa	29° S	Simmons 1986a
Sharp-shinned Hawk	(Te)	50 g/chick/day		Arizona	?	Snyder and Wiley 1976
European Sparrowhawk	(Te)	78 g/chick/day		Scotland	55° N	Newton 1978
Black shouldered Kite	(S)	**43 min/kill**	56	South Africa	25° S	Mendelsohn and Jaksic 1989
Black shouldered Kite	(Te)	19 min/kill		Chile/California	33° S/38° N	Mendelsohn and Jaksic 1989
Snail Kite	(Tr)	**1–1.5 prey/brood/h**	68	Venezuela	8° N	Beissinger 1990
Snail Kite	(S)	3.2–4.7 prey/brood/h		Florida	25° N	Beissinger and Snyder 1987
Rock Kestrel	(S)	**170 g/nest/day**	48	South Africa	32° S	Van Zyl 1993
European Kestrel	(Te)	325 g/nest/day		The Netherlands	53° N	Dijkstra *et al.* 1990
African Marsh Harrier	(S)	**1.7 items/h**	51	South Africa	34° S	This study
Hen Harrier	(Te)	3.5 items/h		W. Europe	53° N	Table 9.12

[a] Matched for brood size, and where possible, maximum effort rates.

Population density comparisons: north versus south

The third testable prediction of Ashmole's hypothesis is that the population density of species in northern climes should be lower (reduced by mortality due to winter weather and migration) than those more sedentary populations closer to the equator. This counter-intuitive prediction is supported by data from the world's harriers. Breeding density data is available for five species of temperate harriers and four subtropical species (Table 9.14). I have excluded the tree-nesting Spotted Harrier because it may be more limited in its nest sites than the other ground-nesting species. The mean breeding density of temperate harriers is 7.7 nests/10 km^2 when the colonial Montagu's Harrier is included. Subtropical species average higher nest densities at 10.9 pairs/10 km^2. The difference becomes greater if we compare only those species which are not colonial (i.e. excluding the Montagu's Harrier). Temperate harriers then average almost half the breeding density found closer to the equator (Table 9.14) although the difference is not statistically significant.

From this comparison it seems that Ashmole was correct in at least two predictions: the prey capture yield is greater for temperate harriers implying greater food resources, and the breeding density is generally lower, implying winter mortality may reduce populations below what the food resources can support. This may the first demonstration that both aspects of Ashmole's hypothesis are supported by any species.

This chapter has assessed five hypotheses accounting for the difference in reproduction. Both experimental and comparative data support Ashmole's density-dependent hypothesis for African Marsh Harriers for the following reasons. Clutch size increases in supplemented birds (independent of laying date), suggesting either that food is limited to pre-laying harriers or that females were making predictive adjustments based on the future food availability. More strikingly, the food gathering abilities of African males before breeding were approximately half that determined for temperate congeners, hence differences can be attributed to ecological constraints on foraging return. These results both point to the conclusion that, because harriers are constrained in their food gathering abilities and provisioned biomass is related to clutch size, African birds lay smaller clutches because of food constraints.

It is revealing, however, that on receiving food supplements African birds did not increase their clutches ($\bar{x}$ = 4.0) to that reported for European Marsh Harriers ($\bar{x}$ = 4.75: Table 9.4), nor the similar-sized Northern/Hen harriers ($\bar{x}$ = 4.65). The lower clutches imply that other limiting factors set a ceiling to which clutches can rise, one of which may be related to latitude. Among these, the inability to rear more than four nestlings is probably crucial. No more than three fledglings per nest were found in 17 records of successful breeding in southern Africa (South African nest record scheme, unpublished) and only one fledged brood of four in 11 successful nests occurred in this study. Experiments with increased broods clearly showed that although some males did increase their provisioning rate in response to enlarged broods, as often they did not. Whether they did or not, the biomass per chick was always reduced in enlarged

Table 9.14 Population densities of breeding harriers throughout the world. Where more than one study was available an average density is given. Subtropical species are given in **bold**.

Species		Nests	Study area size[a]	Density (nests/10 km^2) (n = no. of studies)	Source
Northern	T	3–35	3–210 km^2	5.9 ± 6.6 (n = 10)	MacWhirter and Bildstein 1996
Hen	T[b]	3–71	5–500 km^2	3.7 ± 4.3 (n = 18)	Potts 1998
European Marsh	T	15–72	16–284 km^2	11.0 ± 11.2 (n = 5)	Divis 1982; Altenburg *et al*., 1987; Burneleau *et al*. 1989; Witkowski 1989
European Marsh (*harterti*)		10	20.6 km^2	4.8	Giraud-Audine and Pineau 1974
Montagu's	T	3–54	3–300 km^2	19.5 ± 29 (n = 10)	Arroyo 1995
Pied	T	3	19.6 km^2	1.5	Neufeldt 1964
African Marsh	S	8	10.0 km^2	8.0	This study
Australasian	S	18	12.0 km^2	15.0	Baker-Gabb 1982
Longwinged	S	4	6.8 km^2	5.9	Bó *et al*. 1996
Cinereous	S	10	6.8 km^2	14.7	Bó M.S. unpublished
Means:				5.4 ± 3.5 nests/10 km^2	– excluding colonial Montagu's Harriers
Temperate				[7.7 ± 6.6 nests/10 km^2]	– including colonial Montagu's Harriers
Subtropical				10.9 ± 4.6 nests/10 km^2	c

[a] Study sites smaller than 1 km^2 were omitted since population densities derived from them are artificially high. Similarly, study sites over 500 km^2 were discarded because they give artificially low densities (MacWhirter and Bildstein 1996; Potts 1998).

[b] T = temperate species; S = subtropical species.

[c] Means (temperate vs. subtropical) are not significantly different (t = 1.76, 7 df, $P > 0.05$).

broods and growth rate declined accordingly. Pairs were thus unable to rear more chicks than their planned brood size. Thus clutches may be viewed as adaptively limited to the number of young that parents can rear, yet constrained within proximate limits to food available from the male at the time females are forming eggs.

Is clutch size limited to the number of nourishable young or by the risk of predation?

In challenging Lack's day length hypothesis, Skutch (1949, 1985) cited evidence that tropical species could nourish more young than they normally rear. His evidence was based on (i) the rearing of similar broods by single-parents and two-parent families, the assumption being that if two parents were to help rear broods, they could rear twice as many; (ii) an experiment in which a tanager increased its provisioning rate in response to an enlarged brood; and (iii) experiments in Mexico by Wagner (1957) in which two species reared more young than usual under experimental conditions. The experimental evidence of both Skutch and Wagner, however, suffers from either singularly small samples ($n = 1$), or no quantitative evidence at all (Wagner 1957). Moreover, others (Ward 1965; Lill 1974) have cited contrary evidence from other tropical passerines, and the studies do not take account of the possible costs to survival of the parents, or the survival of their offspring. At best the experimental evidence is poor.

In contradiction to Skutch's predation-reduction hypothesis, harriers increased their clutches when they were food supplemented. This indicates that harrier clutches were partly limited by pre-breeding food and were therefore the product of constraint, not adaptive responses to high predation levels as posited by Skutch (1949). It was also clear that parents (i) did not always respond to increased nestling demands by provisioning more prey, and (ii) nestlings always received less food and consequently grew slower. Again, these lines of evidence point towards a constraint scenario: harriers may be constrained to lay few eggs by pre-breeding food, but given more nestlings, they could not adequately nourish them all. The latter result was confirmed in a review of nine other experimentally increased broods of tropical species and costs to the parents attempting to rear them (Table 9.15). In seven studies, added young could not be reared or grew at much reduced rates. In two studies of passerines, future costs were apparent in reduced clutches or costs to the parents. Indeed, in the study of Wahlberg's Eagle, most of the pairs given extra young to rear did not even attempt to breed the following year. Hence clutches are small initially because food availability limits them and secondarily because of the parent's inability to rear larger brood sizes or because of significant costs to themselves when they do. Do other species conform to these general principles?

Table 9.15 Brood manipulation experiments and their outcome in terrestrial (sub) tropical species.

Species	Broods increased	Extra young	Effect (costs) on young and parents	Source
Quelea	6	(+1, 2)	Fewer young reared	Ward 1965
Hooded Vulture	10	(+1)	Unable to rear more young	Mundy and Cook 1975
Savanna Hawk	10	(+1)	Unable to rear more young	Mader 1982
Redbreasted Sparrowhawk	1	(+2)	Unable to rear more young	Simmons 1986a
White-rumped Swiftlet	11	(+1)	Poor chick survival and reduced body mass	Tarburton 1987
Snail Kite	12	(+1, 2)	No broods of +2 reared, only 40% of +1 reared	Beissinger 1990
Wahlberg's Eagle	8	(+1)	Unable to rear more young, next year's breeding curtailed	Simmons 1993a, 1997b
Bell Miner	11	(+1)	High maternal costs, loss of body mass	Poiani 1993
Tropical House Wren	21	(+2)	Future clutches reduced by 1/3 (next) to 1/2 (next year)	Young 1996
African Marsh Harrier	4	(+2)	Reduced chick growth	This study

Table 9.16 Reproductive responses to experimental food supplements from 32 studies of 25 temperate and subtropical species. Seven species increased clutches independent of laying date, and all were single brooded.

Species	Laying date	Clutch size	Egg size	Hatch success	Fledge success	Date indep?	Broods per year	Temperate subtropical	Sources
Parus ater	N	N	—	—	—	—	DB	T	Krebs 1971
P. major	N	N	—	—	—	—	DB	T	Jones 1973
P. major	A	—	—	—	—	—	DB	T	Kallander 1974
P. caeruleus	A	N	I	N	N	—	DB	T	Ramsay and Houston 1997
Fulica atra	N	N	N	—	—	—	DB	T	Horsfall 1984
F. americana	N	N	I	—	—	—	DB	T	Hill 1988
Pandion haliaetus	N	N	N	{R}	—	—	SB	T	Poole 1985
P. haliaetus	N	N	—	N	—	—	SB	T	Green and Krebs 1995
Aquila wahlbergi	N	N	N	N	N	—	SB	S	Simmons 1993a
Corvus corone	A	N	—	—	I	—	?	T	Yom-Tov 1974
Parus cristatus	A	N	—	—	N	—	DB	T	von Bromsson and
P. montanus	A	N	—	—	N	—	DB	T	Jansson 1980
Melospiza melodia	A	N	—	—	N	—	DB	T	Smith *et al.* 1980
M. melodia	A	I (0.5)	N	N	I	{I}	DB	T	Arcese and Smith 1988
Agelaius phoeniceus	A	N	—	N	{R}	{I}	DB	T	Ewald and Rohwer 1982
A. phoeniceus	A	N	—	—	—	—	DB	T	Wimberger 1988
A. phoeniceus	—	—	—	—	I	—	DB	T	Langston *et al.* 1997
Prunella modularis	A	N	—	—	N	N	DB	T	Davies and Lundberg 1985
P. collaris	N	N	—	—	—	—	DB	T	Nakamura 1995
Sturaus vulgaris	A	N	{I}	—	N	—	DB	T	Kallander and Karlson 1993
Corvus monedula	A	I (0.6)	—	N	I	I	SB	T	Soler and Soler 1996
Pica pica	N	I (0.6)	I	I	I	{I}	SB	T	Högstedt 1981b
P. pica	N	N	I	—	I	—	SB	T	Hochachka and Boag 1987
P. pica	A	N	—	—	—	—	SB	T	Knight 1988
P. pica	N	—	I	—	—	—	SB	T	Clarkson 1984
Aegolius funereus	A	I (0.9)	N	N	I	I	SB	T	Korpimäki 1989

Table 9.16 Continued

Species	Laying date	Clutch size	Egg size	Hatch success	Fledge success	Date indep?	Broods per year	Temperate subtropical	Sources
A. funereus	A	I (0.9)	—	—	—	I	SB	T	Hornfeldt and Eklund 1990
Falco tinnunculus	A	I (0.7)	—	—	—	—	SB	T	Dijkstra *et al.* 1982
F. tinnunculus	A	I (0.2)	—	—	—	N	SB	T	Meijer *et al.* 1988
F. sparverius	N	N*	I	R	—	—	SB/DB	T	Wiebe and Bortolotti 1994
Accipiter nisus	N	I (0.8)	—	N	N	I	SB	T	Newton and Marquiss 1981
Circus ranivorus	N	I (0.5)	N	N	N	I	SB	S	Simmons, this study
Larus fuscus	N	{I} (0.3)	I	{I}	N	I	SB	T	Bolton *et al.* 1992
Ardea herodias	N	{I} (0.2)	—	N	I	I	SB	S	Powell 1983
Lanius collurio	N	I (0.9)	N	N	—	I	SB	T	Carlson 1989

A = advanced; N = no significant effect; I = increased; R = reduced; SB = single brood; DB = double brood; — = no relevant data; { } = marginal (non-significant) response. * = a determinate layer and double brooded in southern latitudes, so no increase expected.

Comparisons with other food supplementation experiments

To date, no consistent trends have been identified from results of food supplementation experiments across avian species. Thus, when certain species are food supplemented prior to breeding, their clutch size either (i) does not increase significantly (53% of 32 studies, mainly temperate passerines: Table 9.16); (ii) increases, but not beyond what the earlier laying date would predict (Kestrels); or (iii) increases independent of laying date (song sparrows, magpies (one study), sparrowhawks, harriers, owls). The latter studies provide the strongest evidence that food has a direct effect on clutch sizes and not through its effect on laying dates. How can we explain the wide variability in response to manipulations and are trends hidden that tell us how subtropical birds apportion their reproductive investment? Ewald and Rohwer (1982) suggested that the variability in response may be related to the quality of the food provided. This would not explain differences between the raptorial species studied: each was fed its principle food type, yet two species laid larger clutches, and two species did not respond. Davies and Lundberg (1985) proposed that small passerines may be limited in their clutch size by their inability to rear more young, and not by pre-egg laying food availability. Egg formation in larger species, on the other hand, may be chiefly limited by food availability. This, too, can be rejected because of the increased clutches found in a small passerine, the Song Sparrow *Melospiza melodia* (Arcese and Smith 1988) and because of the failure by large Ospreys and eagles to increase clutch size (Table 9.16).

Looking to explain clutch size in relation to laying date, and not food per se, Meijer *et al.* (1988) argued that previous studies (e.g. sparrowhawks, Newton and Marquiss 1981) had not provided sufficient proof that supplemented clutches were in fact larger than that predicted by the laying date. They proposed that most species were responding by laying earlier, and that laying date determined clutch size. Their reanalysis of the sparrowhawk data used inappropriate controls (I. Newton pers. comm.) and clutch size did indeed increase to levels higher than that expected by the advance in laying date. Similar conclusions are apparent for seven species now (Table 9.16) and the trend is apparent in both large and small, and temperate and tropical species. So while laying date is clearly important in determining the clutch size of many birds, food alone can increase clutch size independent of laying date. Repeat experiments on the same species have also produced different results (for example Magpies, Coots, Table 9.16). This result alone suggests that individuals may respond differentially, depending on local resources, population density, and perhaps different selection pressures on their current and future survival.

Arcese and Smith (1988) proposed a more general hypothesis of the variations in response to supplemental food: constraint imposed by genetic or developmental factors. Both their results and those of Ewald and Rohwer (1982) indicated that clutch size or early laying rose to a ceiling, above which further increases were probably prevented by proximate factors such as population density or photoperiod and temperature. Hence, once one limiting factor is removed, laying or clutch size is advanced only up to a level at which another factor limits further increases. The corollary of this proposal is that those populations that responded by laying earlier, but not larger, clutches may already be close to their maximum clutch, and can only increase fitness by laying earlier. In such species the inability to rear larger broods (without substantial costs) may be limiting their clutches. This was evident among Dunnocks *Prunella modularis* (Davies and Lundberg 1985), Wahlberg's Eagles *Aquila wahlbergi* (Simmons 1993a), and harriers in this study. Whether these are genetically controlled differences are unproven.

Two-baskets hypothesis

In summary, no single satisfactory answer has emerged from reviews of such experiments. However, there is a fairly obvious difference in breeding strategy is inherent in the 'non-responders' relative to the 'responders' (species that laid larger clutches). As predicted above and mentioned by Martin (1996), a single clutch may not be the most appropriate measure of reproductive effort per season if the season is particularly long or the breeding 'turn-over' time is short, as it is for most small passerines. Reproductive investment must be measured as the total number of eggs attempted per season, and passerines frequently produce two or more clutches. Thus a passerine suddenly experiencing favourable resources (the supplemented pairs) should begin earlier in order to lay a second clutch to increase its reproductive output. This is precisely what Svensson (1995) found in a review of supplementation studies to passerines—those that laid earlier were the multi-brooded species whereas single brood species rarely laid earlier.

Single vs. multi-brooded species

For multi-brooded species very little is gained by increasing individual clutch size if this impairs their ability to lay a second clutch before the season becomes unfavourable. Thus we would predict that multi-brooded species *will not increase their clutch size but will lay earlier.* A larger clutch may also increase predation or reduce survival of young hatching from it. A second clutch will usually double the seasons output, whereas increased (single) clutches are never able to double output. Inspection of the data (Table 9.16) reveals that clutches are typically increased by less than one egg for birds that did respond to supplementary feeding. Those species unable to exploit enhanced food resources by laying additional clutches (single-brooded) are therefore predicted to *increase their clutch size (or egg size), because this is the only means they have of exploiting the rich food resources.*

Thus the predictions of the 'two-baskets' hypothesis are: (i) birds with short egg-to-fledge times and who can rear second broods should lay earlier, but not larger clutches; (ii) species unable to lay a second clutch (single brooders—SB in Table 9.16) before the season becomes unfavourable, should respond by laying larger clutches; (iii) all species in which breeding seasons are extended or relatively undefined (mainly tropical ones), should respond by laying second clutches with no increase in the size of any individual clutch; (iv) some species may be incapable of rearing larger clutches, and no response is expected under any conditions. Thus the well-defined breeding season of most temperate breeding species dictates that single brooders should respond by laying larger clutches.

This prediction was supported: in 11 of 18 studies of single-brooded species target pairs laid larger clutches. This represents larger clutches in 9 (75%) of the 12 species tested because some species were experimentally fed in more than one study. The three exceptions were all raptors (Wahlberg's Eagle, Osprey, and American Kestrel), and at least two of these are known to be incapable of rearing more young, hence a larger clutch is of no advantage to them. This has been experimentally verified in Wahlberg's Eagles (Simmons 1997b) and is likely in Ospreys because enlarged broods experience decreased food per individual, grow slower, fight more, and maternal weight can decrease by up to 40% in a month (Poole 1989). Unlike many other birds of prey, Ospreys are relatively insensitive to food fluctuations and lay three eggs in populations spanning many latitudes (Poole 1989). The more variable character in their breeding strategy is a change in hatching asynchrony with food resources: in good-food years broods hatch synchronously, in poor years they hatch up to ten days apart (Green and Krebs 1995). They, too, may enjoy no benefit in laying larger clutches because costs outweigh any benefits. The third species, the American Kestrel, is reported to be an indeterminate layer by Balgooyen (1976) and double brooded in southern climes (Cade 1982), and for these reasons may not be expected to increase its clutch size.

The second prediction, that those species able to rear a second brood should rather respond by laying earlier but not larger clutches, was strongly supported. Of 12 such species, 92% laid clutches that were no larger than expected. Of these, 9 species laid earlier. The single exception was found in one of two Song Sparrow populations in which extreme density-dependent reduction in clutch size was apparent (Arcese and Smith 1988). Thus when the experiments where performed, females were merely increasing their clutch size (0.5 eggs) to the level typical for lower density populations, and a clutch size increase over normal levels was not evident. They also conform to the expectations of the two-baskets hypothesis.

If we now compare the two expected strategies for single-brooded versus double-brooded species, we find a highly significant difference in their responses in the predicted direction (Fig. 9.7). This includes all species whose non-response to supplementation was explicable. This result is not apparent if we compare passerines with non-passerines, or temperate with subtropical species. It supports the idea, therefore, that each species reacts in the most appropriate manner to a sudden increase in food resources. The definitive experiment, however, has yet to be done: supplemental feeding of tropical passerines with highly extended breeding seasons. The two-basket hypothesis predicts no increase in clutch size, little benefit to laying earlier (because seasons are long anyway), but more clutches.

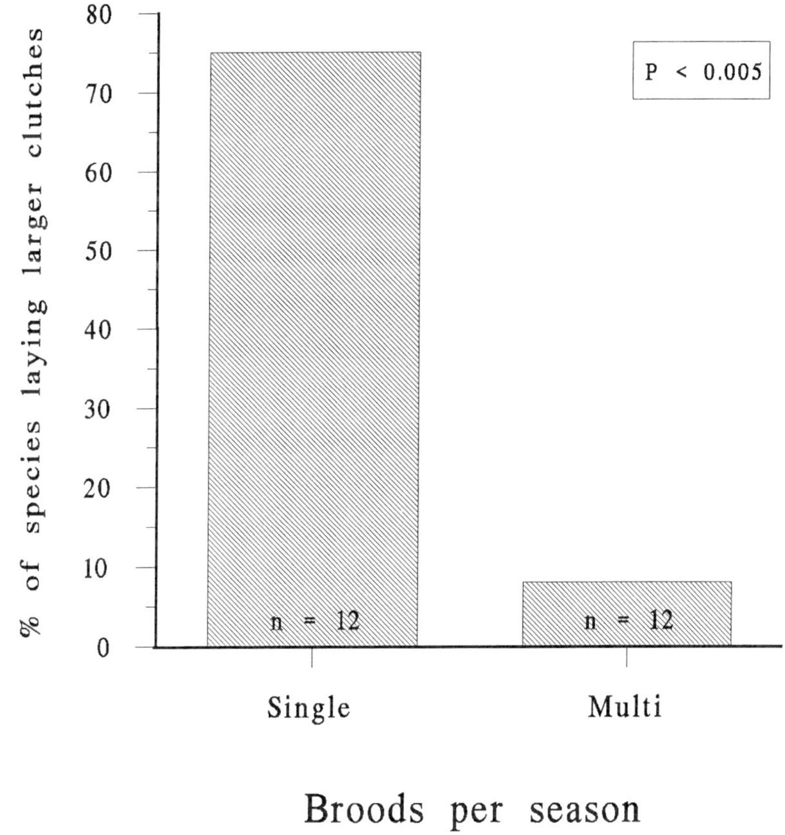

Fig. 9.7 The differing clutch size responses to food supplementation experiments of species producing one brood or multiple broods per year. Single-brooded species (n = 12) put the increased resources into larger clutches while multi-brooded species (n = 12) generally did not, but laid earlier. $X^2 = 8.40$, 1 df, $P < 0.005$ (Yates corrected).

Whether such experiments are undertaken or not, there is a need for future analyses of food supplement responses (the benefit side) to also include brood-enlargement experiments (the costs). A species response can thus be gauged in relation to a food surplus and a food deficit. This helps clarify if those species that do not respond, or respond by laying eggs up to a set ceiling, are constrained by other factors: they may not lay larger clutches because they are incapable of rearing more young. The hypothesis may also show that Alexander Skutch was ultimately correct: neotropical passerines may be able to nourish more young than they can rear in each individual clutch—because they normally rear several clutches in a year (Simmons 1998). Thus experimenting with just one clutch by adding young is an inappropriate test of their abilities, and the effect on their entire breeding season output should be assessed. Possible outcomes may include a reduction in the usual number of broods in the year, rather than an inability to rear the extra number in the present nest.

Table 9.17 Predation levels and clutch size for tree-nesting Spotted Harriers in relation to samples from temperate and tropical ground-nesting species.

Species	Predation level	Clutch size	Source
Spotted Harrier	**13% of 24 nests**	**3.0 ± 0.5**	**Baker-Gabb 1982**
African Marsh Harrier	52% of 25 nests	3.5 ± 0.1	this study
Australasian Swamp Harrier	18% of 25 nests	3.7	Baker-Gabb 1983
Northern Harrier	21% of 99 nests	4.3 ± 1.6	Simmons *et al.* 1986b
Hen Harrier	18% of 480 nests	4.7 ± 0.9	Picozzi 1984a
Montagu's Harrier	21% of 217 nests	3.6 ± 0.4	Butet and Leroux 1993
	9% of 165 nests	4.0 ± 0.1	Arroyo 1995
European Marsh Harrier	11% of 118 nests	4.9 ± 0.7	Witkowski 1989
	13% of 54 nests	4.6 ± 1.5	Underhill-Day 1984

Effect of predation on clutch size

I have dwelt little on predation and its possible role in promoting small clutches, principally because larger species are less likely to show such effects (because they have fewer predators). Nevertheless, we can assess the effect of reduced predation on clutch size by comparing a tree-nesting harrier with ground-nesting ones at the same latitude. Since tree-nesting species would be expected to suffer less predation because they are inaccessible to terrestrial predators, we predict that the tree-nester clutch size should be larger, with body size and latitude held constant. First, it is apparent that for the ground-nesting harriers predation levels were twice the rate on my African study area (despite trapping for mammalian predators) that they were in my eastern North America one (Table 9.17). Other predation levels varied from 21% of *pygargus* nests in western Europe (Butet and Leroux 1993) to 11–13% of *aeruginosus* nests in Poland and England. For the tree-nesting Spotted Harrier breeding at the same latitude (34° S) in south-eastern Australia as African Marsh Harriers in Africa, pairs suffered much lower predation (13%: Table 9.17), as expected for a tree-nester (Baker-Gabb 1982). However, contrary to predictions of the Skutch–Snow hypothesis, clutch size (3.0) was not higher than that predicted (3.86 eggs) by the latitude equations given above for a ground-nesting harrier of this size (725 g).

The same result is apparent if we compare Spotted Harrier clutch size and predation levels with a congener on the same continent: Swamp Harriers in south-east Australia suffered a maximum of 18% nest predation (Table 9.17). This is higher than the Spotted Harrier (13%) but their clutch was also marginally higher (3.7 eggs) than the Spotted Harrier (3.0 eggs) at study sites at equivalent latitudes. This is also contrary to the predation idea.

However, Spotted Harriers are unusual among the harriers in that they lay particularly large eggs for their body size. Close correlations were evident for

nine other species (Fig. 7.3, R^2 = 98%) except the Spotted Harrier, which lays eggs averaging 39.53 cc. This is almost 14% larger than the egg volume (34.77 cc) that its body mass (671 g) predicts. Could it be that its relatively safe nesting sites in trees in the arid lands of southern Australia have led not to an increase in clutch as predicted, but to an increase in egg size instead? Like single-brooded species unable to exploit rich food resources by laying more clutches, do Spotted Harriers instead put resources into larger eggs because, despite having a safer nest site, its elevated position and flimsy structure (Baker-Gabb 1984) precludes larger clutches and broods? Experiments would be fruitful in testing this and other aspects of this ancient harrier's fascinating breeding strategy.

Neither this analysis of harriers, nor the previous finding that clutch size increased following supplemental food (above), offers support for the idea that predation may limit clutch sizes in subtropical harriers. Another way we can assess this clutch size—predation idea is to look for higher predation levels in experimentally enlarged broods of other species in tropical/subtropical areas. Only two studies, to my knowledge, have reported on the effects—the prediction being that enlarged broods should suffer higher predation. No significant increase in predation levels was found in enlarged broods of either Australian Bell Miners (Poiani 1993b) or tropical House Wrens (Young 1996), relative to controls. This suggests larger broods are not limited because predators are more likely to detect them, contrary to the clutch–predation idea.

However, an intriguing result is apparent from an analysis of New Zealand Swamp Harriers and their clutch size. New Zealand has the distinction of historically having no snakes and, until recently, no mammalian predators of ground-nesting birds. A lack of predators should result in a higher clutch size than in adjacent areas at equivalent latitudes. Baker-Gabb (1983) reports that this is indeed the case: clutch size on New Zealand is 4.6 eggs, which is 1.0 eggs larger than those at similar latitudes in his study area on Australia, with its full complement of predators. This is an interesting result, supportive of the clutch–predation idea. However, Baker-Gabb explained it, in a proximate manner, as arising from increased levels of carrion and a lack of competitors for Swamp Harriers in New Zealand. While a viable alternative is that a lack of predators has promoted the large clutch size, we must also remember that clutch size is a fairly plastic trait which varies with ecological factors, and birds there have been exposed to predators for perhaps a thousand years. This is sufficient time for a response to increased levels of predation, and the clutch should have adjusted to present day levels. If this is so, then the larger clutch in the face of predation is poor evidence for predator-mediated small clutches and data from the harriers does not generally support the idea.

I conclude that experiments and comparisons between northern and southern hemisphere harriers strongly support the idea that fluctuating food resources interact with population density levels resulting in more food resources for individuals in temperate populations than in subtropical ones.

Individuals react by laying a clutch to match the number of young they can subsequently nourish—in both hemispheres. Some confusion in the testing of these hypotheses may have arisen by researchers not taking account of breeding strategy differences between single-brooded and multiple-brooded species (common in tropical latitudes). Neither higher predation levels nor greater longevity appear to cause reductions in harrier clutches. Each result supports Ashmole's hypothesis confirming earlier analyses that employed indirect methods to assess the hypothesis. This appears to be the first such direct confirmation of all aspects of the hypothesis.

The experimental food supplements reported here indicate that the clutch size of African harriers is partially limited by early-season food availability (independent of laying date), but that it rises to a ceiling below that of temperate harriers. The difference can be explained by the inability of African harriers to rear more young, which is itself best ascribed to ecological limits (constraint), rather than to the withholding of reproductive effort. This is because African Marsh Harriers, (i) acquired half as much food biomass per unit foraging time as temperate congeners, (ii) could not provide the same biomass per chick when brood size was increased, and (iii) African nestlings typically received less biomass than that delivered to temperate harrier nestlings. The differing results apparent in 32 food supplementation experiments cannot be explained by differences between passerines and non-passerines, or subtropical and temperate species. However, birds laying larger clutches are typically single-brooded, while those showing no clutch size increase were typically multi-brooded and capable of laying more clutches. These trends are sufficient to explain why Skutch found that neotropical passerines can apparently rear more young than they typically nourish—tropical birds often invest in two or more nests, and not in a single breeding attempt. In this respect tropical species differ most from temperate-breeding ones. In most other respects the rules that govern breeding in temperate species are identical to those in subtropical areas. Predation appeared to play no role in selecting for small clutches in subtropical harriers or other species.

10 *Synthesis*

The aim of this book has been twofold: to review current and new research on the evolution and ecology of the world's harriers, and to provide a behavioural ecologist's perspective on old questions about raptors, and harriers in particular. The behavioural ecologist is generally programmed to ask the most difficult question, which is *why*?—why do organisms behave as they do, why have some traits been favoured over others, and so on. To even begin to answer the 'why' question for any behaviour or ecological trait one cannot be content with observing and recording; it also requires perturbing an organism's behaviour in some non-obtrusive way to test its ability to respond to unusual circumstances. The cleverly crafted experiment not only creates endless possibilities for creative thinking on the part of the researcher, it also has the potential to solve numerous puzzles that have not yielded their secrets to the patient observer, even in a lifetime's work. While I have not been very creative in this regard—the experiments reported here are often carbon copies of ones that have been tried in temperate zones—the difference lies in the fact that they have been used on species in southern climes where experimentation is much rarer and where they can provide insights into questions (on clutch size, for example) that have only been tackled from the north, down. In this concluding chapter we will take a look at some of the answers that have emerged from these observations and experiments, draw conclusions from the most interesting findings, and most important of all, highlight the questions that have arisen which may provide research avenues for future evolutionary ecologists.

Evolution

Thanks to the close collaboration of harrier enthusiasts around the globe (in sending feathers and blood) and Michael Wink's genetics lab, we now know that 16 species of harrier occur worldwide. Previously 13 species, and before that (in 1973) 10 species and several subspecies were recognized. The greatest diversity occurs in India where six species may occur post-breeding. The Reunion Harrier (*C. maillardi*), a small dark bird-catching harrier, is now separated from the Madagascar Marsh Harrier (*C. macrosceles*) on morphological and genetic evidence. Speciation probably occurred following its specialization on birds in forested island habitat, about 760 000 years ago, resulting in birds about 30% smaller than found on Madagascar. With the Black Harrier, the Reunion species shares the smallest range and is the rarest of all the harriers. While the Northern Harrier differs little from the Hen Harrier on ecological and behavioural grounds, a genetic distance of 0.8% between the two forms suggests they are indeed two separate species. The third species that is given full species status is the Papuan Harrier of New Guinea based on its sedentary nature in Papua New Guinea. Like the Longwinged Harrier, the oldest living species and coincidentally the one with the highest size dimorphism, the Papuan Harrier occurs in two morphs, one with a clear pale breast and the other more streaked. The genetics and consequences of these differences are unstudied, and indeed the Far Eastern harriers are the least known of the genus, with no ecological studies of the Papuan bird and only two of the Pied Harrier of Mongolia and Russia. The Montagu's Harrier, expected to have arisen from the Hen/Northern/Pallid group of grey harriers, is instead an ancient harrier which gave rise to the marsh harrier about 3 million years ago. The African Marsh Harrier is ancestral to the Eastern Marsh Harrier, while the Australian Swamp Harrier gave rise to the African island harriers and ultimately the European Marsh Harrier. The evidence now suggests a southern origin for the harrier genus, with possibly South America their ancestral home, some 77 million years ago.

Distribution and peculiarities

The Hen and Northern Harriers exhibit the greatest world distribution, but like other harriers do not breed below about 30° N. These breeding distribution

gaps occur in North America, Africa, and Asia for different species, and a paucity of available vertebrate prey in these subtropical areas appears to be the reason that birds do not breed, but can over-winter by exploiting insect prey (Chapter 2). This is well illustrated by the propensity of far-northern Montagu's Harriers to exploit mainly small mammal populations, and those farther south to rely more and more on birds and insects (Arroyo 1997).

The peculiarities of the harriers, which include aerial food-passing, acute auditory abilities, long distance foraging, ground-nesting, and polygyny in certain species, may all have a common origin in the birds' evolutionary switch from the arboreal habits of their Accipiter ancestors to open marsh and savannah. For example, aerial transfers are more common among Northern and Montagu's Harriers, which inhabit relatively dry habitats, than African Marsh Harriers, which frequent more aquatic habitats. Because of the greater likelihood of terrestrial predators being able to follow scent in dry marshes, it may be that aerial transfers evolved in the genus because it is a more efficient (> 96% of all prey is caught) means of avoiding predators than ground passing. The long distance foraging, in which birds may cover over 200 km per day hunting for dependent young, is clearly a response to a paucity of perch sites in an open landscape. The alternative—hovering—is infrequently used by harriers but it is an energetically expensive method compared with the exploitation of winds that frequently blow across open marshes. Because mice and voles are common in these landscapes, acute hearing to locate audible, if not always visible, mice would be of great advantage to foraging birds and all harriers exhibit some specialized facial feathering. That the mouse specialists such as the Hen and Northern Harriers show a complete facial mask, and other less-specialized harriers such as the Montagu's and Pallid show reduced feathering, supports the notion that auditory abilities arose to aid harrier mouse-catching abilities. Further investigations of the size of the ruff and the physical limitations to its sound-catching properties are required to determine if harriers do indeed use the ruff in the manner that is assumed.

Predation behaviour

While the physiological means by which harriers catch small mammals may be in question, the fact that some species hunt them all the time is not in doubt. This leads to some clear predictions about what spatial and temporal tactics prey employ to avoid predators. We saw that prey should be under greater selection to avoid being caught than are the predators attempting to catch them because unsuccessful prey lose their lives, unsuccessful predators merely lose their dinner. Two predictions arise from this Life–Dinner Principle: (i) prey should live in habitats where they are least vulnerable to predators (i.e. there should be an inverse correlation between predation pressure and prey abundance), and (ii) prey should be most active when their main predators are least active. The tests of these ideas with African Marsh Harriers and their main small-mammal prey indicated that (i) that small-mammal abundance was inversely related to harrier predation success in four habitats: that is, mice were more abundant where harriers were least successful. This arose because harriers were least successful in dense vegetation types. In support of prediction (ii), highest mouse activity occurred at dawn and dusk when neither harriers nor other aerial predators hunted mice. Thus the mice were indeed exploiting a short predation-free window. Since harriers were aware of, and exploited, small-mammal activity peaks at other times of day, the absence of predation at dawn was best explained by the negative effects of clear skies and low wind speeds on foraging return. The lack of foraging at dusk, when another mouse activity peak took place, was, however, inexplicable. The synchronization of activity by mice (and voles) may be a 'selfish-herd' effect to dilute the probability of predation, although this is only likely when predators don't precisely match foraging patterns with mouse activity. The seasonal analyses of increasing and decreasing vegetation cover (Chapter 4) indicated a complex temporal relationship between predatory success and mouse availability and abundance. However, in support of other analyses, trends in predation success were best explained by the *availability* of mice in the most mammal-rich habitats and not their *abundance* in these habitats. It seems that small-mammal habitat preferences and activity patterns were strongly influenced by the predation tactics of harriers in this study, confirming results from other studies of predators and their prey.

A second major question in predator–prey studies is whether predators have a major enough influence on their prey to regulate prey populations. Two studies of harriers in Australia and Canada found little evidence of this but the latter did note a greater influence by generalist predators on prey populations. Nevertheless, in the most comprehensive study, on grouse moors in the United Kingdom, harrier predation was sufficient to remove 37% of all grouse chicks and with adult predation reduced future breeders by about 50%. This particular study of Redpath and Thirgood (1997) was motivated by concerns that harriers were regulating and reducing the standing crop of grouse used by

the hunting industry. It could not be conclusively stated, however, that removing harriers would increase grouse populations because wide-scale reduction in heather had reduced the cover available for grouse broods and alternative prey for harriers.

In natural systems relatively little perturbed by man, predators and prey numbers are likely to oscillate about an equilibrium in which neither party gains an advantage. Where predators are introduced or habitat altered, predators often have an advantage and can drive down the populations of their prey below typical levels, which may result in local extinction. Much remains to be learned about predator–prey systems, especially in more complex systems in tropical areas where one expects a greater array of predators and thus greater selection on the prey.

Polygyny

Of the traits that distinguish the harriers from other raptorial birds, perhaps polygyny most clearly sets the harrier genus (*Circus*) apart. Given that polygyny is most prevalent among other marsh-nesting birds, the evolution of this trait in harriers can also be ascribed to their marsh-nesting habitat. Many authors have argued that the high productivity and patchiness of food in marshy landscapes has promoted polygyny, allowing some males to provide large quantities of food to more than one female. I argue that an equally important factor is the defendability of both the (food) resource and the females (from stolen copulations) in open habitats. In semi-closed or closed woodland habitats one can imagine similar patchiness in resources but the detection of intruders and the defence of resources must surely be more difficult in denser, three-dimensional habitats.

Many ideas have been proposed explaining why females join already-mated males. They revolve around the idea that by joining a high quality male, females can leave more recruits than by pairing with a bachelor. Because many species, both passerine and non-passerine, have been found to violate this simple rule, ingenious alternatives have been proposed including: the sexy son hypothesis (sexy sons will become polygynous, compensating for their mothers losses); that females join harems to enhance their own survival through cooperative warning and defence; that females are fooled by the polyterritorial nature of mates they believe to be unmated; and the skewed sex ratio of the adult population which precludes any other choice than pairing with a mated male. Tests of each of these ideas indicated that harrier polygyny cannot be explained by any of them or by the additional possibility that polygynous females join harems to inherit the α female's pole position. Female mate choice cues in Northern Harriers appeared to be based on male provisioning performance and once hierarchies within the harems were investigated it became clear that first-settling females were favoured by males and later-settling ones were neglected, causing them to reproduce very poorly. The obvious conclusion is that males were deceiving these females with high courtship rates and later concentrating their efforts on favoured α females. Where males cannot control good quality territories, (e.g. colonial Montagu's Harriers) or occur in areas where vole populations are apparently lower (Hen, Northern, and European Marsh Harriers in their southern range), polygyny is less likely because the threshold provisioning rate to more than one female cannot be achieved. These conclusions may be unsatisfactory to European-based workers who have favoured skewed sex ratios in promoting polygyny, but the strong association with food resources, the lack of association with skewed sex ratios in North America, and the fact that female Northern Harriers do not fill up all the monogamous situations before polygynous ones arise, argues against such a scenario.

While the evidence suggested that male-provisioned food is the principal choice cue for female Northern Harriers, I do not claim that it is the only criteria on which females choose their mates. Male courtship displays and harem size are significantly correlated in harriers (Chapter 3), suggesting that females initially assess male display vigour, then provisioning, possibly plumage type, and other elements in a hierarchy of choice factors. A female harrier's checklist during the pairing period might be: look for displaying males and choose the most vigorous. If no males are displaying, choose the greyest (oldest) bird, and assess the speed with which the first prey item is delivered, and its size. If food parcels are small, late, or do not arrive, seek mates elsewhere. Experiments can elucidate choice cues by manipulating the ability of males to provide food to prospective mates (i.e. handicapping those most likely to become polygynous). Alternatively, birds can be food supplemented at this time and their mate attraction abilities determined. Choosing young birds which seldom achieve polygyny status, or males in areas where polygyny is uncommon, might be the best targets for such experiments. How females actually sample males as they arrive in a prospective area is of equal interest, and following females introduced into an area in which they have no prior experience, as Bensch and Haselquist (1992) did with warblers, would be an ideal start to such investigations. Additionally very few studies have followed the fate of sons (and daughters) of polygynous pairings to determine if the offspring of polygynous males themselves achieve polygyny status or enhanced viability. In areas where site fidelity is high (some areas of Spain and France for *aeruginosus*), this would be most likely to reap rewards. We additionally do not know if Orkney vole populations cycle as do other voles, which may explain the 5–6-year fluctuations apparent in the reproductive success of Orkney's Hen Harriers.

Displays and RSD

Harriers perform particularly striking aerial displays, which are at their most frequent in early spring just before breeding, and at other times during territorial take-overs. Of all the harriers studied only the ancient Spotted Harrier does not perform skydancing displays. Both male and female harriers perform them although female participation appears to differ between sedentary and migratory populations and between species.

Two forms of aerial display were evident: typical skydancing in which birds soar to great heights before performing U-shaped undulations; and skyspiralling in which birds perform vertical displays, dropping in a series of descending U-shaped undulations. These displays were interpreted as mainly territorial in function because (i) African Marsh Harriers have long-term pair bonds and thus a reduced need for mate attraction, (ii) most male displays were performed in the presence of other male harriers, particularly invading floaters, and (iii) females also display. However, because skyspiralling frequently ended on the nest, and females that were attracted followed males into the sites, spiralling is probably a mate attraction display.

Among Northern Harriers, males that displayed most often had larger harems than males displaying less often. We can be fairly sure that these birds were in better condition (and proving it) from the increased display activity of male African Marsh Harriers that were food supplemented, and the increased display activity of Northern Harriers in food-rich years. Once again food supplements to polygynous species would be ideal in testing whether target birds display more and *then* attract more females. (This would be best undertaken in populations where polygyny is absent or rare.) In addition, we do not know if the males displaying most are the smallest and most agile birds, as predicted from theories of RSD. This is easily determined but has never been tested. From the female side, RSD also predicts that the larger females should be better competitors for small agile males and thus larger females should predominate in the α female position in the harem where competition is intense and the reproductive rewards high. Again this prediction is easily tested but remains unstudied.

The results of the analyses on why female harriers are larger than their mates favoured a combination of natural selection and sexual selection hypotheses. Jehl and Murray's (1986) review of RSD in shorebird families and other groups cautioned, and indeed castigated, raptor biologists for focussing on *raptor-specific* hypotheses to explain RSD in raptors. Because of the high incidence of RSD across the board among raptors this has not been surprising. Fortunately, Penny Olsen's clear thinking on the sexual selection ideas that explain normal dimorphism in other species has helped bridge the gap between raptor-specific ideas and those that are universally applicable. The harriers provide a unique chance to test her hypothesis—that females must be large to compete successfully with other females in contests for essential males. The polygynous species of harriers were found to be more dimorphic than their bird diet suggested, which is predicted for species in which females must compete intensely for access to the best quality males and the males' full services once in the harem. Monogamous harriers by contrast were found to be less dimorphic than expected. The attraction of this hypothesis is that, unlike others, it can explain both the direction and the degree of RSD in the raptors because it explains why females and not males become larger.

An unexpected spin-off of this analysis was that the Longwinged Harrier, whose mating patterns are unstudied, was found to lie above the regression line that linked RSD and a passerine diet. As such it lies with the polygynous species suggesting it, too, is polygynous. An independent assessment of body size variation with latitude suggested exactly the same thing. It was found that both male and female Longwinged Harriers from Nieboer's (1973) sample increased in body size in more southerly populations. However, because the female's body size increased faster than the male's, the bird's RSD values also increased. A plausible reason for this increase in RSD is that more southerly populations are polygynous resulting in greater female–female competition. A third line of reasoning also points to the same possibility: in the latitudinal analysis of polygyny in Northern Harrier populations (Fig. 5.9) it was apparent that populations above about 40° often exhibited some form of polygyny. The Longwinged Harrier breeds as far as 40° S and it is occasionally recorded as far south as Tierra del Fuego (54° S del Hoyo *et al.* 1994). Empirically and theoretically, therefore, this ancient harrier may prove to be polygynous and as such would repay further study.

It is my belief that the reason that RSD is widespread in the raptors and that some of the most dimorphic species occur in this group arises because selection occurs on both males *and* females. The enhanced correlations with RSD in harriers when the proportion of passerines in the diet was substituted for all bird prey, strongly suggested that small agile birds were selecting for small size in males. Natural selection on males therefore plays a role in harriers and other raptors. Talon-grappling and cartwheeling behaviour, chiefly among female harriers and other raptors (Simmons and Mendelsohn 1993), indicates that selection on female size and competitive ability also occurs. However, in the skuas only selection on large female size was identified by Catry *et al.* (1999), and they found no evidence for directional selection on small males. It is of little surprise, therefore, that the RSD indices for the skuas are generally lower than those for bird-eating raptors. Assessing other dimorphic genera for both natural and sexual selection on males and females would prove a fruitful avenue for research. However, there are some groups in which males are clearly essential to female reproductive success—and even their survival—yet exhibit no RSD at all. Among the *Tockus* hornbills, for example, females are sealed into the nest for more than a month and become flightless (Kemp 1995). These females are entirely reliant on their mates and mortality may occur on the loss of their mate. Contrary to expectation, male hornbills are larger than their mates, not smaller as Olsen's idea predicts. However, there is no selection for agility in this species since they are insectivorous and capture their prey by searching on the ground or in trees. There is good reason to believe, therefore, that selection on male raptors is as important as selection on females in explaining the evolution of RSD in raptorial birds.

Manipulation of sex ratio and egg size

The ratio of daughters to sons at birth or in adult populations is generally 1:1 for many vertebrates. Darwin and Fisher reasoned that if parents invest equally in the two sexes, then at the end of the dependency period the least expensive sex should predominate. In human societies this is often the female, but as one sex becomes rarer, so natural selection should favour parents rearing the rarer sex because the rewards in terms of future fitness should increase. Population sex ratios should thus oscillate between a predominance of the cheaper sex and the most rewarding sex. Because female raptors are larger than their brothers, males should predominate, but in some species this prediction was not upheld. This begged the question: are females in fact more expensive to rear? This was verified in one study in which females took 20% more energy to rear to the same age as males—a result not seen in field studies of food ingestion by nestling harriers and sparrowhawks. Nevertheless, sex ratios in four studies of harriers each showed a predominance of males in the fledged young. In the longest data set, from Orkney, Hen Harrier sex ratios changed from a predominance of females to a bias towards males over 30 years. It is tempting to conclude that, because this occurred in parallel with a similar trend towards polygyny over the same time period, selection for more rewarding sons was taking place. However, monogamous populations show the same male bias, and it seems likely that sons predominate because they are less expensive.

Strong sex ratio trends in relation to season, clutch size, laying order, and even colony size (for Montagu's Harriers) were apparent in the harriers but few consistent patterns emerge from comparisons between species. From studies of European Marsh Harriers and European Kestrels Daan *et al.* proposed the early-maturation hypothesis. Sons are favoured early in the season in kestrels because their early maturation increases their (but not their sisters') chances of breeding as yearlings in the very next season. While several species, including the marsh harriers, fit the predictions well, Northern Harriers hatch more sons early in the season when more daughters would be more advantageous for early maturation. Further research on the potential advantages to more early-season sons in this species would be particularly rewarding. One trend that was common to all (three) harrier species investigated was that first eggs were female biased, followed by a preponderance of males. Whether this is specific to the genus will emerge as other harriers are investigated. Its adaptive (or physiological) significance is unknown. Sex ratios clearly showed a great deal of plasticity and appear to be under the control of parent birds. How this is achieved, and whether all such sex ratios need an adaptive explanation, is a particularly fertile field ripe for further experimental manipulation.

In the experimental food supplements with eggs, some surprises were apparent. First, many species, including the harriers, do not show any significant increase in egg size when such an increase could have been beneficial. This result suggests that a ceiling is put on egg size and, from the close correlation apparent between harrier body mass and egg volume, that ceiling is most probably the female's body size. However, a curious relationship emerged when females within a population were assessed: in both Montagu's and African Marsh Harriers it was the smaller females (based on wing chord) who were laying larger eggs (Chapter 7). Since these birds were not trading large eggs for smaller clutches, it is possibly best interpreted as smaller females being better able to mobilize the male-provisioned food into eggs than large females. This result, however, needs to be confirmed in other harriers and other species. A second curious result emerging from this evaluation of egg size was the particularly large eggs of two species, which were larger than their body size predicted. Both the Spotted and Longwinged Harriers are somewhat ancient harriers and phylogenetic peculiarities may explain their large eggs. But both species lay small clutches, and equally likely is the possibility that they have traded flexible clutches for a small-clutch, large-egg 'brood-survival' strategy. One way of testing this is to determine if they lay particularly large final eggs as found in other brood survival species (Slagsvold *et al.* 1984). Food supplements to these species similar to those which changed the energy apportioned to African Marsh Harrier eggs would prove particularly interesting, as would an analysis of egg composition in both species.

The possible link between changes in egg size within the clutch and the manipulation of sex ratio with season, or egg position, suggests that female African Marsh Harriers may alter the sex ratios of their brood when food supplemented. This possibility exists because of the link Witkowski (1989) found between egg size and the sex of the egg (males came from larger eggs). This was not examined in this study but it may occur in both African Marsh Harriers and Montagu's Harriers, in which egg sizes vary within the clutch with changing food resources (Arroyo *et al.* 1999).

Clutch size and food

Clutch size declines with latitude are typically less pronounced in raptors than other species but the analyses of average clutches from Montagu's, European, and Hen Harriers in Europe (Chapter 9) indicated significant declines when samples were limited to the western parts of Europe. This was because longitude also influences clutches, increasing them from east to west. Similar declines are evident from the southern hemisphere, including the Australasian Swamp and African Marsh Harrier. From the experimental food supplements reported here it is apparent that clutch sizes of African harriers are partially limited by early-season food availability (independent of laying date), but they rise to a ceiling (4.0 eggs) below that recorded for temperate harriers. The difference can be ascribed to the inability of African harriers to rear more than four young, and the depressed growth rates of nestlings in experimental nests with increased broods. In determining whether the inability to rear more young was due to birds withholding reproductive effort (adaptation) or because of ecological limits (constraint), three lines of evidence suggested constraint: (i) African Marsh Harriers foraged as long as European birds yet acquired half as much food biomass per unit foraging time; (ii) they rarely increased their provisioning rate to increased broods and could not provide the same biomass per chick even when provisioning did increase; and (iii) African nestlings typically received less biomass than that delivered to temperate harrier nestlings (ruling out adaptive clutch reduction to increase chick quality). So neither the withholding of reproductive effort nor decreasing brood size to increase provisioning to individual chicks was apparent.

Two predictions of Ashmole's fluctuating-food hypothesis were supported: (i) resources were lower in subtropical regions than temperate ones as evidenced by subtropical harriers spending twice as long capturing each prey item as temperate species. This result was common to several other raptors from subtropical and temperate regions. And (ii) there was a trend for subtropical species to breed at higher densities as predicted by the lower mortality expected at lower latitudes.

A major alternative, the Skutch–clutch hypothesis, posits that parents reduce the number of foraging trips and the number of young begging in the nest by building small nests and laying fewer eggs to reduce the probability of predation in tropical areas. Hence clutch size is reduced in tropical passerines. This was tested by assessing results apparent from 32 food supplementation experiments. Some species responded by laying extra eggs while others did not. Because the differences were not explained by different responses from passerines vs. non-passerines or subtropical vs. temperate species, and some tropical species do lay larger clutches when supplemented, there was little support for this hypothesis. The trend that did explain the difference between the responders (clutch size increased) and non-responders (clutch size unaffected) was that birds that did not lay larger clutches were double-brooded, while those that responded by laying larger clutches were single-brooded. If a species can only rear one brood per year, then it is unsurprising that this species puts the increased food resources into laying more eggs. One would expect double-brooded species, however, to put the increased resources into second or third clutches. These trends are sufficient to explain why Skutch found that neotropical passerines can apparently rear more young than they typically nourish—tropical birds put resources into two or more nests, and not into a single breeding attempt (the two-basket hypothesis). In this respect tropical species differ most from temperate-breeding ones. In most other respects the rules that govern breeding in temperate species are identical to those in subtropical areas. A prediction of this two-basket hypothesis is that supplemented tropical passerines should not lay larger clutches, and may not even lay earlier (as found in temperate species with shorter breeding seasons), but the number of clutches should increase significantly over non-fed controls. This has not been tested, even though it is known that, given larger broods, subtropical species do not suffer higher predation levels. Numerous experiments are possible with tropical and subtropical birds to test the conclusion that breeding season length and reduced food resources, rather than predation, is selecting for the reduced clutches of low latitude birds.

There are numerous aspects of harrier and raptor behavioural ecology that have not been explored in this book, but I hope the sampler offered has given an idea of what riches, hinted at by Darwin, may be found by exploring the natural laws and selection pressures acting on any group of species. In some cases I have explored behavioural and ecological traits that I find intrinsically interesting in the harriers, while at other times I have used the harriers as models to test ideas and hypotheses of general applicability. Many questions have been raised in the preceding chapters and a number of them have been summarized here. I hope that some of these, and the many aspects of harrier ecology that I have *not* covered, will stimulate other ecologists, north and south of the equator, to spend more time untangling this particularly rich bank.

References

Alatalo, R. V., Lundberg, G. A., Stahlbrandt, K., and Ulfstrand, S. (1981). The conflict between male polygamy and female monogamy: the case of the pied flycatcher, *Ficedula hypoleuca*. *American Naturalist* **117**, 738–753.

Alatalo, R. V., Lundberg, G. A., and Stahlbrandt, K. (1984). Female mate choice in the Pied Flycatcher *Ficedula hypoleuca*. *Behavioural Ecology and Sociobiology* **14**, 253–261.

Alatalo, R., Hoglund, J., and Lundberg, A. (1991). Lekking in the Black grouse—a test of male viability. *Nature* **352**, 155–156.

Ali, S. and Ripley, S. D. (1968). *Handbook of the birds of India and Pakistan*. Vol. 1. Divers to hawks. Oxford University Press, London.

Altenburg, W., Daan, S., Starkensberg, J., and Zijlstra, M. (1982). Polygamy in the marsh harrier, *Circus aeruginosus*: individual variations in hunting performance and number of mates. *Behaviour* **79**, 272–312.

Altenburg, W., Bruinenberg-Rinsma, J., Wildschut, P., and Zijlstra, M. (1987). Colonization of a new area by the Marsh Harrier. *Ardea* **75**, 213–220.

Altmann, S. (1974). Observational study of behaviour: sampling methods. *Behaviour* **49**, 227–267.

Altmann, S. A., Wagner, S. S., and Lenington, S. (1977). Two models for the evolution of polygyny. *Behavioural Ecology and Sociobiology* **2**, 397–410.

Alvarez, F. (1993). Alertness signalling in two rail species. *Animal Behaviour* **46**, 1229–1231.

Amadon, D. (1959). The significance of sexual differences in size among birds. *Proceedings of the American Philosophical Society* **103**, 531–536.

Amadon, D. (1978). Remarks on the taxonomy of some Australian raptors. *Emu* **78**, 115–118.

Amadon, D. and Bull, J. (1988). Hawks and owls of the world, a distributional and taxonomic list. *Proceedings of the Western Foundation of Vertebrate Zoology* **3**, 295–327.

Anderson, C., Kok, O. B., and Erasmus, J. (1999). Diet, body mass and body condition of Lesser Kestrels in South Africa. *Ostrich*. (In press.)

Andersson, M. (1982a). Female choice selects for extreme tail length in a widow bird. *Nature* **299**, 818–820.

Andersson, M. (1982b). Sexual selection, natural selection and quality advertisement. *Biological Journal of the Linnean Society* **17**, 375–393.

Andersson, M. (1994). *Sexual selection*. Princeton University Press.

Andersson, M. and Eriksson, M. O. G. (1982). Nest parasitism in goldeneyes *Bucephala clangula*: some evolutionary aspects. *American Naturalist* **120**, 1–16.

Andersson, M. and Norberg, R. A. (1981). Evolution of reversed sexual dimorphism and role partitioning among predatory birds, with a size scaling of flight performance. *Biological Journal of the Linnean Society* **15**, 105–130.

Ankney, C. D. and McInnes, C. D. (1978). Nutrient reserves and reproductive performance of female Lesser Snow Geese. *Auk* **95**, 459–471.

Arcese, P. and Smith, J. N. M. (1988). Effects of population density and supplemental food on reproduction in Song Sparrows. *Journal of Animal Ecology* **57**, 119–136.

Armstrong, E. A. (1955). *The Wren.* Collins, London.

Armstrong, W. A. (1947). *Bird Display and Behaviour.* Drummond, London.

Arroyo, B. E. (1995). *Breeding ecology and nest dispersion of Montagu's Harrier Circus pygargus in central Spain.* Unpublished D.Phil. thesis, University of Oxford.

Arroyo, B. (1996a). Reproductive success of Montagu's Harrier (*Circus pygargus*) and Hen Harrier (*Circus cyaneus*) in an agricultural habitat of central Spain. In *Biology and conservation of Mediterranean raptors.* SEO/Birdlife (Monografia No. 4), Madrid.

Arroyo, B. (1996b). A possible case of polyandry in Montagu's Harrier. *Journal of Raptor Research* **30**, 100–102.

Arroyo B. E. (1997). Diet of Montagu's Harrier *Circus pygargus* in central Spain: analysis of temporal and geographic variation. *Ibis* **139**, 664–672.

Arroyo, B. E. (1999). Copulatory behaviour of semi-colonial Montagu's Harriers. *Condor.* (In press.)

Arroyo. B. E., King, J. R., and Palomares, L. E. (1995). Observations on the ecology of Montagu's and Marsh Harriers wintering in north-west Senegal. *Ostrich* **66**, 37–39.

Arroyo, B., Bretagnolle, V., and Mougeot, F. (1998a). Nest defence in Montagu's Harrier *Circus pygargus*: assessing the effects of coloniality, predation risk and offspring survival prospects. Abstract V World Conference on Birds of Prey and Owls. World Working Group on Birds of Prey and Owls, South Africa.

Arroyo, B., Leroux, A., and Bretagnolle, V. (1998b). Patterns of egg and clutch size variation in the Montagu's Harrier. *Journal of Raptor Research.* (In press.)

Arroyo, B., Salamolard, M., and Bretagnolle, V. (in press). Environmental and parental components in egg and clutch size variation in a raptor. *Condor.* (In press.)

Ashmole N. P. (1963). The regulation of numbers of tropical oceanic birds. *Ibis* **103b**, 458–473.

Askenmo C. (1977). Effects of addition and removal of nestlings on nestling weight, nestling survival, and female weight loss in the Pied Flycatcher *Ficedula hypoleuca* (Pallas). *Ornis Scandinavica* **8**, 1–8.

AOU. (1982). Thirty-fourth supplement to the AOU checklist of North American birds. *Auk* **99**, 1–16.

Baker, J. A. and Brooks, R. J. (1981). Distribution patterns of raptors in relation to density of voles. *Condor* **83**, 42–47.

Baker-Gabb, D. J. (1979). Remarks on the taxonomy of the Australasian Harrier (*Circus approximans*). *Notornis* **26**, 325–329.

Baker-Gabb, D. J. (1981). Breeding behaviour and ecology of the Australasian Harrier (*Circus approximans*) in the Manawatu-Rangitikei sand country, New Zealand. *Notornis* **28**, 103–119.

Baker-Gabb, D. J. (1982). *Comparative ecology and behaviour of Swamp Harriers Circus approximans and Spotted Harriers C. assimilis and other raptors in Australia and New Zealand.* Unpublished D.Phil. thesis, Monash University, Victoria.

Baker-Gabb, D. J. (1983). Observations on the mating system and breeding success of Marsh Harriers in coastal south-eastern Australia. *Corella* **7**, 109–113.

Baker-Gabb, D. J. (1984). The evolution of tree-nesting and the origin of the Spotted Harrier. *Corella* **8**, 67–69.

Baker-Gabb, D. J. (1986). Ecological release and behavioural and ecological flexibility in marsh harriers on islands. *Emu* **86**, 71–81.

Bakst, M. R. and Bird, D. M. (1987). Localization of oviductal sperm-storage tubules in the American kestrel (*Falco sparverius*). *Auk* **104**, 321–324.

Balfour, E. (1957). Observations on the breeding biology of the hen harrier in Orkney. Il. *Bird Notes* **27**, 216–224.

Balfour, E. (1962–63). The Hen Harrier in Orkney. Part 3. Courtship, display and sociability. *Bird Notes* **30**, 145–153.

Balfour, E. and Cadbury, C. J. (1975). A population study of the Hen Harrier, *Circus cyaneus* in Orkney. In *The natural environment of Orkney* (ed. R. Goodier). Pp.122–128. Nature Conservancy Council, Edinburgh.

Balfour, E. and Cadbury, C. J. (1979). Polygyny, spacing and sex ratio among Hen Harriers *Circus cyaneus* in Orkney, Scotland. *Ornis Scandinavica* **10**, 133–141.

Balfour, E. and Macdonald, M. A. (1970). Food and feeding behaviour of the Hen Harrier in Orkney. *Scottish Birds* **6**, 157–166.

Balgooyen, T. G. (1976). Behaviour and ecology of the American Kestrel (*Falco sparverius L.). University of California Publications in Zoology* **103**,1–85.

Barnard, P. E. (1983). *Foraging behaviour and energetics of northern harriers* Circus cyaneus (L.). Unpublished BSc (Hons) thesis, Acadia University.

Barnard, P. (1987). Foraging site selection by three raptors in relation to grassland burning in a montane habitat. *African Journal of Ecology* **25**, 35–45.

Barnard, P. E. (1990). Male tail length, sexual display intensity and female sexual response in a parasitic African finch. *Animal Behaviour* **39**, 652–656.

Barnard, P. E. (1995). Scientific research tradition and collaboration in tropical ecology. *Trends in Ecology and Evolution* **10**, 38–39.

Barnard, P. and Markus, M. B. (1989). Male copulation frequency and female competition for fertilizations in a promiscuous brood parasite, *Vidua macroura. Ibis* **131**, 421–425.

Barnard, P. and Simmons, R. (1986). The function of leg-lowering in territorial predatory birds. *Ostrich* **57**, 107–109.

Barnard, P., MacWhirter, R. B., Simmons, R. E., Hansen, G. L., and Smith, P. C. (1987). Timing of breeding and seasonal importance of passerine prey to Northern Harriers (*Circus cyaneus*). *Canadian Journal of Zoology* **65**, 1942–1950.

Barth, E. (1964). Supplement til fokstumyras fuglefauna. *Sterna* **6**, 49–74.

Bavoux, C. Burneleau, G. Leroux, A. & Nicolau-Guillaumet, P. (1988). Le Busard des Roseax *Circus a. aeruginosus* en Charente-Maritime (France). I—Presentation—Determination biometrique du sexe. *Alauda* **56**, 246–260.

Bavoux, C., Burneleau, G., Leroux, A., and Nicolau-Guillaumet, P. (1989). Le Busard des Roseax *Circus a. aeruginosus* en Charente-Maritime (France). *Alauda* **57**, 247–262.

Bavoux, C, Burneleau, G., Nicolau-Guillaumet, P, and Picard, M. (1995). Le Busard des Roseax *Circus a. aeruginosus* en Charente-Maritime (France). VIII Attachment au site de reproduction et appatements. *Alauda* **61**, 273–280.

Bavoux, C., Burneleau, G., Nicolau-Guillaumet, P., and Picard, M. (1994). Le Busard des Roseaux *Circus a. aeruginosus* en Charente-Maritime (France). VII—Deplacements et activite journaliere des adultes en hiver. *Alauda* **62**, 281–288.

Beacham, T. D. (1979). Selectivity of avian predation in declining populations of the vole *Microtus townsendii. Canadian Journal of Zoology* **57**, 1767–1772.

Bechard, M. J. (1982). Effect of vegetative cover on foraging site selection by Swainson's Hawk. *Condor* **84**, 153–159.

Bednarz, J. C. (1987). Pair and group reproductive success, polyandry, and cooperative breeding in Harris' Hawks. *Auk* **104**, 393–404.

Beecher, M. D. and Beecher, I. M. (1979). Sociobiology of bank swallows: reproductive strategy of the male. *Science* **205**, 1282–1285.

Beehler, B. M., Pratt T. K., and Zimmerman, D. A. (1986). *Birds of New Guinea.* Princeton University Press.

Beissinger, S. R. (1990). Experimental brood manipulations and the mono-parental threshold in Snail Kites. *American Naturalist* **136**, 20–38.

Beissinger, S. R. and Snyder, N. F. R. (1987). Mate desertion in the Snail Kite. *Animal Behaviour* **35**, 477–487.

Belanger, R. (1981). *The influence of rain on the behaviour of small mammals, muskrats (Onadatra zibethica) and their predators.* Unpublished PhD thesis, McGill University, Montreal.

Bengtson, S.-A. (1967). Observations on the reproductive success in 26 nests of the Marsh Harrier (*Circus aeruginosus*) in Ska[o]ne Province, Sweden. *The Oologists' Record* **61**, 23–28.

Bensch, S. and Hasselquist, D. (1992). Evidence for active choice in a polygynous warbler. *Animal Behaviour* **44**, 301–311.

Benson, C. W., and Benson F. M. (1977). *The birds of Malawi.* Mountfort Press. Limbe, Malawi.

Benson, C. W., Brooke, R. K., and Vernon, C. J. (1964). Bird breeding data for the Rhodesias [Zambia, Zimbabwe] and Nyasaland [Malawi]. *Occasional Papers of the Natural History Museum of Southern Rhodesia.*

Berger, J. (1996). Animal behaviour and plundered mammals: is the study of mating systems a scientific luxury or a conservation necessity? *Oikos* **77**, 207–216.

Bider, J. R. (1968). Animal activity in uncontrolled terrestrial communities as determined by a sand transect technique. *Ecological Monographs* **8**, 69–308.

Bildstein, K. L. (1979). Behavior of a pair of Northern Harriers on the day of nest abandonment. *Inland Bird Banding News* **51**, 63–65.

Bildstein, K. (1987). Behavioural ecology of Red-tailed Hawks (*Buteo jamaicensis*), Rough-legged Hawks (*B. lagopus*), Northern Harriers (*Circus cyaneus*), and American Kestrels (*Falco sparverius*) in south central Ohio. *Ohio Biological Survey* **18**, 1–53.

Bildstein, K. and Collopy, M. W. (1985). Escorting flight and agonistic interactions in wintering Northern Harriers. *Condor* **87**, 398–401.

Bildstein, K. L. and Collopy, M. (MS). Geographic variation in the hunting behaviour of Northern Harriers (*Circus cyaneus*) in North America. Paper presented at Raptor Research Foundation first European meeting 1993. Hawk and Owl Trust, Canterbury, UK.

Bildstein, K. and Gollop, J. B. (1988). Northern Harrier *Circus cyaneus*. In *Handbook of North American birds* (ed. R. S. Palmer), Vol. 4. Yale University Press, New Haven.

Bildstein, K. L. and Hamerstrom, F. (1980). Age and sex differences in the size of Northern Harriers. *Journal of Field Ornithology* **51**, 356–360.

Bird, D. M. and Lague, P. C. (1977). Semen production of the American Kestrel. *Canadian Journal of Zoology* **55**, 1351–1358.

Birkhead, T. R. (1979). Timing and duration of mate guarding in magpies *Pica pica*. *Animal Behaviour* **30**, 277–283.

Birkhead, T. R. (1988.) Behavioural aspects of sperm competition in birds. *Advances in the Study of Behaviour* **18**, 35–72.

Birkhead, T. R. (1991a). Sperm depletion in the Bengalese finch *Lonchura striata*. *Behavioural Ecology* **2**, 267–275.

Birkhead, T. R. and Lessells, C. M. (1988). Copulation behaviour of the Osprey *Pandion haliaetus*. *Animal Behaviour* **36**, 1672–1682.

Birkhead, T. R. and Møller, A. P. (1987). Avian mating games. *New Scientist*, Dec. 1986, 34–36.

Birkhead, T. R. and Møller, A. P. (1992). *Sperm competition in birds*. Academic, London.

Birkhead, T. R., Atkin, L., and Møller, A. P. (1987). Copulation behaviour of birds. *Behaviour* **101**, 101–138.

Birkhead, T. R., Pellat, J., and Hunter, F. M. (1988). Extra-pair copulation and sperm competition in the zebra finch. *Nature, London* **334**, 60–62.

Birkhead, T. R., Stanback, M. T., and Simmons, R. E. 1993. The phalloid organ of Buffalo Weavers *Bubolornis*. Ibis **135**, 326–331.

Birney, E. C., Grant, W. E., and Baird, D. D. (1976). Importance of vegetative cover to cycles of *Microtus* populations. *Ecology* **57**, 1043–1057.

Blanco, G., Harrera, M. A., Fargallo, J. A., and Cuevas, J. A. (1993). Female participation in courtship displays of Western Marsh Harriers (*Circus aeruginosus*) in central Spain. *Journal of Raptor Research* **27**, 165–166.

Blomqvist, D., Johansson, O. C., Unger, O., Larsson, M., and Flodin, L.-A. (1997). Male aerial display and reversed sexual size dimorphism in the Dunlin. *Animal Behaviour* **54**, 1291–1299.

Bó, M. S., Cicchino, S. M., and Martiniz, S. M. (1996). Diet of Long-winged Harrier (*Circus buffoni*) in southeastern Buenos Aires Province, Argentina. *Journal of Raptor Research* **30**, 237–239.

Bó, M. S., Cicchino, S. M., Caceres, R. M and Martinez, S. M. (MS). Diet of breeding Cinereous Harriers (*Circus cinereus*) in southeastern Buenos Aires Province, Argentina. Unpublished, Universidad Nacional de mar del Plata, Argentina.

Bock, W. F. (1979). Zur situation der Rohrweihe (*Circus aeruginosus*) in Schleswig-Holstein. *Journal für Ornithologie* **120**, 416–430.

Boersma, P. D. (1982). Why some birds take so long to hatch. *American Naturalist* **120**, 733–750.

Bolton, M., Houston, D., and Monaghan, P. (1992). Nutritional constraints on egg formation in the lesser black-backed gull: an experimental study. *Journal of Animal Ecology* **61**, 521–532.

Bond, W., Ferguson, M., and Forsyth, G. (1980). Small mammals and habitat structure along altitudinal gradients in the southern Cape mountains. *South African Journal of Zoology* **15**, 34–43.

Bortolotti, G. R. (1986). Evolution of growth rates in eagles: sibling competition vs. energy considerations. *Ecology* **67**, 182–194.

Bortolotti, W. R. (1986). Sex ratios of fledgling Golden Eagles. *Auk* **106**, 520–521.

Bosque and Bosque (1995). Nest predation as a selective factor in the evolution of developmental rates in altricial birds *American Naturalist* **145**, 234–260.

Bowers, M. A. and Dooley, J. L. (1993). Predation hazard and seed removal by small mammals: microhabitat versus patch scale effects. *Oecologia* **94**, 247–254.

Bowland, A. E. (1987). The effect of wind on small mammal trapping. *Lammergeyer* **38**, 35–39.

Boyce, M. S. and Perrins, C. M. (1987). Optimizing Great Tit clutch size in a fluctuating environment. *Ecology* 68, 142–153.

Bray, O. E., Kennelly, J. K., and Guarino, J. L. (1975). Fertility of eggs produced on territories of vasectomized red-winged blackbirds. *Wilson Bulletin* **87**, 187–195.

Breckenridge, W. J. (1935). An ecological study of some Minnestoa Marsh Hawks. *Condor* **37**, 268–276.

Bretagnolle, V., Thiollay, J.-M., and Attie, C. (1998a). Status of Reunion Marsh Harrier *Circus maillardi* on Reunion Island. *Proceedings of the V World Conference on Birds of Prey*. Pretoria.

Bretagnolle, V., Arroyo, B., and Mougeot, F. (1998b). Nest defence in Montagu's Harrier: assessing the effects of coloniality, predation risk and offspring survival prospects. In *Proceedings of 22nd International Ornithological Congress, Durban (*ed. N. J. Adams and R. H. Slotow). *Ostrich* **69**, 243.

Brooker, M. G., Rowley, I., Adams, M., and Baverstock, P. R. (1990). Promiscuity: an inbreeding avoidance mechanism in a socially monogamous species? *Behavioural Ecology and Sociobiology* **26**, 191–199.

Brooks, P. M. (1974). *The ecology of the four-striped mouse Rhabdomys pumilio (Sparmann* **1784***) with particular reference to a population in the van Riebeck Nature Reserve, Pretoria*. Unpublished D.Sc. thesis, Pretoria University.

Brown, C. J. (1988). *A study of the Bearded Vulture* Gypaetus barbatus *in southern Africa*. Unpublished Ph.D. thesis, University of Natal, RSA.

Brown, L. H. (1971). The relations of the Crowned Eagle *Stephanoeatus coronatus* and some of its prey animals. *Ibis* **113**, 240–243.

Brown, L. H. and Amadon, D. (1968). *Eagles, hawks and falcons of the world*. Wellfleet, Secaucus, NJ.

Brown, L. H. and Pomeroy, D. E. (1984). The age structure of populations of wild birds in tropical Africa as demonstrated by plumage characteristics and marking techniques. *Proceedings of the V Pan African Ornithology Conference*, pp. 97–119.

Brown, L. H. and Brown, C. H. (1984). Is food the main determinant of African birds' breeding seasons? *Proceedings of the Vth Pan African Ornithological Congress* 753–761.

Brown, L. H., Urban, E., and Newman, K. (1982). *The birds of Africa.* Vol. 1. Academic, London.

Bryson, B. (1996). *Notes from a small island.* Black Swan, London.

Buczek, T. (1994). An unusual case of breeding cooperation in the Marsh Harrier. *Notatki Ornithologiczne* **35**, 3–4.

Burke, C. J. (1979). *Effect of prey and land use on mating systems of harriers.* Unpublished M.Sc. thesis. University of Wisconsin.

Burley, N. (1986). Sexual selection for aesthetic traits in species with biparental care. *American Naturalist* **127**, 415–445.

Burneleau, B. G., Leroux, A., and Nicolai-Guillaumet, P. (1989). Le Busard des Roseaux *Circus a. aeruginosus* en Charente-Maritime (France). *Alauda* **57**, 247–262.

Butet, A. and Leroux, A. (1993). Effect of prey on a predator's breeding success. A 7-year study on common vole (*Microtus arvalis*) and Montagu's harrier (*Circus pygargus*) in a West France marsh. *Acta Oecologia* **14**, 857–865.

Cáceres, R. M. (1996). *Ecologia troficade dos especies de gavilanes (Circus buffoni y Circus cinereus) en la reserva municipal Mar Chiquita (Buenos Aires Prov.) durante el periodo de cria.* Licenciatura en Ciencias Biologicas. University Nacional de Mar del Plata.

Cade, T. J. (1982). *The falcons of the world.* Comstock/Cornell Press, New York.

Calford, M. B., Wise, L. Z., and Pettigrew, J. D. (1985). Coding of sound location and frequency in the auditory midbrain of diurnal birds of prey, families Accipitridae and Falconidae. *Journal of Comparative Physiology A* **157**, 149–160.

Carlson, A. (1989). Courtship feeding and clutch size in Red-backed Shrikes (*Lanius collurio*). *American Naturalist* **133**, 454–457.

Carlsson, B.-G., Hornfeldt, B., and Lofgren, O. (1987). Bigyny in Tengmalm's Owl *Aegolius funereus*: effect of mating strategy on breeding success. *Ornis Scandinavica* **18**, 237–243.

Caro, T. M. (1986). The functions of stotting in Thomson's Gazelles: some tests of the predictions. *Animal Behaviour* **34**, 663–684.

Castano, J. P. (1997). Fenelogia de puesta y paramestros reproductivos en una poblacion de Aguilucho cenizo (*Circus pygargus*) en el Campo de Montiel. *Ardeola* **44**, 166–175.

Catchpole, C., Leisler, B., and Winkler, H. (1985). Polygyny in the Great Reed Warbler, *Acrocephalus arundinaceus:* a possible case of deception. *Behavioural Ecology and Sociobiology* **16**, 285–291.

Catry, P., Phillips, R. A. and Furness, R.W. (1999). Evolution of reversed size dimorphism in skuas and jaegars. *Auk* **116**, 158–168.

Cave, A. J. (1968). The breeding of the Kestrel *Falco tinnunculus* in the reclaimed area Oostelijk Flevoland. *Netherlands Journal of Zoology* **18**, 313–407.

Chadwick, P. I. (1997). Breeding biology by Black Harriers in the West Coast National Park South Africa. *Journal of African Raptor Biology* **12**, 14–19.

Charnov, E. L. and Krebs, J. R. (1974). On clutch size and fitness. *Ibis* **116**, 217–219.

Cheke, A. S. (1985). The ecology of surviving native land-birds of Reunion. In *Studies of Mascarene island birds* (ed. A. W. Diamond), pp. 301–358. Cambridge University Press.

Cheng, K. M., Burns, S. T., and McKinney, F. (1983). Forced copulations in captive mallards III. Sperm competition. *Auk* **110**, 302–310.

Choate, T. S. (1972). Behavioural studies of some Rhodesian rodents. *Zoologica Africana* 7, 103–118.

Christian, D. P. (1977). Diurnal activity of the four-striped mouse Rhabdomys pumilio. *Zoologica Africana* 12, 238–239.

Clark, A. B. and Wilson, D. S. (1981). Avian breeding adaptations: hatching asynchrony, brood reduction and nest failure. *Quarterly Review of Biology* **56**, 253–277.

Clark, R. J. and Stanley, B. L. (1976). Facial feathers of the harrier (*Circus cyaneus hudsonius*), Longeared Owl (*Asio otus*) and Short-eared Owl (*Asio flammeus*) compared. *Proceedings of the Pennsylvania Academy of Sciences* **50**, 86–88.

Clark, W. S. (1987). The dark morph of the [European] Marsh Harrier. *British Birds* **80**, 61–72.

Clark, W. S. (1999). Display flights of the Black Harrier in the southern Cape South Africa. *Journal of African Raptor Biology*. (In press.)

Clark, W. S. (MS). A dark morph for the African Marsh Harrier. (*Journal of African Raptor Biology*).

Clarke, J. A. (1983). Moonlight's influence on predator/prey interactions between short-eared owls (*Asio flammeus*) and deermice (*Peromyscus maniculatus*). *Behavioural Ecology and Sociobiology* **13**, 205–209.

Clarke, R. (1995). *The Marsh Harrier.* Hamlyn, London.

Clarke, R. (1996). *Montagu's Harrier.* Arlequin Press, Chelmsford, UK.

Clarkson, K. (1984). *The breeding and feeding ecology of the Magpie* Pica pica. Unpublished D.Phil. thesis. University of Sheffield.

Clancey, P. A. (1964). *The birds of Natal and Zululand.* Oliver and Boyd, London.

Clouet, M. (1978). Le busard de Maillard (*Circus aeruginosus maillardi*) de L'ile de la Reunion. *L'Oiseauet R. F. O.* **48**, 95–106.

Clutton-Brock, T. H. (1991). *The evolution of parental care.* Princeton University Press.

Clutton-Brock, T. and Godfray, C. (1991). Parental investment. In *Behavioural ecology: an evolutionary approach* (ed. J. R. Krebs and N. B. Davies), pp. 234–262. Blackwell Scientific, Oxford.

Clutton-Brock, T. and Harvey, P. H. (1980). Primates, brains and ecology. *Journal of Zoology (London)* **190**, 309–323.

Coates, B. J. (1985). *The birds of Papua New Guinea.* Vol. 1. Dove, Queensland.

Cockburn, A. (1992). Maternal infanticide optimises sex allocation in Antechinuses. *Abstract 4th Behavioural Ecology Congress*, Princeton, NJ.

Cody, M. L. (1966). A general theory of clutch size. *Evolution* 20, 174–184.

Cody, M. L. (1971). Ecological aspects of reproduction. In *Avian biology* (ed. D. S. Farner and J. R. King), pp. 461–512. Vol. 1. Academic, New York.

Collopy, M. W. (1986). Food consumption and growth energetics of nestling Golden Eagles. *Wilson Bulletin* **98**, 44–58.

Collopy, M. W. and Bildstein, K. L. (1987). Foraging behaviour of Northern Harriers wintering in southeastern salt and freshwater marshes. *Auk* **104**, 11–16.

Collopy, M. W. and Edwards, T. C. (1989). Territory size, activity budget, and role of undulating flight in nesting Golden Eagles. *Journal of Field Ornithology* **60**, 43–51.

Cooke, F., Findlay, C.S., and Rockwell, R.F. (1984). Recruitment and the timing of reproduction in Lesser Snow Geese (*Chen caerulescens caerulescens*). *Auk* **101**, 451–458.

Corbacho, C., Sanchez, J. M., and Sanchez, A. (1997). Breeding biology of Montagu's Harrier *Circus pygargus* L. in agricultural environments of southwest Spain; comparison with other populations in the western Palearctic. *Bird Study* **44**, 107–114.

Cormier, J. P. (1983). Etude preliminaire de quelques aspects de la biologie de la reproduction du Busard Saint Martin *Circus cyaneus* L. En Anjou. *L'Oiseau et R. F. O.* **53**, 241–249.

Cormier, J. P. (1985). La reproduction du Busard cendre, *Circus pygargus* L. dans deux sites de l'ouest de la France. *L'Oiseau et R. F. O.* **55**, 107–114.

Cormier, J. P. (1989). Selection des sites de reproduction chez *Circus cyaneus* (L.). *L'Oiseaux et R. F. O.* **59**, 45–55.

Cormier, J.-P. (1990). Une cas d'aide a l'elevage des jeunes de le part d'un male de deux aus chez le busard cendre *Circus pygargus* (L.). *AlaudaI* **58**, 203–204.

Cormier, J. P. and Baillon, F. (1991). Concentration de busards cendres *Circus pygargus* dans la region de M'bour (Senegal) durant l'hiver 1988–1989: utilisation du milieu et regime alimentaire. *Alauda* **59**, 163–168.

Coulson, J. C. (1966). The influence of pair bond and age on the breeding biology of the Kittiwake Gull *Rissa tridactyla*. *Journal of Animal Ecology* **35**, 269–279.

Cowling, R. M. and Hilton-Taylor, C. (1994). Patterns of plant diversity and endemism in South Africa: an overview. In *Botanical diversity in southern Africa* (ed. B. J. Huntley). *Strelitzia* **1**, 31–52.

Craig, T. H., Craig, E. H., and Marks, J. S. (1982). Aerial talon-grappling in Northern Harriers. *Condor* **84**, 239.

Craighead, J. J. and Craighead, F. C. (1969). *Hawks owls and wildlife*. Dover, New York.

Cramp, S. and Simmons, K. E. L. (ed.) (1980). *The birds of the western Palearctic*. Oxford University Press.

Cresswell, W. (1994). Song as a pursuit-deterrent signal, and its occurrence relative to other anti-predator behaviours of skylark (*Alauda arvensis*) on attack by merlins (*Falco columbarius*). *Behavioural Ecology and Sociobiology* **34**, 217–223.

Crook, J. H. (1964). The evolution of social organisation and visual communication in the weaver birds (Ploceinae). *Behaviour* **10** (Supplement), 1–178.

Cupper, J. and Cupper, L. (1981). *Hawks in focus*. Jaclin Enterprises, Mildura, Australia.

Curtis, B. A. and Perrin, M. R. (1979). Food preferences of the vlei rat (*Otomys irroratus*) and the four-striped mouse (*Rhabdomys pumilio*). *South African Journal of Zoology* **14**, 224–229.

Daan, S. and Dijkstra, C. (1982). Annual cycles of breeding in the Kestrel: individual tuning to vole availability. *Ibis* **124**, 389–390.

Daan, S. and Raptor Group (1982). Timing of vole hunting in aerial predators. *Mammal Review* **12**, 169–181.

Daan, S. and Slopsema, S. (1978). Short term rhythms in foraging behaviour of the common vole, *Microtus arvalis*. *Journal of Comparative Physiology* **127**, 215–227.

Daan, S., Dijkstra, C., Drent, R., and Meijer, T. (1989). Food supply and the annual timing of avian reproduction. *Acta XIX Ornithological Congress*. University of Ottawa Press.

Daan, S., Dijkstra, C., and Tinbergen, J. M. (1990). Family planning in the kestrel (*Falco tinnunculus*): the ultimate control of covariation of laying date and clutch size. *Behaviour* **114**, 83–116.

Daan, S., Dijkstra, C., and Weissing, F. J. (1996). An evolutionary explanation for seasonal trends in avian sex ratios. *Behavioural Ecology* **7**, 426–430.

Darwin, C. (1871). *The descent of man and selection in relation to sex*. Random House, New York.

David, J. H. M. and Jarvis, J. U. M. (1985). Population fluctuations, reproduction and survival in the Striped field mouse *Rhabdomys pumilio* on the Cape Flats, South Africa. *Journal of Zoology London* **207**, 251–276.

Davies, N. B. (1992). *Dunnock behaviour and social evolution*. Oxford University Press.

Davies, N. B. and Lundberg, A. (1985). The influence of food on time budgets and timing of breeding of the dunnock (*Prunella modularis*). *Ibis* **127**, 100–110.

Davies, R. A. G. and Ferguson, J. W. M. (1998). The influence of predation by Black Eagles *Aquila verreauxii* on Rock Hyrax numbers in the arid Karoo. *Proceedings of Vth World Conference on Birds of Prey*. Pretoria.

Davis, M. R. (1973). The ecology and life history of the Vlei Rat *Otomys irroratus* (Brants, 1827) on the van Riebeck Nature Reserve, Pretoria. Unpublished D.Sc. thesis. University of Pretoria.

Davygora, A. V. and Belik, V. P. (1994). The Pallid Harrier *Circus macrourus* as an endangered species in the Palearctic. In *Raptor conservation today* (ed. B.-U. Meyburg and R. D. Chancellor). pp. 93–96 World Working Group on Birds of Prey/Pica Press Berlin.

Dawkins, R. and Krebs, J. R. (1979). Arms races between and within species. *Proceedings of the Royal Society B* **205**, 489–511.

Decker, M. D., Parker, P. G., Minchella, D. J., and Rabenold, K. N. (1993). Monogamy in black vultures: genetic evidence from DNA fingerprinting. *Behavioural Ecology* **4**, 29–35.

Delannoy, C. A. and Cruz, A. (1988). Breeding biology of the Puerto Rican Sharp-shinned Hawk (*Accipiter striatus venator*). *Auk* **105**, 649–662.

Del Hoyo, J. L., Elliot, and Sargatal, (1994). *The handbook of the birds of the world*. Vol. 2. New World Vultures to Guineafowl. Lynx Edicions, Barcelona.

Desey, B. A., Batzil, G. O., and Liu, J. (1990). Effects of food and predation on behaviour of prairie voles: a field experiment. *Oikos* **58**, 159–168.

Deubbert, H. F. and Lokemoen, J. T. (1977). Upland nesting of American Bitterns, Marsh Hawks and Short-eared Owls. *Prairie Naturalist* **9**, 33–40.

Dewsbury, D. A. (1982). Ejaculate cost and male choice. *American Naturalist* **119**, 601–610.

Diamond, J. (1985). Evolution of ecological segregation in the New Guinea montane avifauna. In *Community ecology* (ed. T. Case and J. M. Diamond), pp. 98–125. Harper and Row, New York.

Dickman, C. R., Predavec, M., and Lynam, A. J. (1991). Differential predation of size and sex classes of mice by the barn owl, *Tyto alba*. *Oikos* **62**, 67–76.

Dijkstra, C. (1988). *Reproductive tactics in the kestrel Falco tinnunculus*. PhD thesis, Rijksuniversitiet Groningen, The Netherlands.

Dijkstra, C., Vuursteen, L., Daan, S., and Masman, D. (1982). Clutch size and laying date in the kestrel, *Falco tinnunculus*: effect of supplementary food. *Ibis* **124**, 210–213.

Dijkstra, C., Daan, S., Meijer, T., Cave, A. J., and Foppen, R. P. B. (1988). Daily and seasonal variations in body mass of the Kestrel in relation to food availability and reproduction. *Ardea* **76**, 127–140.

Dijkstra, C., Daan, S., and Buker, J. B. (1990). Adaptive seasonal variation in the sex ratio of kestrel broods. *Functional Ecology* **4**, 143–147.

Dittami, J. P. (1987). A comparison of breeding and moult cycles and life histories in two tropical starling species. *Ibis* **129**, 69–85.

Divis, T. (1982). [Number of Marsh Harriers and its population density and dynamics (*Circus aeruginosus* L.) in the region of Nachod.] Report on activities of Group for Research and Protection of Birds of Prey and Owls in Czechoslovakia in 1982. [Czech, English summary.]

Donazar, J. A. (1990). Geographic variation in clutch and brood size of the Eagle Owl *Bubo bubo* in the western Palearctic. *Journal of Ornithology* 131, 439–443.

Donazar, J. A., Negro, J. J., and Hiraldo, F. (1992). Functional analysis of mate feeding in the Lesser Kestrel *Falco naumanni*. *Ornis Scandinavica* **23**, 190–194.

Donazar, J. A., Ceballos, O., and Tella, J. L. (1994). Copulation behaviour in the Egyptian Vulture *Neophron percnopterus*. *Bird Study* **41**, 37–41.

Drent, R. and Daan, S. (1980). The prudent parent: energetic adjustments in avian breeding. *Ardea* **68**, 225–252.

Duebbert, H. F. and Lokemoen, J. T. (1977). Upland nesting of American Bitterns, Marsh Hawks and Short-eared Owls. *Prairie Naturalist* **9**, 33–40.

Dunn, P. O. and Hannon, S. J. (1991). Intraspecific competition and the maintenance of monogamy in Tree Swallows. *Behavioural Ecology* **2**, 258–266.

du Plessis, M. A., Siegfried, W. R., and Armstrong, A. J. (1995). Ecological and life-history correlates of cooperative breeding in South African birds. *Oecologia* **102**, 180–188.

Dyrcz, A. (1986). Factors affecting facultative polygyny and breeding results in the Great Reed Warbler (*Acrocephalus arundinaceus*). *Journal für Ornithologie* **127**, 447–461.

Earle, R. A. (1981). Factors governing avian breeding in *Acacia* savanna, Pietermaritzburg, Part 1: extrinsic factors. *Ostrich* **52**, 65–73.

Edwards, A. W. F. (1997). Natural selection and the sex ratio. *Nature* **389**, 780.

Elliot, P. F. (1975). Longevity and the evolution of polygyny. *American Naturalist* **109**, 281–287.

Ellis, D. H. and Powers, L. (1982). Mating behaviour in the golden eagle in non-fertilisation contexts. *Raptor Research* **16**, 134–136.

Emlen, S. T. and Oring, L. W. (1977). Ecology, sexual selection and the evolution of mating systems. *Science* **197**, 215–223.

Endler, J. A. (1986). *Natural Selection in the Wild.* Monographs in population biology. Princeton University Press.

Endler, J. A. (1991). Interactions between predators and prey. In *Behavioural ecology an evolutionary approach* (ed. J. R. Krebs and N. B. Davies), pp. 169–196. Blackwell Scientific, Oxford.

Erickson, M. O. G. (1985). Prey detectability for fish-eating birds in relation to fish density and water transparency. *Ornis Scandinavica* **16**, 1–7.

Erickson, M. O. G. (1986). Fish delivery, production of young, and nest density of Ospreys (*Pandion haliaetus*) in southwest Sweden. *Canadian Journal of Zoology* **64**, 1961–1965.

Erickson, C.J. and Zenone, P.G. (1976). Courtship differences in male ring doves: avoidance of cuckoldry? *Science* **192**, 1353–1354.

Erlinge, S., Goransson, G., Hogstedt, G., Jansson, G., Liberg, O., Loman, I., *et al.* (1984). Can vertebrate predators regulate their prey? *American Naturalist* **123**, 125–133.

Errington, P. L. (1946). Predation and vertebrate populations. *Quarterly Review of Biology* **21**, 144–177 and 221–245.

Estes, R. D. (1991). *The behaviour guide to African mammals.* University of California Press, Berkeley.

Ewald, P. W. and Rohwer, S. A. (1982). Effects of supplemental feeding on timing of breeding, clutch-size and polygyny in Red-winged Blackbirds *Agelaius phoeniceus. Journal of Animal Ecology* **51**, 429–450.

Faaborg, J., de Vries, T., Patterson, C.B. and Griffin, C.R. (1980). Preliminary observations of the occurrence and evolution of polyandry in the Galapagos Hawk (*Buteo galapagoensis*). *Auk* **97**, 581–590.

Falconer, D. S. (1981). *Introduction to quantitative genetics.* 2nd edn. Longman, New York.

Faralli, U. (1994). Breeding biology, habitat selection and conservation of Montagu's Harrier *Circus pygargus* in the northern Apennines, Italy. In *Raptor conservation today* (ed. B.-U. Meyburg and R. D. Chancellor), pp. 97–101. World Working Group on Birds of Prey/Pica Press, Berlin.

Fefelov, I. G. (1996). The Eastern Marsh Harrier *Circus aeruginosus spilonotus* in the Selenga River delta. *Russian Journal of Ornithology* **5**, 41–46. [Russian, English abstract.]

Fernandez, C. and Azkona, P. (1992). Sexual differences in conspecific territorial defence of Marsh Harriers (*Circus aeruginosus*). *Raptor Research* **28**, 23–26.

Festing, M. F. and Nordskog, A. W. (1967). Response to selection for body weight and egg weight in chickens. *Genetics* **55**, 219–231.

Figuerola, J. (1999). A comparative study on the evolution of reversed size dimorphism in monogamous waders. *Biological Journal of the Linnean Society* **67**, 1–18.

Fisher, R. A. (1930). *The genetical theory of natural selection.* Clarendon, Oxford.

Fitch, M. A. and Shugart, G. W. (1983). Comparative biology and behaviour of monogamous pairs and one male-two female trios of Herring Gulls. *Behavioural Ecology and Sociobiology* **14**, 1–7.

Forbes, L. S. (1991). Insurance offspring and brood reduction in a variable environment: the costs and benefits of pessimism. *Oikos* **62**, 325–332.

Ford, N. L. (1983). Variation in mate fidelity in monogamous birds. In *Current ornithology* (ed. R. F. Johnston). Vol. 1. pp. 329–356. Plenum, New York.

Fry, C. H. (1980). Survival and longevity among tropical land birds. *Proceedings IV Pan African Ornithological Congress*, pp. 333–343.

Furness, R. W. and Furness, B. (1981). A technique for estimating the hatching dates of eggs of unknown laying date. *Ibis* **123**, 98–102.

Gans, C. and Huey, R. B. (ed.) (1988). *Biology of the reptilia*. Liss, New York.

Gard, N. W. & Bird, D. M. (1990). Breeding behavior of American Kestrels raising manipulated brood sizes in years of varying prey abundance. *Wilson Bull.* **102**, 605–614.

Gargett, V. (1990). *The Black Eagle, a study.* Acorn and Russel Friedman, Randburg, RSA.

Geer, T. (1981). Factors affecting the delivery of prey to nestling Sparrowhawks (*Accipiter nisus*). *J. Zool., Lond.* **195**, 71–80.

Getz, L. L. (1961). Factors influencing the local distribution of Microtus and Synaptomys in southern Michigan. *Ecology* **42**, 110–119.

Getz, L. L. (1968). Influences of weather on the activity of the Red backed vole. *Journal of Mammalogy* **49**, 565–570.

Getz, L. L. (1971). Microclimate, vegetative cover and local distribution of the Meadow Vole. *Transactions of the Illinois Academy of Science* **64**, 9–21.

Giraud-Audine, M. and Pineau, J. (1974). Nidification du Busard des Roseaux *Circus aeruginosus harterti* en milieu non aquatique dans la region de Tanger. *Alauda* **42**, 281–288.

Gibb, H. L., Weatherhead, P. J., Boag, P. T., White, B. N., Tabak, L. M., and Hoysak, D. J. (1990). Realized reproductive success of polygynous red-winged blackbirds revealed by DNA markers. *Science* **250**, 1394–1397.

Gibson, R. M. and Bradbury, J. W. (1985). Sexual selection in lekking grouse: phenotypic correlates of male strutting success. *Behavioural Ecology and Sociobiology* **18**, 117–123.

Glutz von Blotzheim, U. N., Bauer, K., and Bezzel, E. (1971). Handbuch der Vogel Mitteleuropas. Vol. 4. Akademische Verlagsgesselschaft, Frankfurt am Main.

González, J. L. (1991). *El Aguilucho lagunero* Circus aeruginosus (*L.*, 1748*) en España. Situacion, biologia de la reproduccion, alimentacion y conservacion. Colecc. Tec.* ICONA, Madrid.

Gotmark, F. (1992). The effects of investigator disturbance on nesting birds. *Current Ornithology* **9**, 63–104.

Gould, S. J. (1971). Geometric similarity in allometric growth: a contribution to the problem of scaling in the evolution of size. *American Naturalist* **105**, 113–136.

Gould, S. J. and Lewontin, R. C. (1979). The spandrels of San Marco and the Panglossian paradigm: a critique of the adaptationists programme. *Proceedings of Royal Society London B* **205**, 581–598.

Gowaty, P. A. and Karlin, A. A. (1984). Multiple maternity and paternity in single broods of apparently monogamous eastern bluebirds (*Sialia sialis*). *Behavioural Ecology and Sociobiology* **15**, 91–95.

Gowaty, P. A. and Droge, D. L. (1992). Sex ratio conflict and the evolution of sex-biased provisioning in birds. In *Acta XX Congressus Internationalis Ornithologici*, 932–945.

Grant, B. R. and Grant, P. R. (1989). *Evolutionary dynamics of a natural population. The large Cactus Finch of the Galapagos*. Chicago University Press.

Grant, P. R. (1986). *Ecology and evolution of Darwin's finches*. Princeton University Press.

Graves, J., Whiten, A., and Henzi, P. (1984). Why does the Herring Gull lay three eggs? *Animal Behaviour* **32**, 798–805.

Green, D. J. and Krebs, E. A. (1995). Courtship feeding in Osprey *Pandion haliaetus*: a criterion for mate assessment? *Ibis* **137**, 35–43.

Greene, E. (1990). Individuals in an Osprey colony discriminate between high and low quality information. *Nature* **329**, 239–241.

Gregory, R. D., Keymer, A. E., and Harvey, P. H. (1991). Life history, ecology and parasite community structure in Soviet birds. *Biology Journal of the Linnean Society* **43**, 249–262.

Gustafsson, L. (1987). Lifetime reproductive success and heritability: empirical support of Fisher's Fundamental Theorem. *American Naturalist* **128**, 761–764.

Hagen, Y. (1969). Norwegian studies on the reproduction of birds of prey and owls in relation to micro-rodent population fluctuations. *Fauna* **22**, 73–126.

Hahn, D. C. (1981). Asynchronous hatching in the laughing gull: cutting losses and reducing rivalry. *Animal Behaviour* **29**, 421–427.

Hale, W. G. and Ashcroft, R. P. (1982). Pair formation and pair maintenance in the Redshank *Tringa totanus*. *Ibis* **124**, 471–490.

Hale, W. G. and Ashcroft, R. P. (1983). Studies of the courtship behaviour of the Redshank *Tringa totanus*. *Ibis* **125**, 3–23.

Halle, S. (1993). Diel pattern of predation risk in microtine rodents. *Oikos* **68**, 510–518.

Hamerstrom, F. (1963). The use of Great Horned Owls in catching Marsh Hawks. *Proceedings of the XIII International Ornithological Congress*, pp. 866–869.

Hamerstrom, F. (1968). Ageing and sexing harriers. *Inland Bird Banding News* **40**, 143–146.

Hamerstrom, F. (1969). A harrier population study. In *Peregrine Falcon populations: their biology and decline* (ed. J. J. Hickey), pp. 367–385. University of Wisconsin Press.

Hamerstrom, F. (1979). Effect of prey on predators: voles and harriers. *Auk* **96**, 370–374.

Hamerstrom, F. (1986). *Harrier, hawk of the marshes*. Smithsonian Institute Press, Washington.

Hamerstrom, F., Hamerstrom, F. N., and Burke, C. J. (1985). Effect of voles on mating systems of harriers in a central Wisconsin population of harriers. *Wilson Bulletin* **97**, 332–346.

Hamilton, W. D. (1971). Geometry of the selfish herd. *Journal of theoretical biology* **7**, 1–52.

Hamilton, W. D. and Zuk, M. (1982). Heritable true fitness and bright birds: a role for parasites? *Science* **218**, 384–387.

Hansson, L. and Henttonen, H. (1985). Gradients in density variations of small rodents: the importance of latitude and snow cover. *Oecologia* **67**, 394–402.

Harmata, A. R. (1982). What is the function of undulating flight display of Golden Eagles? *Raptor Research* **16**, 103–109.

Harrison, J. A., Allan, D. G., Underhill, L. G., Herremans, M., Tree, A. J., Parker, V., and Brown, C. J. (ed.) (1997). *The atlas of southern African birds*. Vols 1 and 2. Birdlife South Africa, Cape Town.

Hartley, R. and Hustler, K. (1993). A less-than-annual breeding cycle in a pair of African Bat Hawks *Macheiramphus alcinus*. *Ibis* **135**, 456–458.

Hays, C. (1971). Essai sur la biologie de reproduction du busard cendre (*Circus pygargus*) dans le Morbihan. *Ar Vran* **4**, 1–15.

Hecht, W. R. (1951). Nesting of the Marsh Hawk at Delta, Manitoba. *Wilson Bulletin* **63**, 167–176.

Heisler, I. L. (1981). Offspring quality and the polygyny threshold: a new model for the 'sexy-son' hypothesis. *American Naturalist* **117**, 316–328.

Henschel, J.R, David, J. H. M. and Jarvis, J. U. M. (1982). Age determination and age structure of a striped fieldmouse, *Rhabdomys pumilio*, population from the Cape Flats. *South African Journal of Zoology* **17**, 136–142.

Heredia, R. and Donazar, J. A. (1990). High frequency of polyandrous trios in an endangered population of Lammergeyers *Gypaetus barbatus* in northern Spain. *Biological Conservation* **53**, 163–171.

Herklots, G. A. C. (1961). *The birds of Trinidad and Tobago*. Collins, London.

Hesse, R., Allee, W. C., and Schmidt, K. P. (1937). *Ecological animal geography*. Wiley, New York.

Hill, G. E. (1991). Plumage colouration is a sexually selected indicator of male quality. *Nature* **350**, 337–339.

Hill, W. L. (1988). The effect of food abundance on the reproductive patterns of coots. *Condor* **90**, 324–331.

Hillstrom, L. (1992). *Patterns of hatching and adult body mass change in the Pied Flycatcher Ficedula hypoleuca: adaptations and constraints*. Unpublished D.Phil. thesis, Uppsala University, Sweden.

Hirons, G. J. M., Hardy, A. R., and Stanley, P. I. (1984). Body weight, gonadal development and moult in the Tawny Owl. *Journal of Zoology London* **202**, 145–164.

Hochachka, W. M. (1988). The effect of food supply on the composition of Black-billed Magpie eggs. *Canadian Journal of Zoology* **66**, 692–695.

Hochachka, W. M. and Boag, D. A. (1987). Food shortage for breeding black-billed magpies (*Pica pica*): An experiment using supplemental food. *Canadian Journal of Zoology* **65**, 1270–1274.

Högstedt, G. (1981a). Should there be a positive or negative correlation between survival of adults in a bird population and their clutch size? *American Naturalist* **118**, 568–571.

Högstedt, G. (1981b). Effect of additional food on reproductive success in the Magpie (*Pica pica*). *Journal of Animal Ecology* **50**, 219–229.

Holtzhuijzen, M. A. (1992). Frequency and timing of copulations in the Prairie Falcon. *Wilson Bulletin* **104**, 333–338.

Hornfeldt, B. and Eklund, U. (1990). The effect of food on laying date and clutch size in Tengmalm's Owl. *Ibis* **132**, 395–406.

Horsfall, J. A. (1984). Food supply and egg mass variation in the European Coot. *Ecology* **65**, 89–95.

Horvath, L. (1955). Red-footed falcons in Ohat-woods, near Hortobagy. *Acta Zoologica Academia of Sciences Hungary* 1: 245–287.

Hosking, E. J. (1943). Some observations on the Marsh Harrier. *British Birds* **37**, 2–10.

Houston, D.C. (1990). Control of breeding in African vulture. *Proceedings of the III world conference on birds of prey.*

Howe, H. F. (1976). Egg size, hatching asynchrony, sex and brood reduction in the Common Grackle. *Ecology* **57**, 1195–1207.

Hussell, D. J. T. (1985). Clutch size, day length and seasonality of resources: comments on Ashmole's hypothesis. *Auk* **102**, 632–634.

Il'ichev, V. D., Bogoslovskaya, L. S., Barsova, L. L. (1974) Neuronal organisation and homology of avian auditory centres. *Doklady Biologia Sciences* **218**, 1249–1252.

Iribarren, J. J. (1975). Biologia del Aguila Calzada (*Hieraeetus pennatus*) durante el periodo de nidisicacion en Navarra. *Ardeola* **21**, 305–320.

Jamieson, I.G., McRae, S.B. Simmons, R.E. and Trewby, M. (2000). High rates of conspecific brood parasitism and egg rejection in coots and moorhens breeding in ephemeral wetlands in Namibia. *Auk* **117**, 250–255.

Janes, S. W. (1985). Habitat selection in raptorial birds. In *Habitat selection in birds* (ed. M. L. Cody), pp. 159–188. Academic, London.

Janzen, D. H. (1980). When is it co-evolution? *Evolution* **34**, 611–612.

Jarvinen, A. and Väisanen, R. A. (1983). Egg size and related reproductive traits in a southern passerine *Ficedula hypoleuca* breeding in an extreme northern environment. *Ornis Scandinavica* **14**, 253–262.

Jarvinen, A. and Ylimaunu, J. (1986). Intraclutch egg-size variation in birds: physiological responses of individuals to fluctuations in environmental conditions. *Auk* **103**, 235–237.

Jehl, J. R. and Murray, B. G. (1986). The evolution of normal and reverse sexual size dimorphism in shorebirds and other birds. *Current Ornithology* **3**, 1–86.

Jenkins, A. R. (1991). Latitudinal prey productivity and potential density in the Peregrine Falcon. *Gabar* **6**, 20–24.

Jennions, M. D. and Macdonald, D. W. (1994). Cooperative breeding in mammals. *Trends in Ecology and Evolution* **9**, 89–93.

Jimenez, J. E. and Jaksic, F. M. (1988). Ecology and behavior of southern South American Cinereous Harriers, *Circus cinereus*. *Revista Chilena de Historia Natural* **61**, 199–208.

Jones, C. G. (1989). The aerial display of the Reunion Harrier. *Gabar* **4**, 22–23.

Jones, P. J. (1973). *Some aspects of the feeding ecology of the Great Tit (Parus major* L.). Unpublished D.Phil. thesis, University of Oxford.

Jones, P.J. and Ward, P. (1976). The level of reserve protein as the proximate factor controlling the timing of breeding and clutch size in the Redbilled Quelea, *Quelea quelea*. *Ibis* **118**, 547–574.

Jover, L., Ruiz, X., and Gonzales-Martin, M. (1993). Significance of intra-clutch egg size variation in the Purple Heron. *Ornis Scandinavica* **24**, 127–134.

Kallander, H. (1974). Advancement of laying of Great Tits by the provision of food. *Ibis* **116**, 365–376.

Kallander, H. and Karlsson, J. (1993). Supplemental food and laying date in the European Starling. *Condor* **95**, 1031–1034.

Karlsson, J. (1983). *Breeding of the Starling (Sturnus vulgaris)*. Unpublished D. Phil. thesis, University of Lund, Sweden.

Karr, J.R., Nichols, J.D., Klimkiewicz, M.K. and Brawn, J.D. (1990). Survival rates of birds in tropical and temperate forests: will the dogma survive? *American Naturalist* **136**, 277–291.

Kear, J. and Duplaix-Hall, N. (ed.) (1975). *Flamingos*. T. and A. D. Poyser, Berkhamsted, England.

Keith, L. B., Todd, A. W., Brand, C. J., Adamcik, R. S., and Rusch, D. H. (1977). An analysis of predation during a cyclic fluctuation of snowshoe hares. *Proceedings XIIIth Congress on Game Biology*, pp. 151–175.

Kemp, A.C. (1990). Timing of breeding by Greater Kestrels *Falco rupicoloides*, near Pretoria, South Africa. *Ostrich* **60**, 35–39.

Kemp, A. C. (1995). A comparison of hunting behaviour by each sex of adult Greater Kestrels *Falco rupicoloides* resident near Pretoria, South Africa. *Ostrich* **66**, 21–33.

Kemp, A. C. (1995). *The Hornbills*. Oxford University Press.

Kemp, A. C. and Dean, R. (1988). Diet of African Marsh Harrier from pellets. *Gabar* **3**, 54–55.

Kempenaers, B., Verheyen, G. R., van den Broeck, M., Burke, T., van Broeckhoven, C., and Dhondt, A. A. (1992). Extra-pair paternity results from female preference for high quality males in the Blue Tit. *Nature* **357**, 494–496.

Kempenaers, B., Verheyen, G. R., and Dhondt, A. A. (1997). Extra-pair paternity in the Blue Tit (*Parus caeruleus*): female choice, male characteristics, and offspring quality. *Behavioural Ecology* **8**, 481–492.

Kingdon, J. (1978). *East African Mammals: an atlas of evolution in Africa*. Vol. IIB. Academic Press, London.

Kirkpatrick, M. (1985). Evolution of female choice and male parental care in polygynous species: the demise of the 'sexy-son'. *American Naturalist* **125**, 788–810.

Klomp H. (1970). The determination of clutch-size in birds. A review. *Ardea* **58**, 1–96.

Knight, R. L. (1988). Effect of supplementary food on the breeding biology of the Black-billed Magpie. *Condor* **90**, 956–958.

Knudsen, E. I. and Konishi, M. (1979). Sound localization in the Barn Owl (*Tyto alba*): measures with the search coil technique. *Journal of Comparative Physiology* **133**, 1–11.

Kodric-Brown, A. and Brown, J. H. (1984). Truth in advertising: the kinds of traits favoured by sexual selection. *American Naturalist* **124**, 309–323.

Koenig, W. D. (1982). Ecological and social factors affecting hatchability of eggs. *Auk* **99**, 526–536.

Koga, K. and Shiraishi, S. (1994). Copulation behaviour of the Black Kite *Milvus migrans* in Nagasaki Peninsula. *Bird Study* **41**, 29–36.

Korpimäki, E. (1984). Population dynamics of birds of prey in relation to fluctuations in small mammal populations in western Finland. *Annals Zoologica Fennici* **21**, 287–293.

Korpimäki, E. (1985). Rapid tracking of microtine populations by their avian predators: possible evidence for stabilising selection. *Oikos* **45**, 281–284.

Korpimäki, E. (1987). Timing of breeding in Tengmalm's Owl *Aegolius funereus* in relation to vole dynamics in western Finland. *Ibis* **129**, 58–68

Korpimäki, E. (1988). Factors promoting polygyny in European birds of prey—a hypothesis. *Oecologia* **77**, 278–285.

Korpimäki, E. (1989). Breeding performance of Tengmalm's Owl *Aegolius funereus*: effects of supplementary feeding in a peak vole year. *Ibis* **131**, 51–56.

Korpimäki, E. (1991). Poor reproductive success of polygynously mated female Tengmalm's owls: are better options available? *Animal Behaviour* **41**, 37–47.

Korpimäki, E. (1992). Fluctuating food abundance determines lifetime reproductive success of male Tengmalm's Owls. *Journal of Animal Ecology* **61**, 103–111.

Korpimäki, E. and Hakkarainen, H. (1991). Fluctuating food supply affects the clutch size of Tengmalm's Owls independent of laying date. *Oecologia* **85**, 543–552.

Korpimäki, E. and Hakkarainen, H. (1992). Blood parasites reduce fecundity in female Tengmalm's Owls. Abstract *4th International Behavioral Ecology Congress*, University of Princeton.

Korpimäki, E., Lahti, K., May, C. A., Parkin, D. T., Powell, G. B., Tolonen, P., and Wetton, J. H. (1996). Copulatory behaviour and paternity determined by DNA fingerprinting in kestrels: effects of cyclic food abundance. *Animal Behaviour* **51**, 945–955.

Kotler, B. P., Brown, J. S., and Hasson, O. (1991). Factors effecting gerbil behaviour and rates of owl predation. *Ecology* 72, 2249–2260.

Krackow, S. (1993). Note on falconiform sex ratio given in Olsen and Cockburn 1991: avian raptors exhibit no unique sex ratio bias. *Behavioural Ecology and Sociobiology* **32**, 429–430.

Krebs, J. R. (1971). Territory and breeding density in the Great Tit, *Parus major* L. *Ecology* **52**, 2–22.

Krebs, J. R. and Kacelnik, A. (1991). Decision-making. In *Behavioural ecology an evolutionary approach* (ed. J. R. Krebs and N. B. Davies), pp. 105–136. Blackwell Scientific, Oxford.

Kruuk, H. (1964). Predators and anti-predator behaviour of the Black headed Gull. (*Larus ridibundus*). *Behaviour* **11**, 1–129.

Kuleza, G. (1990). An analysis of clutch size in New World passerine birds. *Ibis* **132**, 407–422.

Lack, D. (1940). Courtship feeding in birds. *Auk* **57**, 169–178.

Lack, D. (1947). The significance of clutch size I. *Ibis* **89**, 302–352.

Lack, D. (1954). *The natural regulation of animal numbers*. Oxford University Press.

Lack, D. (1968). *Ecological adaptations for breeding in birds*. Methuen, London.

Lake, P. E. (1975). Gamete production and the fertile period with particular reference to domesticated birds. *Symposium Zoological Society of London* **35**, 225–244.

Langston, N., Rohwer, S., Gori, D. (1997). Experimental analyses of intra- and inter-sexual competition in red-winged blackbirds. *Behavioral Ecology* **8**, 524–533.

Lendrum, A. (1979). The Augur Buzzard in the Matopos Rhodesia. *Ostrich* **50**, 203–214.

Lenington, S. (1980). Female choice and polygyny in Redwinged Blackbirds. *Animal Behaviour* **28**, 347–361.

Leonard, M. L. (1990). Polygyny in Marsh Wrens: asynchronous settlement as an alternative to the polygyny threshold model. *American Naturalist* **136**, 446–458.

Leroux, A. and Bretagnolle, V. (1996). Sex ratio variations in broods of Montagu's Harriers *Circus pygargus*. *Journal of Avian Biology* **27**, 63–69.

Lessells, K. and Boag, P. T. (1987). Unrepeatable repeatabilities: a common mistake. *Auk* **104**, 116–121.

Lidicker, W. Z. (1975). The role of dispersal in the demography of small mammals. In *Small mammals: their productivity and population dynamics* (ed. F. B. Golley, K. Petrusewicz, and L. Ryszkowski), pp. 103–128. Cambridge University Press.

Lill, A. (1974). The evolution of clutch-size and male 'chauvinism' in the Whitebearded Manakin. *Living Bird* **13**, 211–231.

Lima, S. L. and Dill, L. M. (1990). Behavioural decisions made under the risk of predation: a review and prospectus. *Canadian Journal of Zoology* **68**, 619–640.

Lindén, M and Møller, A. P. (1989). Cost of reproduction and covariation of life history traits in birds. *TREE* **4**, 367–371.

Linden, H. and Wikman, M. (1983). Goshawk predation on Tetraonids: availability of prey and diet of the predator in the breeding season. *Journal of Animal Ecology* **52**, 953–968.

Linner, S. C. (1980). *Resource partitioning in breeding populations of Marsh Hawks and Short-eared Owls*. Unpublished M.Sc. thesis. Utah State University, Logan, Utah.

Locati, M. (1990). Female Chamois defends kids from eagle attack. *Mammalia* **54**, 155–156.

Lockie, J. D. (1955). The breeding habits and food of Short-eared Owls after a vole plague. *Bird Study* **2**, 53–69.

Longland, W. S. and Price, M. V. (1991). Direct observations of owls and heteromyd rodents: can predation risk explain microhabitat use? *Ecology* **72**, 2261–2273.

Lundberg, A. and Alatalo, R. V. (1992). *The Pied Flycatcher*. T. and A. D. Poyser. Calton, England.

Lundevall, C. F. and Rosenberg, E. (1954). Some aspects of the behaviour and breeding biology of the Pallid Harrier (*Circus macrourus*). *Proceedings of XI International Ornithological Congress*, pp. 599–603. Basel, Switzerland.

Maclean, G.L. (1976). Factors governing breeding of African birds in non-arid habitats. *Proceedings 16th International Ornithological Congress* 258–271.

MacWhirter, R. B. (1985). *Breeding ecology, prey selection and provisioning strategies of Northern Harriers Circus cyaneus (L.)*. Unpublished BSc. Honours thesis, Mount Allison and Acadia Universities, Wolfville, Nova Scotia.

MacWhirter, R. B. (1994). *Offspring sex ratios, mortality and relative provisioning of sons and daughters in the Northern Harrier* (*Circus cyaneus*). Unpublished D.Phil. thesis. Ohio State University, Columbus.

MacWhirter, R. B. and Bildstein, K. L. (1996). Northern Harrier *Circus cyaneus*. In *The Birds of North America (*ed. A. Poole and F. Gill), No. 210, The Academy of Natural Sciences, Philadelphia and the American Ornithologists Union, Washington D.C.

Mader, W. J. (1982). Ecology and breeding habits of the Savanna Hawk in the Ilanos of Venezuela. *Condor* **84**, 261–271.

Magrath, R. D. (1989). Hatching asynchrony and reproductive success in the blackbird. *Nature* **339**, 536–538.

Magrath, R. D. (1990). Hatching asynchrony in altricial birds. *Biological Review* **65**, 587–622.

Magrath, R. D. (1992). Seasonal changes in egg-mass within and among clutches of birds: general explanations and a field study of the Blackbirds *Turdus merula*. *Ibis* **134**, 171–179.

Malan, G., Crowe, T. M., Biggs, R., and Herholdt, J. J. (1997). The social system of the Pale Chanting Goshawk *Melierax canorus*; monogamy v polyandry and delayed dispersal. *Ibis* **139**, 313–321.

Marchant, S. and Higgins, P. J. (1993). *Handbook of Australian, New Zealand and Antarctic birds*. Vol. 2. Oxford University Press.

Martelli, D. and Sandri, V. (1991). Status ed ecologia dell'Abanella minore (*Circus pygargus*) in Emilia Romagna. Analisi conclusiva. In: SROPU (ed.) Atti V Conv. Ital. Ornit. *Ricerche di Biologia della Selvaggina*, *XVII*, 49–52.

Martin, T. E. (1987). Food as a limit on breeding birds: a lifehistory perspective. *Annual Review of Ecology and Systematics* **18**, 453–487.

Martin, T. E. (1995). Avian life history evolution in relation to nest sites, nest predation and food. *Ecological Monographs* **65**, 101–127.

Martin, T. E. (1996). Life history evolution in tropical and south temperate birds: what do we really know? *Oikos* **27**, 263–272.

Masman, D. (1986). The annual cycle of the kestrel *Falco tinnunculus*. Unpublished D.Phil. thesis, University of Groningen, The Netherlands.

Masman, D., Daan, S., and Dijkstra, C. (1988). Time allocation in the Kestrel (*Falco tinnunculus*), and the principle of energy minimization. *Journal of Animal Ecology* **57**, 411–432.

Mather, M. H. and Robertson, R. J. (1992). Honest advertisement in flight displays of Bobolinks (*Dolichonyx oryzivorus*). *Auk* **109**, 869–873.

McKinney, F., Cheng, K. M., and Bruggers, D. J. (1984). Sperm competition in apparently monogamous birds. In *Sperm competition and the evolution of animal mating systems (*ed. R. L. Smith), pp. 523–545. Academic, New York.

Mclleny, E. A. (1940). Sex ratio in birds. *Auk* **57**, 85–93.

Meijer, T., Daan, S., and Dijkstra, C. (1988). Female condition and reproduction: effects of food manipulation in free-living and captive Kestrels. *Ardea* **76**, 141–154.

Meijer, T., Masman, D., and Daan, S. (1989). Energetics of reproduction in female Kestrels. *Auk* **106**, 549–559.

Mendelsohn, J. M. (1982). The feeding ecology of the Blackshouldered Kite *Elanus caeruleus*. *Durban Museum Novitates* **13**, 75–116.

Mendelsohn, J. M. (1986). Why are females big, or males small? *Gabar* **1**, 22–26.
Mendelsohn, J. M. and Jaksic, F. M. (1989). Hunting behaviour of Blackshouldered Kites in the Americas, Europe, Africa and Australia. *Ostrich* **60**, 1–12.
Meyburg, B.-U. (1974). Sibling aggression and mortality among nestling eagles. *Ibis* **116**, 224–228.
Mock, D. W. and Parker, G. A. (1997). *The evolution of sibling rivalry.* Oxford University Press.
Mock, D. W., Drummond, H., and Stinson, C. H. (1990). Avian siblicide. *American Scientist* **78**, 438–449.
Møller, A. P. (1982). Clutch size in relation to nest size in the swallow *Hirundo rustica*. *Ibis* **124**, 339–343.
Møller, A. P. (1986). Mating systems among European passerines: a review. *Ibis* **128**, 234–250.
Møller, A. P. (1987a). House sparrow, *Passer domesticus*, communal displays. *Animal Behaviour* **35**, 203–210.
Møller, A. P. (1987b). Copulation behaviour in the goshawk, *Accipiter gentilis*. *Animal Behaviour* **35**, 755–763.
Møller, A. P. (1988). Female choice selects for male sexual tail ornaments in the monogamous swallow. *Nature* **332**, 640–642.
Møller, A. P. (1990a). Nest predation selects for small nest size in the Blackbird. *Oikos* **57**, 237–240.
Møller, A. P. (1990b). Effects of a Haematophagous mite on the Barn Swallow (*Hirundo rustica*): a test of the Hamilton and Zuk hypothesis. *Evolution* **44**, 771–784.
Møller, A. P. (1991a). Influence of wing and tail morphology on the duration of song flight in skylarks. *Behavioural Ecology and Sociobiology* **28**, 309–314.
Møller, A. P. (1991b). Viability is positively related to degree of ornamentation in male swallows. *Proceedings of the Royal Society of London B* **243**, 145–148.
Møller, A. P. (1993). Parasites increase the cost of reproduction in their hosts. *Journal of Animal Ecology* **62**, 309–322.
Møller, A. P. (1996). *Sexual selection and the Barn Swallow.* Oxford University Press.
Møller, A. P. and Birkhead, T. R. (1993a). Certainty of paternity covaries with paternal care in birds. *Behavioural Ecology and Sociobiology* **33**, 261–268.
Møller, A. P. and Birkhead, T. R. (1993b). Cuckoldry and sociality: a comparative study of birds. *American Naturalist* **142**, 118–140.
Møller, A. P. and de Lope, F. (1993). Differential costs of a secondary sexual character: an experimental test of the handicap principle. *Evolution.* (In press.)
Monaghan, P., Bolton, M., and Houston, D. C. (1995). Egg production constraints and the evolution of avian clutch size. *Proceedings of the Royal Society of London B* **259**, 189–191.
Monaghan, P., Nager, R. G., and Houston, D. C. (MS). The price of eggs: increased investment in egg production reduces the offspring rearing capacity of parents.
Moreau, R. E. (1944). Clutch-size: a comparative study, with special reference to African Birds. *Ibis* **86**, 286–347.
Moss, R. and Watson, A. (1982). Heritability of egg size, hatch weight, body weight, and viability in Red Grouse (*Lagopus lagopus scoticus*). *Auk* **99**, 683–686.

Moss, W. W. and Camin, J. H. (1970). Nest parasitism, productivity and clutch size in Purple Martins. *Science* **168**, 1000–1003.

Mougeot, F., Arroyo, B., and Bretagnolle, V. (1998). Testing the 'mate-guarding' and 'frequent copulation' paternity assurance strategies in three semi-colonial raptors. In *Proceedings of the 22nd International Ornithological Congress, Durban (*ed. N. J. Adams and R. H. Slotow). *Ostrich* **69**, 261.

Mougeot, F., Arroyo, B. and Bretagnolle, V. (MS). An experimental test of paternity assurance strategies in a semi-colonial raptor, the Montagu's Harrier *Circus pygargus*. CNRS, France.

Mueller, H. C. and Meyer, K. (1985). The evolution of reversed sexual dimorphism in size: a comparative analysis of the Falconiformes of the Western Palearctic. In *Current Ornithology (*ed. R. F. Johnston) **2**, pp. 65–101. Plenum, New York.

Munoff, J. A. (1963). Food habits, growth and mortality in nesting marsh hawks. *Kingbird* **13**, 67–74.

Mundy, P.J. and Cook A.W. (1975). Hatching and rearing of two chicks by the Hooded Vulture. *Ostrich* **46**, 45–50.

Mundy, P. J., Piper. S. E., Ledger, J., and Butchart, D. (1992). *The vultures of Africa.* Russel Friedman, Johannesburg.

Murray, B. G. (1985). Evolution of clutch size in tropical species of birds. In *Neotropical Ornithology*, Ornithological Monographs, No. 36 (ed. M. S. Foster *et al.*), pp. 505–519. American Ornithologists Union, Washington D.C.

Murray, G. A. (1976). Geographic variation in the clutch sizes of seven owl species. *Auk* **93**, 602–613.

Murphy, E. C. and Haukioja, E. (1986). Clutch size in nidicolous birds. *Current Ornithology* **4**, 141–210.

Murphy, M. T. (1986). Body size and condition, timing of breeding, and aspects of egg production in Eastern Kingbirds. *Auk* **103**, 465–476.

Murton, R. K. and N. J. Westwood. (1977). *Avian breeding cycles.* Clarendon, Oxford.

Narayan, G. and Rosalind, L. (1990). New record of the Pied Harrier *Circus melanoleucos* (Pennant) breeding in Assam duars, with a brief review of its distribution. *Journal of the Bombay Natural History Society* **88**, 28–34.

Negro, J. J. and Hiraldo, F. (1991). Sex ratios in broods of Lesser Kestrels *Falco naumanni. Ibis* **134**, 190–191.

Negro, J. J., Donazar, J. A. and Hiraldo, F. (1992). Copulatory behaviour in a colony of Lesser Kestrels: sperm competition and mixed reproductive strategies. *Animal Behaviour* **43**, 921–930.

Negro, J. J., Villarroel, M., Tella, J. L., Kuhnlein, U., Hiraldo, F., Donazar, J. A., and Bird, D. M. (1996). DNA fingerprinting reveals a low incidence of extra-pair fertilizations in the Lesser Kestrel. *Animal Behaviour* **51**, 935–943.

Nelson, J. B. (1978). *The Sulidae: gannets and boobies.* Oxford University Press.

Neufeldt, I. A. (1964). Notes on the nidification of the Pied Harrier, *Circus melanoleucos* (Pennant), in Amurland, U. S. S. R. *Journal of the Bombay Natural History Society* **64**, 284–306.

Newton, I. (1976b). Breeding of Sparrowhawks (*Accipiter nisus*) in different environments. *Journal of Animal Ecology* **45**, 831–849.

Newton, I. (1976a). Population limitation in diurnal raptors. *Canadian Field Naturalist* **90**, 274–300.

Newton, I. (1978). Feeding and development of sparrowhawk *Accipiter nisus* nestlings. *Journal of Zoology (London)* **184**, 465–487.

Newton, I. (1979). *Population ecology of raptors.* T. and A. D. Poyser, Berkhamsted, England.

Newton, I. (1985). Lifetime reproductive output of female sparrowhawks. *Journal of Animal Ecology*, **54**, 241–253.

Newton, I. (1986). *The Sparrowhawk.* T. and A. D. Poyser, Berkamsted, England.

Newton, I. (1989). *Lifetime reproduction in birds.* Academic Press, London.

Newton, I. (1993). Predation and limitation of bird numbers. *Current Ornithology* **11**, 143–198.

Newton, I. and Marquiss, M. (1979). Sex ratio among nestlings of the European Sparrowhawk. *American Naturalist* **113**, 309–315.

Newton, I. and Marquiss, M. (1981). Effects of additional food on laying dates and clutch sizes of sparrowhawks. *Ornis Scandinavica* **12**, 224–229.

Newton, I. and Marquiss, M. (1982). Food, predation and breeding season in Sparrowhawks. *Journal of Zoology London* **197**, 221–240.

Newton, I. and Marquiss, M. (1984). Seasonal trend in breeding performance of Sparrowhawks. *Journal of Animal Ecology* **53**, 809–829.

Newton, I., Marquiss, M., and Moss, D. (1981). Age and breeding in sparrowhawks. *Journal of Animal Ecology* **50**, 839–854.

Newton, I., Marquiss, M. and Village, A. (1983). Weights, breeding and survival in European Sparrowhawks. *Auk* **100**, 344–354.

Nieboer, E. (1973). *Geographical and ecological differentiation in the genus* Circus. Unpublished D.Phil. thesis. Free University, Amsterdam.

Niebuhr, V. (1981). An investigation of courtship feeding in Herring Gulls *Larus argentatus*. *Ibis* **123**, 218–223.

Nilsson, J.-Å. (1991). Clutch size determination in the Marsh Tit (*Parus palustris*). *Ecology* **72**, 1757–1762.

Nilsson, J.-Å. and Svensson, E. (1993a). Causes and consequences of egg mass variation between and within Blue Tit clutches. *Journal of Zoology London* **230**, 469–481.

Nilsson, J.-Å. and Svensson, E. (1993b). Energy constraints and ultimate decisions during egg-laying in the Blue Tit. *Ecology* **74**, 244–251.

Nisbet, I. C. T. (1973). Courtship-feeding, egg size and breeding success in Common Terns. *Nature* **241**, 141–142.

Norberg, Å. (1981). Temporary weight decrease in breeding birds may result in more fledged young. *American Naturalist* **118**, 838–850.

Norrdahl, K. and Korpimaki, E. (1993). Predation and interspecific competition in two Microtus voles. *Oikos* **67**, 149–158.

Nur, N. (1988a). The consequences of brood size for breeding Blue Tits. III Measuring the cost of reproduction: survival, future fecundity, and differential dispersal. *Evolution* **42**, 351–362.

Nur, N. (1988b). The cost of reproduction in birds: an examination of the evidence. *Ardea* **76**, 155–168.

O'Connor, R. J. (1978). Egg weights and brood reduction in the Swift (*Apus apus). Condor* **81**, 133–145.

O'Connor, R. J. (1982). Habitat occupancy and the regulation of clutch size in the European Kestrel *Falco tinnunculus. Bird Study* **29**, 17–26.

Ojanen, M., Orell, M., and Vaisanen, R. A. (1979). Role of heredity in egg size variation in the Great Tit *Parus major* and the Pied Flycatcher *Ficedula hypoleuca. Ornis Scandinavica* **10**, 17–21.

Ollason, J. C. and Dunnet, G. M. (1978). Age experience and other factors affecting the breeding success of the Fulmar, *Fulmarus glacialis*, in Orkney. *Journal of Animal Ecology* **47**, 961–976.

Olsen, P. & Olsen, J. (1984). Book review of the Falcons of the world. *Australian Wildlife Research* **11**, 205–206.

Olsen, P. D. and Cockburn, A. (1991). Female-biased sex allocation in Peregrines Falcons and other raptors. *Behavioural Ecology and Sociobiology* **28**, 417–423.

Olsen, P. and Olsen, J. (1987). Sexual size dimorphism in raptors: intrasexual competition in the larger sex for a scarce breeding resource, the smaller sex. *Emu* **87**, 59–62.

Onofre, N. (1992). Bio-ecologia da aguia cacadeira *Circus pygargus* (L.) Numa area de agricultura cerealifera em Castro Verde. Dados preliminaires. I Conferencia Nacional sobre Aves de Rapina, Gaia, Portugal.

Orians, G. H. (1969). On the evolution of mating systems in birds and mammals. *American Naturalist* **103**, 589–603.

Owen, D. F. (1977). Latitudinal gradients in clutch size: an extension of Lack's hypothesis. In *Evolutionary ecology* (ed. B. Stonehouse and C. Perrins), pp. 171–179. University Park Press, Baltimore.

Palmer, R. S. (1988). *Handbook of North American birds*. Vol. 4. Yale University Press, New Haven.

Palokangas, P., Alatalo, R. V., and Korpimaki, E. (1992). Female choice in the kestrel under different availability of mating options. *Animal Behaviour* **43**, 659–665.

Palokangas, P., Korpimaki, E., Hakkarainen, H., Huhta, E., Tolenen, P., and Alatalo, R. V. (1994). Female kestrels gain reproductive success by choosing brightly ornamented males. *Animal Behaviour* **47**, 443–448.

Pandolfi, M. and Giacchini, P. (1991). Distribuzione e successo riproduttivo di albanella minore *Circus pygargus* nell Marche. *Riv. Italian Ornitologi* **61**, 25–32.

Pandolfi, M. and Pino D'Astore, P. R. (1990a). Analysis of breeding behaviour in Montagu's Harrier *Circus pygargus* in a site of central Italy. *Avocetta* **14**, 97–102.

Pandolfi, M. and Pino D'Astore, P. R. (1990b). 'Food pass' behaviour pattern in *Circus pygargus. Ethology, Ecology and Evolution* **2**, 321.

Pandolfi, M., Pagliarani, R., and Olivetti, G. (1998). Pair copulations, extra-pair copulations and females refusing mating in the Montagu's Harrier. *Journal of Raptor Research* **32**, 269–277.

Parsons, J. (1970). Relationship between egg-size and post hatch survival in the Herring Gull *Larus argentatus. Nature* **228**, 1221–1222.

Parsons, J. (1976). Factors determining the number and size of eggs laid by the Herring Gull. *Condor* **78**, 481–492.

Partridge, L. and Farquar, M. (1981). Sexual activity reduces lifespan of male fruitflies. *Nature* **294**, 580–582.

Payne, R. S. (1971). Acoustical location of prey by the Barn Owl (*Tyto alba*). *Journal of Experimental Biology* **56**, 535–573.

Payne, R. (1984). Sexual selection, lek behaviour and sexual size dimorphism in birds. *Ornithological Monographs* **33**, 1–52.

Perrin, M. R. (1981). Notes on the activity patterns of 12 species of southern African rodents and a new design of activity monitor. *South African Journal of Zoology* **16**, 248–258.

Perrin, M. R. (1986). Some perspectives on the reproductive tactics of southern African rodents. *Cimbebasia* **8**, 63–77.

Perrins, C.M. (1970). The timing of birds' breeding seasons. *Ibis* **112**, 242–255.

Perrins, C. M. (1979). *British Tits*. Collins, London.

Perrins, C. M. and Birkhead, T. R. (1983). *Avian ecology*. Blackie, Glasgow.

Petit, L. J. (1991). Experimentally induced polygyny in a monogamous bird species: prothonotary warblers and the polygyny threshold. *Behavioural Ecology and Sociobiology* **29**, 177–187.

Pettifor, R. A., Perrins, C. M., and McCleery, R. H. (1988). Individual optimization of clutch size in great tits. *Nature* **336**, 160–162.

Picozzi, N. (1971). Wing tags. A method of marking birds. *Birds* **3**, 310–311.

Picozzi, N. (1978). Dispersion, breeding and prey of the Hen Harrier *Circus cyaneus* in Glen Dye, Kincardineshire. *Ibis* **120**, 498–509.

Picozzi, N. (1980). Food, growth, survival and sex ratio of nestling Hen Harriers *Circus c. cyaneus* in Orkney. *Ornis Scandinavica* **11**, 1–11.

Picozzi, N. (1981). Weight, wing length and iris colour of Hen Harriers in Scotland. *Bird Study* **28**, 159–161.

Picozzi, N. (1983). Two hens but a single nest: an unusual case of polygyny by Hen Harriers in Orkney. *British Birds* **76**, 122–128.

Picozzi, N. (1984a). Breeding biology of polygynous Hen Harriers *Circus c. cyaneus* in Orkney. *Ornis Scandinavica* **15**, 1–10.

Picozzi, N. (1984b). Sex ratio, survival and territorial behaviour of polygynous Hen Harriers *Circus c. cyaneus* in Orkney. *Ibis* **126**, 356–365.

Pietiainen, H., Saurola, P., and Vaisanen, R. A. (1986). Parental investment in clutch size and egg size in the Ural Owl *Strix uralensis*. *Ornis Scandinavica* **17**, 309–325.

Poiani, A. (1993a). Small clutch sizes as a possible adaptation against ecto-parasitism: a comparative analysis. *Oikos* **68**, 455–462.

Poiani, A. (1993b). Effects of clutch size manipulations in reproductive behaviour and nesting success in the cooperatively breeding Bell Miner (*Manorina melanophrys*). *Evolutionary Ecology* **7**, 329–356.

Poole, A. (1985). Courtship feeding and Osprey reproduction. *Auk* **102**, 479–492.

Poole, A. (1989). *Ospreys: a natural and unnatural history*. Cambridge University Press.

Poole, K. G. 1988. Feeding responses by Gyrfalcons to brood size manipulation. *J. Rapt. Res.***22**, 67–70.

Potts, G. R. (1998). Global dispersion of nesting Hen Harriers *Circus cyaneus*; implications for grouse moors in the U.K. *Ibis* **140**, 76–88.

Poulin, R. (1991). Group-living and infestation by ectoparasites in passerines. *Condor* **93**, 418–423.

Powell, G. V. N. (1983). Food availability and reproduction by Great White Herons *Ardea herodias*: a food addition study. *Colonial Waterbirds* **6**, 139–147.

Preston, C. R. (1990). Distribution of raptor foraging in relation to prey biomass and habitat structure. *Condor* **92**, 107–112.

Price, T., Kirkpatrick, M. and Arnold, S. J. (1988). Directional selection and the evolution of breeding date in birds. *Science* **240**, 798–799.

Pyke, H. G., Pulliam, H. R., and Charnov, E. L. (1977). Optimal foraging theory: a selective review of theory and tests. *Quarterly Review of Biology* **52**, 137–154.

Quarmen, D. (1994). *Song of the Dodo*. Viking Press, New York.

Quay, W. B. (1985). Sperm release in migrating wood-warblers (Parulinae) nesting at higher latitudes. *Wilson Bulletin* **97**, 283–295.

Rahn, H. and Ar, A. (1974). The avian egg: incubation time and water loss. *Condor* **76**, 147–152.

Rahn, H., Pagannelli, C. V. and Ar, A. (1975). Relation of avian egg weight to body weight. *Auk* **92**, 750–765.

Ramsay, S. and Houston, D. C. (1997). Nutritional constraints on egg production in the Blue Tit: a supplementary feeding study. *J. Anim. Ecol.* **66**, 649–657.

Rasa, O. A. E. (1986). Coordinated vigilance in dwarf mongoose family groups: the watchman's song hypothesis and the costs of guarding. *Ethology* **71**, 340–344.

Rautenbach, I. L. (1976). A survey of the mammals occurring in the Golden Gate Highlands National Park. *Koedoe* **19**, 133–144.

Redmond, G. W., Keppie, D. M., and Herzog, P. W. (1982). Vegetative structure, concealment and success at nests of two races of Spruce Grouse. *Canadian Journal of Zoology* **60**, 670–675.

Redmond, R. L. (1986). Egg size and laying date of long-billed curlews *Numenius americanus*: implications for female reproductive tactics. *Oikos* **46**, 330–338.

Redpath, S. M. (1989). *The effect of Hen Harriers and other predators on red grouse populations in Scotland*. Unpublished D.Phil. thesis, University of Leeds.

Redpath, S. M. (1992). Behavioural interactions between Hen Harriers and their moorland prey. *Ornis Scandinavica* **23**, 73–80.

Redpath, S. M. and Thirgood, S. (1993). Provisioning rates at Hen Harrier nests. Paper presented at first European Meeting of Raptor Research Foundation and Hawk and Owl Trust, University of Kent, Canterbury.

Redpath, S. M. and Thirgood, S. J. (1997). *Birds of prey and red grouse*. HMSO, London.

Rhan, H., Paganelli, C. V., and Ar, A. (1975). Relation of egg weight to body weight. *Auk* **92**, 750–765.

Rice, W. R. (1982). Acoustical location by the Marsh Hawk: adaptation to concealed prey. *Auk* **99**, 403–413.

Ricklefs, R. E. (1976). Growth rates of birds in the humid new world tropics. *Ibis* **118**, 179–207.

Ricklefs, R. E. (1980). Geographical variation in clutch size among passerine birds: Ashmole's hypothesis. *Auk* **97**, 38–49.

Ricklefs, R. E. (1993). Sibling competition, hatching asynchrony, incubation period, and lifespan in altricial birds. *Current Ornithology* **11**, 199–276.

Ricklefs, R. E. and Hussell, D. J. T. (1984). Changes in adult mass associated with the nesting cycle in the European Starling. *Ornis Scandinavica* **15**, 155–161.

Riedstra, B., Dijkstra, C., and Daan, S. (1998). Daily energy expenditure of male and female marsh harrier nestlings. *Auk* **115**, 635–641.

Rijnsdorp, A., Daan, S., and Dijkstra, C. (1981). Hunting in the Kestrel, *Falco tinnunculus*, and the adaptive significance of daily habits. *Oecologia* **50**, 391–406.

Robertson, A. S. (1986). Copulations throughout breeding in a colonial Accipitrid vulture. *Condor* **88**, 535–539.

Robinson, W. (1950). Montagu's Harriers (*Circus pygargus*). *Bird Notes* **24**, 103–114.

Romanoff, A. L. and Romanoff, A. J. (1949). *The avian egg*. Wiley, New York.

Rood, J. P. (1983). Banded mongoose rescues pack member from eagle. *Animal Behaviour* **31**, 1261–1262.

Rose, M. and Charlesworth, B. (1980). A test of evolutionary theories of senescence. *Nature* **287**, 141–142.

Roskaft, E. (1985). The effect of enlarged brood size on the future reproductive potential of the Rook. *Journal of Animal Ecology* **54**, 255–260.

Rowan, M. K. (1966). Some observations on reproduction and mortality in the Cape Sparrow *Passer melanurus*. *Ostrich Supplement* **6**, 425–434.

Rowe, L., Ludwig, D., and Schluter, D. (1999). Time, condition and the seasonal decline of avian clutch size. *American Naturalist*. (In press.)

Rowley, I. and Russell, E. (1991). Demography of passerines in the temperate southern hemisphere. In *Bird population studies* (ed. C. M. Perrins, J.-D. Lebreton, and G. J. M Hirons), pp. 22–44. Oxford University Press.

Royama, T. (1966). A re-interpretation of courtship feeding in birds. *Bird Study* **13**, 116–129.

Rudebeck, G. (1950). The choice of prey and modes of hunting of predatory birds with special reference to their selective effect. *Oikos* **2**, 65–88.

Saether, B-E. (1985). Variation in reproductive traits in European passerines in relation to nesting site: allometric scaling to body weight or adaptive variation? *Oecologia* **68**, 7–9.

Saether, B-E. (1988). Pattern of covariation between life-history traits of European birds. *Nature* **331**, 616–617.

Safina, C. (1984). Selection for reduced male size in raptorial birds: the possible roles of female choice and mate guarding. *Oikos* **43**, 159–164.

Saitou, N. and Nei, M. (1987). The neighbour-joining method: a new method for reconstructing phylogenetic trees. *Molecular Biology and Evolution* **4**, 406–425.

Salomolard, M. (1997). Utilisation de L'espace par le Busard Cendré *Circus pygargus*. *Alauda* **65**, 307–320.

Scharf, W. C. and Balfour, E. (1971). Growth and development of nestling Hen Harriers. *Ibis* **113**, 323–339.

Scharf, W.C. and Hamerstrom, F. (1975). A morphological comparison of two harrier populations. *Raptor Research* **9**, 27–32.

Schipper, W. J. A. (1973). A comparison of prey selection in sympatric harriers (*Circus*) in western Europe. *Le Gerfaut* **63**, 17–120.

Schipper, W. J. A. (1977). Hunting in three European Harriers (Circus) during the breeding season. *Ardea* **65**, 53–72.

Schipper, W. J. A. (1978). A comparison of breeding ecology in three European harriers (*Circus*). *Ardea* **66**, 77–102.

Schipper, W. J. A., Buurma, L. S., and Bossenbroek, P. H. (1975). Comparative study of hunting behaviour of wintering Hen harriers *Circus cyaneus* and Marsh Harriers *Circus aeruginosus*. *Ardea* **63**, 1–29.

Scott, D., Clarke, R., and Shawyer, C. (1993/94). Tree-nesting Hen Harriers—evolution in the making? *The Raptor* **21**, 53–56.

Sealy, S. G. (1967). Notes on the breeding biology of Marsh Hawks in Alberta and Saskatchewan. *Blue Jay* **25**, 63–69.

Searcy, W. A. (1979). Female choice of mates: a general model for birds and its application to Red-winged Blackbirds. *American Naturalist* **114**, 77–100.

Seyfarth, R. M. and Cheyney, D. L. (1986). Vocal development in Vervet Monkeys. *Animal Behaviour* **34**, 1640–1658.

Sharland, M. S. (1932). Notes on the Swamp Harrier. *Emu* **32**, 87–90.

Sheldon, B. C., Rasanen, K., and Dias, P. (1997). Certainty of paternity and paternal effort in the Collared Flycatcher. *Behavioral Ecology* **8**, 421–428.

Sherman, P. W. (1985). Alarm calls of Belding's ground Squirrel to aerial predators: nepotism or self preservation? *Behavioural Ecology and Sociobiology* **17**, 313–323.

Sherrod, S. K., White, C. M., and Williamson, F. S. L. (1977). Biology of the Bald Eagle (*Haliaeetus leucocephalus alascanus*) on Amchitka Island. *Living Bird* **15**, 143–182.

Sherrod, S.K. (1983). *Behavior of fledgling Peregrines*. Peregrine Fund, Ithaca, New York.

Siegfried, W. R. and Frost. P. G. H. (1975). Continuous breeding and associated behaviour in the Moorhen *Gallinula chloropus*. *Ibis* **117**, 102–109.

Sih, A. (1980). Optimal behaviour: can foragers balance two conflicting demands? *Science* **210**, 1041–1043.

Simmons, R. E. (1983). *Polygyny, ecology and mate choice in Northern Harriers* Circus cyaneus, (L.). Unpublished Masters thesis. Acadia University, Wolfville, Nova Scotia.

Simmons, R. E. (1984a). Do Northern Harriers lay replacement clutches? *Raptor Research* **18**, 103–106.

Simmons, R. E. (1984b). Pre-independence behaviour, morphometrics, and trapping of fledgling Redbreasted Sparrowhawks. *Ostrich* **55**,158–162.

Simmons, R. (1986a). Food provisioning, nestling growth and experimental manipulation of brood size in the African Redbreasted Sparrowhawk *Accipiter rufiventris*. *Ornis Scandinavica* **17**, 31–40.

Simmons, R. E. (1986b). Ecological segregation of the Redbreasted Sparrowhawk *Accipiter rufiventris* and six coexisting Acciptrine raptors in southern Africa. *Ardea* **74**, 137–149.

Simmons, R. E. (1986c). A tagged African Marsh Harrier. *Bee-eater* **37**, 26–27.

Simmons, R. E. (1988a). Offspring quality and the evolution of cainism. *Ibis* **130**, 339–357.

Simmons, R. E. (1988b). Food and the deceptive acquisition of mates by polygynous male harriers. *Behavioural Ecology and Sociobiology* **23**, 83–92.

Simmons, R. E. (1988c). Honest advertising, sexual selection, courtship displays, and body condition of polygynous male harriers. *Auk* **105**, 303–307.

Simmons, R. E. (1988d). Notes on foraging European Marsh Harriers in the Nyl River Floodplain, South Africa. *Gabar* **3**, 35–37.

Simmons, R. E. (1989). *Adaptation and constraint in the breeding of subtropical harriers and eagles*. Unpublished D.Phil. thesis, University of the Witwatersrand, Johannesburg.

Simmons, R. (1990). Copulation patterns of African Marsh Harriers: evaluating the paternity assurance hypothesis. *Animal Behaviour* **40**, 1151–1157.

Simmons, R. E. (1991a). Comparisons and functions of aerial skydancing displays of *Circus* harriers: untangling the Marsh Harrier complex. *Ostrich* **62**, 45–51.

Simmons, R. E. (1991b). The evolution and efficiency of food passing in harriers. *Gabar* **6**, 51–56.

Simmons, R. E. (1992). Brood adoption and deceit among African Marsh Harriers *Circus ranivorus*. *Ibis* **134**, 32–34.

Simmons, R. E. (1993a). Effects of supplementary food on density-reduced breeding in an African eagle: adaptive restraint or ecological constraint? *Ibis* **135**, 394–402.

Simmons, R. E. (1993b). Chance, choice and habitat-related breeding success in dense population of Wahlberg's Eagles *Aquila wahlbergi*. *Proceedings of VIII Pan-African Ornithological Congress*. Burundi. pp. 133–139.

Simmons, R. E. (1994). Supplemental food alters egg size hierarchies among harrier clutches. *Oikos* **71**, 341–348.

Simmons, R. E. (1997a). African Marsh Harrier/Black Harrier. In *The atlas of southern African birds* (ed. J. A. Harrison, *et al.*), pp. 236–237/241–243. Birdlife South Africa, Johannesburg.

Simmons, R. E. (1997b). Why don't all siblicidal eagles lay insurance eggs? The egg quality hypothesis. *Behavioral Ecology* **8**, 544–550.

Simmons, R. E. (1998). The Skutch–clutch dilemma: nourishable young and predation in the tropics—a solution. *Proceedings of the International Ornithological Congress*, Durban, South Africa.

Simmons, R. E. (MS). Allometric and ecological predictors of breeding and growth in eagles: Is lifespan or body size best?

Simmons, R. E. (MSb). Brood reduction does provide food benefits for raptor chicks. Submitted

Simmons, R. E. and Mendelsohn, J. M. (1993). A critical review of cartwheeling flights in raptors. *Ostrich* **64**, 13–24.

Simmons, R. E. and Smith, P. C. (1985). Do Northern Harriers choose nest sites adaptively? *Canadian Journal of Zoology* **63**, 494–498.

Simmons, R. E., Smith, P. C., and MacWhirter, R. B. (1986a). Hierarchies among Northern Harrier harems and the costs of polygyny. *Journal of Animal Ecology* **55**, 755–771.

Simmons, R. E., MacWhirter, R. B., Barnard, P. E., and Hansen, G. (1986b). The influence of microtines on polygyny, productivity, age, and provisioning of breeding Northern Harriers: a 5-year study. *Canadian Journal of Zoology* **64**, 2447–2456.

Simmons, R. E., Barnard P., and Smith, P. C. (1987). Reproductive behaviour of *Circus cyaneus* in North America and Europe: a comparison. *Ornis Scandinavica* **18**, 33–41.

Simmons, R. E., Avery, D. M., and Avery, G. (1991). Biases in diets determined from pellets and remains: correction factors for a mammal and bird-eating raptor. *Journal of Raptor Research* **25**, 63–67.

Simmons, R. E., Griffin, M., Griffin, R. E., Marais, E., and Kolberg, H. (1998). Endemism in Namibia: patterns, processes and predictions. *Biodiversity and Conservation* **7**, 513–530.

Skutch, A. F. (1949). Do tropical birds rear as many young as they can nourish? *Ibis* **91**, 430–455.

Skutch, A. F. (1985). Clutch size, nesting success, and predation on nests of neotropical birds, revisited. In *Neotropical Ornithology*, Ornithological Monographs No. 36 (ed. M. S. Foster *et al.*), pp. 575–594. American Ornithologists Union, Washington D.C.

Slagsvold, T. (1982). Clutch size, nest size, and hatching asynchrony in birds: experiments with Fieldfares (*Turdus pilaris*). *Ecology* **63**, 1389–1399.

Slagsvold, T. and Amundsen, T. (1992). Do Great Tits adjust hatching spread, egg size and offspring sex ratio to changes in clutch size? *Journal of Animal Ecology* **61**, 249–258.

Slagsvold, T. and Lifjeld, J. T. (1989). Constraints on hatching asynchrony and egg size in pied flycatchers. *Journal of Animal Ecology* **58**, 837–849.

Slagsvold, T. and Lifjeld, J. T. (1994). Polygyny in birds: the role of competition between females for male parental care. *American Naturalist* **143**, 59–94.

Slagsvold, T., Sandvik, J., Rofstad, G., Lorentsen, O., and Husby, M. (1984). On the adaptive value of intraclutch egg-size variation in birds. *Auk* **101**, 685–697.

Smith, J. N. M., Montgomerie, R. D., Taitt, M. J., and Yom-Tov, Y. (1980). A winter-feeding experiment on an island Song Sparrow population. *Oecologia* **47**, 101–106.

Smith, J. N. M., Yom-Tov, Y., and Moses, R. (1982). Polygyny, male parental care, and sex ratio in Song Sparrows: an experimental study. *Auk* **99**, 555–564.

Smithers, R. H. N. (1971). *The mammals of Botswana.* Trustees of the National Museum of Rhodesia, Salisbury.

Smithers, R. H. N. (1983). *The mammals of the southern African subregion.* University of Pretoria Press, Pretoria.

Snow, D. W. (1976). *The web of adaptation; bird studies in the American tropics.* Quadrangle/New York Times Book Co., N.Y.

Snow, D. W. (1978). The nest as a factor determining clutch size in tropical birds. *Journal of Ornithology* **119**, 227–230.

Snow, D. W. and Lill, A. (1974). Longevity records for some neotropical land birds. *Condor* **76**, 262–267.

Snyder, N. F. R. & Snyder, H. A. (1973). Experimental study of feeding rates of nesting Cooper's hawks.*Condor* **75**, 461–463.

Snyder, N. F. R. and Wiley, J. W. (1976). Sexual size dimorphism in hawks and owls of North America. *Ornithological Monograph* **20**, 1–96.

Sodhi, N. S. (1991). Pair copulations, extra-pair copulations, and intra-specific nest intrusions in merlin. *Condor* **93**, 433–437.

Sodhi, N. S., Oliphant, L. W., James, P. C., and Warkentin, I. G. (1993). Merlin (*Falco columbarius*). In *The birds of North America*, No. 44 (ed. A. Poole and F. Gill), Philadelphia, The Academy of Natural Sciences; Washington DC.

Soler, M. and Soler, J. J. (1992). Latitudinal trends in clutch size in single brooded hole nesting bird species: a new hypothesis. *Ardea* **80**, 293–300.

Sotherland, P. R. and Rahn, H. (1987). On the composition of bird eggs. *Condor* **89**, 49–65.

Spurrier, M. F., Boyce, M. S., and Manly, B. F. J. (1994). Lek behaviour in captive Sage Grouse *Centrocercus urophasianus*. *Animal Behaviour* **47**, 303–310.

Stacey, P. B. and Koenig, W. D. (ed.). (1990). *Cooperative breeding in birds: long term studies of ecology and behaviour.* Cambridge University Press.

Statistical Graphics Corporation. (1986). Statgraphics, version 3.0. STSC, Inc. USA.

St Clair, C. C., Waas, J. R., St Clair, R. C., and Boag, P. T. (1995). Unfit mothers? Maternal infanticide in royal penguins. *Animal Behaviour* **50**, 1177–1185.

Stearns, S. C. (1992). *The evolution of life histories.* Oxford University Press.

Stenmark, G., Slagsvold, T., and Lifjeld, J. T. (1988). Polygyny in the Pied Flycatcher, *Ficedula hypoleuca:* a test of the deception hypothesis. *Animal Behaviour* **36**, 1646–1657.

Steyn, P. (1982). *Birds of prey of southern Africa.* Croom Helm, Beckenham, Kent.

Stockland, J. N. and Amundsen, T. (1988). Initial size hierarchy in broods of the shag: relative significance of egg size and hatching asynchrony. *Auk* **105**, 308–315.

Stonehouse, B. (1969). Environmental temperatures of tertiary penguins. *Science* **163**, 673–675.

Storer, R. W. (1966). Sexual dimorphism and food habits in three North American accipiters. *Auk* **83**, 423–426.

Streseman, E. (1924). Der Formenkries der Rohrweihe, *Circus aeruginosus*. *Journal für Ornithologie* **72**, 262–269.

Stronach, N. R. H. (1991). Wintering harriers in Serengeti National Park, Tanzania. *African Journal of Ecology* **29**, 90–92.

Struhsaker, T. T. and Leakey, M. (1990). Prey selectivity by Crowned Eagles on monkeys in the Kibale Forest, Uganda. *Behavioural Ecology and Sociobiology* **26**, 435–443.

Svensson, E. (1995). Avian reproductive timing: when should parents be prudent? *Animal Behaviour* **49**, 1569–1575.

Swofford, D.L. (1993). PAUP: Phylogenetic analysis using parsimony, version 3.1.1. *Illinois Natural History Survey,* Champaign Illinois.

Tarboton, W. R. (1978). Hunting and energy budget of the Blackshouldered Kite. *Condor* **80**, 88–91.

Tarboton, W. R. and Allan, D. G. (1984). The status and conservation of birds of prey of the Transvaal. *Transvaal Museum Monograph* No. 3, Pretoria.

Tarburton, M. K. (1986). An experimental manipulation of clutch and brood size of White-rumped Swiftlets *Aerodrammus spodiopygius* of Fiji. *Ibis* **129**, 107–114.

Tasker, C. R. and Mills, J. A. (1981). A functional analysis of courtship-feeding in the Red-billed Gull (*Larus novaehollandiae*). *Behaviour* 77, 222–241.

Taylor, K. D. and Green, M. G. (1976). The influence of rainfall on diet and reproduction in four African rodent species. *Journal of Zoology London* **180**, 367–389.

Tella, J. L. (1993). Polyandrous trios in a population of Egyptian Vultures. *Journal of Raptor Research* **27**, 119–120.

Tella, J. L., Donazar, J. A., Negro, J. J., and Hiraldo, F. (1996). Seasonal and interannual variations in the sex-ratio of Lesser Kestrel *Falco naumanni* broods. *Ibis* **138**, 342–345.

Temeles, E. J. (1980). *Sexual size dimorphism in the New World Falconiformes.* Unpublished Master's thesis. Louisiana State University, Baton Rouge.

Temeles, E. J. (1985). Sexual size dimorphism of bird-eating hawks: the effect of prey vulnerability. *American Naturalist* **125**, 485–499.

Temeles, E. J. (1987). The relative importance of prey availability and intruder pressure in feeding territory regulation by harriers, *Circus cyaneus. Oecologia* 74, 286–297.

Temple, S. A. 1987. Do predators always capture substandard individuals disproportionately from prey populations? *Ecology* **68**, 669–674.

Temrin, H. (1989). Female pairing options in polyterritorial wood warblers *Phylloscopus sibilatrix*: are females deceived? *Animal Behaviour* **37**, 579–586.

Thiollay, J.-M. (1970). Observations sur l'ecologie d'une population de Busards des Roseaux, *Circus aeruginosus*, en Camargue. *Nos Oiseaux* **30**, 214–229.

Thiollay, J.-M. (1988). Comparative foraging success of insectivorous birds in tropical and temperate forests: ecological implications. *Oikos* **53**, 17–30.

Thomas, C. (1983). The relationship between breeding experience, egg volume and reproductive success of the kittiwake *Rissa tridactyla. Ibis* **125**, 567–574.

Toland, B. (1986). Hunting success of some Missouri raptors. *Wilson Bulletin* **98**, 116–125.

Trail, P. W. (1985). Courtship disruption modifies mate choice in a lek-breeding bird. *Science* **227**, 778–780.

Trivers, R. and Willard, D. E. (1973). Natural selection of parental ability to vary the sex ratio of offspring. *Science* **179**, 90–92.

Underhill-Day, J. C. (1984). Population and breeding biology of Marsh Harriers in Britain since 1900. *Journal of Applied Ecology* **21**, 773–787.

Underhill-Day, J. C. (1985). The food of breeding Marsh Harriers *Circus aeruginosus* in East Anglia. *Bird Study* **32**, 199–206.

Underhill-Day, J. C. (1990). *The status and breeding biology of Marsh Harrier and Montagu's Harrier in Britain since 1900.* Unpublished Ph.D. thesis, CNAA, London.

Underhill-Day, J. (MS). Hunting behaviour and success of British Marsh Harriers foraging over arable farmland. Paper presented at Raptor Research Foundation first European meeting. Hawk and Owl Trust, Canterbury, UK, 1993.

Urano, E. (1985). Polygyny and breeding success of the Great Reed Warbler *Acrocephalus arundinaceus. Researches on Population Ecology (Kyoto)* **27**, 393–412.

Van der Kraan, C. and van Strien, (1969). Polygamie bij de Blaue Kiekendief (*Circus cyaneus*). *Limosa* **42**, 34–35.

Van der Merwe, F. (1981). Review of the status and biology of the Black Harrier. *Ostrich* **52**, 193–207.

Vander Werfe, E. (1992). Lack's clutch size hypothesis: an explanation using meta-analysis. *Ecology* **73**, 1699–1705.

van Noordwijk, A. J. (1984). Quantitative genetics in natural populations of birds illustrated with examples from the Great Tit *Parus major*. In *Population biology and evolution* (ed. K. Worhmann and V. Loeschke), pp. 67–79. Springer, Berlin.

van Noordwijk, A. J., Keizer, L. C. P., van Balen, J. H., and Scharloo, W. (1981). Genetic variation in egg dimensions in natural populations of the Great Tit. *Genetica* **55**, 221–232.

Van Zyl, A. J. (1993). *Aspects of the foraging and breeding ecology of the southern African Kestrel*, Falco tinnunculus rupicolus. Masters thesis. University of Cape Town.

Van Zyl, A. J. (MS). Can the Common Kestrel (*Falco tinnunculus*) provision increased broods in a mid-latitude environment? FitzPatrick Institute, University of Cape Town.

Van Zyl, A. J. and Amutete, G. (1998). Time-energy budgets of Common Kestrels over latitudes: energy minimisation or maximization of intake rates. In *Proceedings of 22nd International Ornithological Congress, Durban* (ed. N. J. Adams and R. H. Slotow). *Ostrich* **69**, 224.

Veit, R. and Petersen, W. (1993). *Birds of Massachusetts.* Massachusetts Audobon Society, Lincoln, MA.

Vermeij, G. J. (1982). Unsuccessful predation and evolution. *American Naturalist* **120**, 701–720.

Verner, J. (1964) The evolution of polygamy in the long-billed marsh wren. *Evolution* **18**, 252–261.

Verner, J. and Willson, M. (1966). The influence of habitats on mating systems of North American passerine birds. *Ecology* **47**, 143–147.

Village, A. (1985). Spring arrival times and assortative mating of kestrels in south Scotland. *Journal of Animal Ecology* **54**, 857–868.

Village, A. (1990). *The Kestrel.* T and A. D. Poyser, Berkhamsted, England.

Von Bromssen, A. and Jansson, C. (1980). Effects of food addition to Willow Tit *Parus montanus* and Crested Tit *P. cristatus* at the time of breeding. *Ornis Scandinavica* **11**, 173–178.

von Haartman, L. (1969). Nest site and evolution of polygamy in European passerine birds. *Ornis Fennica*, **46**, 1–12.

Wagner, H. O. (1957). Variation in clutch size at different latitudes. *Auk* **75**, 243–250.

Wagner, R. H. (1998). Hidden leks: mechanisms of sexual selection and colony formation. In *Proceedings of 22nd International Ornithological Congress, Durban* (ed. N. J. Adams and R. H. Slotow). *Ostrich* **69**, 70.

Wakeley, J. S. (1978). Factors affecting the use of hunting sites by Ferruginous Hawks. *Condor* **80**, 316–326.

Ward, P. (1965). The breeding biology of the Black-faced Dioch *Quelea quelea* in Nigeria. *Ibis* **107**, 326–349.

Ward, P. and Zahavi, A. (1973). The importance of certain assemblages of birds as 'information centres' for food-finding. *Ibis* **115**, 517–534.

Warham J. (1975). The crested penguins. In *Biology of penguins* (ed. B. Stonehouse), pp. 189–269. Macmillan, London.

Wasser, J. S. (1986). The relationship of energetics of Falconiform birds to body mass and climate. *Condor* **88**, 57–62.

Watson, D. (1977). *The Hen Harrier.* T. and A. D. Poyser, Berkhamsted, England.

Watson, J. (1997). *The Golden Eagle.* T. and A. D. Poyser, London.

Watson, R. T. (1989). Aggressive display and territoriality of the Bateleur *Terathopius ecaudatus. South African Journal of Zoology* **24**, 146–150.

Weatherhead, P. J. and Robertson, R. J. (1979). Offspring quality and the polygyny threshold: the 'sexy-son' hypothesis. *American Naturalist*, **113**, 201–208.

Weatherhead, P. J. and Robertson, R. J. (1981). In defence of the 'sexy-son' hypothesis. *American Naturalist* **117**, 349–356.

Webster, M. S. (1997). Extreme sexual size dimorphism, sexual selection, and the foraging ecology of Montezuma Oropendulas. *Auk* **114**, 570–580.

Weis, H. (1923). *The life of the harrier in Denmark.* Wheldon and Wesley, London.

Westneat, D. F. (1987). Extra-pair fertilizations in a predominantly monogamous bird: genetic evidence. *Animal Behaviour* **35**, 877–886.

Wetton, J.H. and Parkin, D.T. (1991). An association between fertility and cuckoldry in the House Sparrow *Passer domesticus. Proceedings of the the Royal Society Series B* 245.

Wicklund, C. G. (1982). Fieldfare (*Turdus pilaris*) breeding success in relation to colony size, nest position and association with Merlins (*Falco columbarius*). *Behavioural Ecology and Sociobiology* **11**, 165–172.

Widén, P. (1984). Reversed sexual dimorphism in birds of prey: revival of an old hypothesis. *Oikos* **43**, 259–263.

Wiebe, K. L. and Bortolotti, G. R. (1992). Facultative sex ratio manipulation in American Kestrels. *Behavioural Ecology and Sociobiology* **30**, 379–386.

Wiebe, K. L. and Bortolotti, G. R. (1994). Food supply and hatching spans of birds: energy constraints or facultative manipulation? *Ecology* **75**, 813–823.

Wiehn, J. (1997). Plumage characteristics as an indicator of male parental quality in the American Kestrel. *Journal of Avian Biology* **28**, 47–55.

Wiggins, D. A. (1990). Sources of variation in egg mass of Tree Swallows *Tachycineta bicolor. Ornis Scandinavica* **21**, 157–160.

Wiggins, D. A. and Morris, R. D. (1986). Criteria for female choice of mates: courtship feeding and paternal care in the Common Tern. *American Naturalist* **128**, 126–129.

Wiklund, C. G. (1982). Fieldfare (*Turdus pilaris*) breeding success in relation to colony size, nest position and association with Merlins (*Falco columbaris*). *Behavioral Ecology and Sociobiology* **11**, 165–172.

Wiklund, C. G., Karlsson, B., and Forsberg, J. (1987). Adaptive versus constraint explanations for egg-to-body size relationships in two butterfly families. *American Naturalist* **130**, 828–838.

Williams, A. J. (1980). Variation in weight of eggs and its effect on the breeding biology of the Great Skua. *Emu* **80**, 198–202.

Williams, G. C. (1966). *Adaptation and natural selection.* Princeton University Press

Williams, T. D. (1993). Intraspecific variation in egg size and egg composition in birds: effects on offspring fitness. *Biological Review* **68**, 35–59.

Willis, E. O., Wechsler, D., and Oniki, Y. (1978). On behaviour and nesting of McConnell's Flycatcher (*Pipromorpha macconnelli*): does female rejection lead to male promiscuity? *Auk* **95**, 1–8.

Willson, M. F. and Pianka, E. R. (1963). Sexual selection, sex ratio, and mating system. *American Naturalist*, **97**, 405–407.

Wimberger, P. H. (1988). Food supplement effects on breeding time and harem size in the Red winged Blackbird (*Agelaius phoeniceus*). *Auk* **105**, 799–802.

Wink, D., Ristow, D., and Wink, C. (1985). Biology of Eleonora's Falcon (*Falco eleonorae*): 7. Variability of clutch size, egg dimensions and egg colouring. *Raptor Research* **19**, 8–14.

Wink, M. (1998). Advances in the molecular phylogeny of vultures, eagles, hawks and falcons. *Proceedings of Vth World Conference on Birds of Prey and Owls*. Pretoria, South Africa.

Winkler, D. W. and Walters, J. R. (1985). The determination of clutch size in precocial birds. In *Current ornithology* (ed. R. F. Johnstone), pp. 33–67. Plenum Press, N.Y.

Winterbottom, M., Burbe, T. and Birkhead, T.R. (1999). A stimulatory phalloid organ in a weaver bird. *Nature* 399, 28.

Witkowski, J. (1989). Breeding biology and ecology of the marsh harrier *Circus aeruginosus* in the Barcycz valley, Poland. *Acta Ornithologica* **25**, 223–320.

Wittenberger, J. F. (1976). The ecological factors selecting for polygyny in altricial birds. *American Naturalist* **110**, 779–799.

Wittenberger, J. F. (1980). Group size and polygamy in social mammals. *American Naturalist* **115**, 197–222.

Woets, D. (1989). Grauwe Kiekendieven in De Weerribben. *Het Vogeljaar* **37**, 16–19.

Wyndham, E. (1986). Length of birds' breeding seasons. *American Naturalist* **128**, 155–164.

Ydenberg, R. C. and Bertram, D. (1989). Lack's clutch size hypothesis and brood enlargement studies on colonial seabirds. *Colonial Waterbirds* **12**, 134–137.

Ydenburg, R. C. and Forbes, L. S. (1991). The survival-reproduction selection equilibrium and reversed size dimorphism in raptors. *Oikos* **60**, 115–120.

Yom-Tov, Y. (1974). The effect of food and predation on breeding density and success, clutch size and laying date of the Crow (*Corvus corone* L.). *Journal of Animal Ecology* **4**, 479–498.

Yom-Tov, Y. (1987). The reproductive rate of Australian passerines. *Australian Wildlife Research* **14**, 319–330.

Yom-Tov, Y., Christie, M. I., and Iglesias, G. J. (1994). Clutch size in passerines of southern South America. *Condor* **96**, 170–177.

Young, B.E. (1996). An experimental analysis of small clutch size in tropical House Wrens. *Ecology* **77**, 472–488.

Zijlstra, M. Daan, S. and Bruinenberg-Rinsma, J. (1992). Seasonal variation in the sex ratio of marsh harrier *Circus aeruginosus* broods. *Functional Ecology* **6**, 553–559.

Index

Species are entered under their common names. Latin names which occur frequently in the text are given in brackets after the common name.